About Island Press

Island Press is the only nonprofit organization in the United States whose principal purpose is the publication of books on environmental issues and natural resource management. We provide solutions-oriented information to professionals, public officials, business and community leaders, and concerned citizens who are shaping responses to environmental problems.

In 2003, Island Press celebrates its nineteenth anniversary as the leading provider of timely and practical books that take a multidisciplinary approach to critical environmental concerns. Our growing list of titles reflects our commitment to bringing the best of an expanding body of literature to the environmental community throughout North America and the world.

Support for Island Press is provided by The Nathan Cummings Foundation, Geraldine R. Dodge Foundation, Doris Duke Charitable Foundation, Educational Foundation of America, The Charles Engelhard Foundation, The Ford Foundation, The George Gund Foundation, The Vira I. Heinz Endowment, The William and Flora Hewlett Foundation, Henry Luce Foundation, The John D. and Catherine T. MacArthur Foundation, The Andrew W. Mellon Foundation, The Moriah Fund, The Curtis and Edith Munson Foundation, National Fish and Wildlife Foundation, The New-Land Foundation, Oak Foundation, The Overbrook Foundation, The David and Lucile Packard Foundation, The Pew Charitable Trusts, The Rockefeller Foundation, The Winslow Foundation, and other generous donors.

The opinions expressed in this book are those of the author(s) and do not necessarily reflect the views of these foundations.

About Future Harvest

The Future Harvest Foundation is a global nonprofit organization that builds awareness and support for food and environmental research for a world with less poverty, a healthier human family, well-nourished children, and a better environment. Future Harvest supports research, promotes partnerships, and sponsors projects that bring the results of research to rural communities, farmers, and families in Africa, Latin America, and Asia. It is an initiative of the sixteen food and environmental research centers that are primarily funded through the Consultative Group on International Agricultural Research. Future Harvest, PMB 238, 2020 Pennsylvania Avenue, NW, Washington, DC 20006, USA; tel: +1-202-473-1142; email: info@futureharvest.org; web: http://www.futureharvest.org.

About the IUCN

The World Conservation Union (IUCN, formally known as the International Union for Conservation of Nature) was founded in 1948 and brings together 78 states, 112 government agencies, 735 nongovernmental organizations (NGOs), 35 affiliates, and some 10,000 scientists and experts from 181 countries in a unique worldwide partnership. Its mission is to influence, encourage, and assist societies throughout the world to conserve the integrity and diversity of nature and to ensure that any use of natural resources is equitable and ecologically sustainable. Within the framework of global conventions, the IUCN has helped over seventy-five countries to prepare and implement national conservation and biodiversity strategies. IUCN/The World Conservation Union, Rue Mauverney 28, 1196 Gland, Switzerland; tel: +41 (22) 999-0001; email: reception@iucn.org; web: http://www.iucn.org.

ECOAGRICULTURE

ECOAGRICULTURE

Strategies to Feed the World and Save Biodiversity

Jeffrey A. McNeely and Sara J. Scherr

FUTURE HARVEST AND
IUCN (WORLD CONSERVATION UNION)

ISLAND PRESS
Washington • Covelo • London

Library of Congress Cataloging-in-Publication Data
McNeely, Jeffrey A.
  Ecoagriculture : strategies to feed the world and save wild biodiversity /
Jeffrey A. McNeely and Sara J. Scherr.
      p.   cm.
Includes bibliographical references (p.   ).
    ISBN 1-55963-644-0 (hardcover : alk. paper) — ISBN 1-55963-645-9 (pbk. :
alk. paper)
    1. Agrobiodiversity. 2. Agrobiodiversity conservation. I. Scherr, Sara J. II. Title.
    S494.5.A43 M37 2002
    333.95'16—dc21   2002005949
British Cataloguing-in-Publication Data available

Book design by Brighid Willson

Printed on recycled, acid-free paper ♲

Manufactured in the United States of America
10 9 8 7 6 5 4 3 2 1

*This book is dedicated to all those innovators around the
world—farmers, scientists, and environmentalists—who are actively
seeking and finding ways to conserve wild biodiversity in agricultural
lands, while increasing our food supply and improving the livelihoods
of the rural poor. May they inspire others to action.*

The fate of birds, mammals, frogs, fish, and all the rest of biodiversity depends not so much on what happens in parks but what happens where we live, work, and obtain the wherewithal for our daily lives. To give biodiversity and wildlands breathing space, we must reduce the size of our own imprint on the planet.

—John Tuxill (1998)

Contents

List of Cases, Maps, Figures, Tables, and Boxes

Cases

Maps

Figures

Tables

Boxes

Preface

When we began our collaboration in the year 2000, we did not intend to write a book on this topic—indeed, we had not even conceived of "ecoagriculture" at that time. Barbara Rose, then executive director of Future Harvest, had asked us simply to write a piece that would illustrate the relevance of agriculture and agricultural research for biodiversity conservation. Jeff had worked for decades on biodiversity conservation in the tropics; Sara, on agricultural development in the tropics.

What emerged from this partnership went well beyond what either of us had anticipated. As an anthropologist and ecologist, Jeff had long been interested in the role of extensive agricultural systems (such as shifting cultivation) as wildlife habitat. As an agricultural and natural resource economist, Sara had long studied the potential for intensification of so-called marginal lands while protecting ecosystem services such as watershed functioning and carbon sequestration. But both of us started out highly skeptical that healthy wild species populations could be compatible on a large scale with the agricultural intensification needed to meet growing food and livelihood needs in the developing world.

As we began to share and integrate information and uncover new case material, however, that skepticism began to fade. On the one hand, when we examined the most recent global data on agricultural systems and wildlife habitats, it became crystal clear that the scale of agriculture's impacts on ecosystems was so massive that without addressing these directly, other efforts at biodiversity conservation—such as many critical protected areas, especially in the so-called biodiversity hotspots—were likely doomed to failure. At the same time, we experienced a growing excitement on discovering the potentials for coexistence that are emerging from new scientific understanding and new resource management systems being developed in different parts of the world. We coined the term *ecoagriculture* to reflect such systems. At first we were a little uncomfortable with the term and wondered whether it would resonate with the production-oriented farmers and development planners who needed to support these systems. But none of the other terms in use fully reflected our vision of land use systems that were intentionally designed and managed *both* to increase food production and farmer incomes, and to conserve wild biodiversity and other ecosystem services.

Meanwhile, our experience in writing this book has reinforced our commitment to interdisciplinary collaboration and analysis. Learning the languages, concepts, and perspectives of new disciplines takes time and an open mind, but it reaps a rich harvest. We hope that the concept of ecoagriculture can serve as an umbrella for the many different groups that are experimenting with new ways of managing agroecosystems. It is as yet far too early to tell which approaches will be successful in the long term; our aim is to help foster greater experimentation and cross-fertilization.

The ecoagriculture concept has prompted several initial reactions. An unexpected degree of interest and support has come from diverse actors involved in sustainable agricultural development, from those conservationists already actively engaged in projects with farming projects and farming communities to policy-makers who have already begun to grapple with the challenges of maintaining, or even augmenting, ecosystem services of all types in regions dominated by agricultural land use. Indeed, we have discovered that far more initiatives to combine agriculture and biodiversity conservation have arisen over the past five years than almost anyone realizes, in all types of land use systems.

At the same time, some conservationists focused on protecting globally unique or highly threatened biodiversity have expressed concern that highlighting the potentials for biodiversity in areas of production agriculture and forestry could undermine policy-maker commitment and financial support for protected areas. We believe that, on the contrary, ecoagriculture can help raise public consciousness of the value of biodiversity conservation and mobilize greater support to expand and protect these reserves, while restoring biodiversity in ecologically degraded areas in the 90 percent of land areas outside the protected areas.

On the other hand, some agriculturalists concerned mainly with raising farm productivity in the developing world have assumed that ecoagriculture is relevant only for the minority of farmers who can find a price premium in niche markets of ecologically conscious consumers. But a careful look at the successful ecoagriculture case studies highlighted in this book will reveal that only a few depend on such premia. We strongly support consumer education and certification strategies, where feasible, to accelerate the transition to ecoagriculture. But ecoagriculture fundamentally depends on technical and institutional innovation in resource management, drawing as much on scientific advances as on indigenous ecological knowledge.

We hope that the global community, particularly those who recently met for the World Summit on Sustainable Development in Johannesburg, will renew their commitment to poverty reduction through agricultural development *and* biodiversity conservation. Mobilizing ecoagriculture strategies offers a key to achieve those mutually reinforcing goals.

Jeffrey A. McNeely and Sara J. Scherr, July 2002

Acknowledgments

The authors would like to express their appreciation to the many generous colleagues who contributed materials used in this book, including Brian Belcher, Steve Franzel, Naoya Furoda, Dennis Garrity, David Kaimowitz, Ted Lefroy, Erik Lichtenberg, Ruth Meinzen-Dick, Peter Neuenschwander, Michele Pena, Ruth Raymond, Colin Rees, Tom Simpson, and Stanley Wood. We benefited greatly from thoughtful comments provided by Weber Amaral, Lukas Brader, Gretchen Daily, Toby Hodgkin, Roger Leakey, Jules Pretty, Jeff Sayer, Meine van Noordjik, and three anonymous reviewers on earlier drafts. For sharing their perspectives on the current state of biodiversity-friendly agriculture, we thank Bruce Boggs, George Boody, Kate Clancy, Randy Curtis, Dana Jackson, Kathy MacKinnon, Gunnars Platais, Barbara Russmore, and all the members of the Katoomba Group, particularly Carl Binning and Ken Chomitz.

Barbara Rose, then the executive director of Future Harvest, provided the inspiration for this paper and valuable feedback all along the way. We drew heavily from materials produced by IUCN and all the Future Harvest Centers: CIAT, CIFOR, CIMMYT, CIP, ICARDA, ICLARM, ICRAF, ICRISAT, IFPRI, IITA, ILRI, IPGRI, IRRI, ISNAR, IWMI, and WARDA. Special thanks go to Kate Sebastian for preparing many of our maps. We also appreciate Population Action International's help in reproducing some of their maps from *Nature's Place* (Cincotta and Engelman 2000). Thanks to Shannon Allen, Ben Dappen, Nathan Dappen, Sandra Gagnon, Joseph McNeely, Michael McNeely, Uday Mohan, Yolanda Palis, Arthur Rosenberg, and Jason Wettstein for their excellent assistance in research and manuscript preparation, and especially to Sue Rallo for her steadfast secretarial support at IUCN. We are grateful to our editors Todd Baldwin, James Nuzum, Cecilia González, and Randy Baldini for their help in finalizing the manuscript for Island Press. Any errors that remain are ours alone.

Jeff thanks Pojanan Suyaphan McNeely for her kind suffering in silence while her husband was occupied in writing this book. Sara warmly thanks her husband, Alan Dappen, and sons, Ben and Nathan Dappen, for their encouraging support of her work on this project and for making sure she still took some time off to enjoy the biodiversity of Virginia.

List of Acronyms

AVHRR	Advanced very high resolution radiometer
CABI	Commonwealth Agricultural Bureau International
CATIE	Centro Agronómico Tropical de Investigación y Enseñanza (Tropical Center for Agricultural Research and Education)
CBD	Convention on Biological Diversity
CGIAR	Consultative Group on International Agricultural Research
CIAT	Centro Internacional de Agricultura Tropical (International Center for Tropical Agriculture)
CIBC	Commonwealth Institute of Biological Control
CIFOR	Center for International Forestry Research
CIMMYT	Centro Internacional de Mejoramiento de Maíz y Trigo (International Maize and Wheat Improvement Center)
CIP	Centro Internacional de la Papa (International Potato Center)
CITES	Convention on International Trade in Endangered Species
CSIRO	Commonwealth Scientific and Industrial Research Organization
DDT	Dichloro-diphenyl-trichloro-ethane
DNA	Deoxyribonucleic acid
EID	Emerging infectious disease
FAO	Food and Agriculture Organization of the United Nations
FSC	Forest Stewardship Council
FUG	Forest User Group

GDP   Gross domestic product

GEF   Global Environment Facility

GISP   Global Invasive Species Programme

GM(O)   Genetically modified (organism)

IBA   Important Bird Areas (Europe)

ICARDA   International Center for Agricultural Research in the Dry Areas

ICDP   Integrated conservation and development project

ICLARM   International Center for Living Aquatic Resources Management

ICRAF   International Centre for Research in Agroforestry

ICRISAT   International Crops Research Institute for the Semi-Arid Tropics

IFAD   International Fund for Agricultural Development

IFPRI   International Food Policy Research Institute

IITA   International Institute of Tropical Agriculture

ILRI   International Livestock Research Institute

IPGRI   International Plant Genetic Resources Institute

IPM   Integrated pest management

IRRI   International Rice Research Institute

ISNAR   International Service for National Agricultural Research

IUCN   World Conservation Union

IWMI   International Water Management Institute

MMC   Marine Management Committee

MVP   Minimum viable population

NGO   Nongovernmental organization

NTFP   Nontimber forest products

OECD   Organization for Economic Cooperation and Development

PAGE   Pilot Assessment of Global Ecosystems

POP   Persistent organic pollutant

TDR   Tradeable development right

UNDP   United Nations Development Programme

UNEP   United Nations Environment Programme

UNESCO   United Nations Educational, Scientific, and Cultural
          Organization

USAID   United States Agency for International Development

WARDA   West Africa Rice Development Association

WCMC   World Conservation Monitoring Centre

WRI   World Resources Institute

WWF   World-Wide Fund for Nature (World Wildlife Fund in the
          United States and Canada)

Part I

The Challenge: Agricultural Intensification, Rural Poverty, and Biodiversity

Many ecologists fear that the world is poised on the brink of the largest wave of wild species extinctions since the dinosaurs disappeared 65 million years ago. If current trends continue, we could lose or greatly reduce populations of 25 percent of the world's species by the middle of this century. Since global awareness of this crisis emerged in the late 1970s, conservationists have focused on protecting endangered species and endangered habitats primarily through the establishment of protected areas. Nearly 10 percent of the earth's land is now officially protected, and land purchases to create private reserves are expanding such areas. Agricultural production areas, by contrast, have been largely ignored by conservationists. These areas were assumed to have habitat conditions so radically modified from their original state that their potential contribution to biodiversity conservation could only be marginal. Permanent croplands were estimated in the early 1980s to account for only 12 percent of global land area, so conservation efforts were understandably focused elsewhere (apart from widespread efforts to limit farmland conversion).

Part I draws on new global data to argue that in this new century food and fiber production—both that produced by agriculture (domesticated crops, livestock, trees, and fish) and harvested from natural systems (forests, grasslands, and fisheries) has come to be the dominant influence on rural habitats outside the arctic, boreal, high mountain, and desert ecoregions. Growing

human populations, increasing demand for food and fiber products, and growing concern about rural poverty mean that agricultural output must necessarily expand for at least several more decades until the rate of human population
growth begins to stabilize, or even begins to decline (as it already has in some
eastern European countries). Adequate growth in supply is by no means
assured, especially in areas where productivity is limited by poor soils, difficult
climates, and insufficient water. Indeed, the World Bank says that billions of
people are at risk of serious food insecurity and deepening poverty.

Future economic development in the poorest and most biodiversity-rich
countries will depend heavily on agriculture and natural resource management
that continue to enhance productivity and adapt to changing conditions. Agriculture will remain economically and socially important. Even industrialized
countries cannot reasonably expect to save biodiversity at the expense of agricultural output and incomes, much less the developing countries of the tropics. Rather, the challenge is to conserve biodiversity while maintaining or
increasing agricultural production. Protected areas will remain a critical element
of any conservation strategy, but this book stresses that it is essential to focus
greater conservation effort on the large areas under agricultural use.

Chapter 1 presents an overview of the issues the book will address. Chapter
2 summarizes the value and global distribution of biodiversity and identifies
some of the places where it is most threatened. In Chapter 3 is an overview of
agricultural production systems, followed by a demonstration of why continuing increases in agricultural production are so important to food security and
economic development in the tropics. Chapter 4 documents the historically
large negative impacts of agricultural expansion and intensification on wild biodiversity; it argues that we can have little hope of conserving wild biodiversity
without major changes in the way we farm.

Introduction

During the twentieth century we humans witnessed momentous economic, social, and technological changes. New technologies such as automobiles, airplanes, container ships, telephones, and computers profoundly affected our way of life, enabling us to escape reliance on local ecosystems and become part of a global economy. Radio, movies, and television transformed the way we related to one another and to the world. Public health systems and education became much more widespread, and material wealth—even in the poorest of countries—reached levels inconceivable at the beginning of the century. Our population more than quadrupled, from 1.4 billion in 1900 to more than 6 billion in 2000. As a species, we had a very good century in many ways.

Our twentieth-century prosperity was fueled in part by a constantly growing supply of food, enabling us not only to feed a rapidly growing population, but also to amass food surpluses on a scale never before reached. Based on improved seeds, widespread use of agricultural chemicals, modern farm machinery, and better transportation systems, agricultural production soared. In the past decade alone, production of cereal crops increased by 17 percent, roots and tubers by 13 percent, meat by 46 percent, and marine fish by 17 percent (World Resources Institute 2000). With such impressive gains on so many fronts, why should we worry about the twenty-first century?

First, although more people are consuming more food than ever before, inequity is increasing as well: some parts of the world suffer from growing overconsumption while others go hungry. The World Bank estimates that some 800 million people remain undernourished, in large part because they cannot access the food that is produced. That number is likely to grow because the world's population increases by 75 to 85 million people each year. Some experts suggest that in thirty years we will need at least 50 to 60 percent more food than we produce now, in order to meet global food demand and enjoy at least a modest degree of greater affluence. If that food is to be accessible to the rural poor, then much of it must be produced where they live, and in ways that increase both their consumption and income. Yet food-producing systems throughout the world are already stressed by eroding soils, declining freshwater reserves, declining fish populations, deforestation, desertification, natural disasters, and global climate change. These and various other factors are making it increasingly difficult to maintain, much less increase, food production in many areas of the world.

What is more, the impressive gains for our species have often come at the expense of other species with whom we share our planet. The main victim of our affluence has been wild biodiversity—the nondomesticated portion of our planet's wealth of genes, species, and ecosystems. Agricultural production has converted highly diverse natural ecosystems into greatly oversimplified ecosystems, led to pollution of soils and waterways, and hastened the spread of invasive alien species. According to Heywood and Watson (1995), "overwhelming evidence leads to the conclusion that modern commercial agriculture has had a direct negative impact on biodiversity at all levels: ecosystem, species, and genetic; and on natural and domestic diversity."

While major investments continue to improve agricultural productivity in centers of surplus commercial production, the needs of the rural poor tend to be ignored. As a result, the poor struggle to survive, managing their resources to meet immediate needs rather than invest in a more secure future. Many of these poor people live in areas remote from modern agricultural development but close to habitats supporting the greatest wild biodiversity. Often they have little choice but to exploit these habitats for survival.

Without urgent action to develop the right kind of agriculture, wild biodiversity will be further threatened. The resulting destruction of natural habitats will deprive both local people and the global community of important benefits such as food, fodder, fuel, construction materials, medicines, and genetic resources, as well as services such as watershed protection, clean air and water, protection against floods and storms, soil formation, and even human inspiration.

These threats to biodiversity pose a major dilemma for modern society. On the one hand, modern intensive agriculture has made it possible for the expanding human population to eat more food. On the other hand, agriculture is now spreading into the remotest parts of the world, often in destructive forms that further reduce wild biodiversity and undermine the sustainability of the global food production system. At the same time, reducing biodiversity and simplifying ecosystems can undermine local livelihoods by destabilizing ecosystem services. Recent mudslides in several Latin American countries, floods in Bangladesh, and droughts in southern Africa are all "natural" phenomenon made into a disaster for local people due at least in part to loss of biodiversity.

This situation has led many in the environmental community and the general public to promote the establishment of protected areas where human use—in particular agricultural use—is supposed to be greatly restricted. While such management measures clearly are needed to preserve many types of wild biodiversity, they face many challenges. Some centers of the greatest or most valued wild biodiversity are being surrounded by areas of intensive agricultural production and high rural population densities. In some areas, large human populations preclude the establishment of extensive reserves, so the protected areas tend to be too small to support viable populations of the species they are designed to protect. In these human-dominated ecosystems, conservation action in isolated protected areas is doomed to fail, unless fundamental changes also take place in the adjacent agricultural landscape. Moreover, some types of wild biodiversity, such as some species of birds and butterflies, actually thrive best in farmed and populated landscapes. Farming is a practice that extends at least 10,000 years back into human history, and many species of plants and animals have evolved in concert with the development of agriculture. Some species of large mammals (especially wild cattle in Asia) may even depend on shifting cultivation (Wharton 1968).

Aggressive efforts to conserve wild biodiversity have sometimes reduced the livelihood security of rural people, especially the poor in developing countries (Pimbert and Toledo 1994). But this need not be the case (McNeely 1999). Rural populations historically have established conservation practices to protect environmental services important to their own food production, water supply, and spiritual values (see, for example, Western and Wright 1994; Singh et al. 2000). Examples from this book will show that managing biodiversity through a combination of conservation measures and improved and diversified agricultural systems can increase incomes and household nutrition, reduce livelihood risks, and provide collateral benefits such as increased freshwater reserves and fewer mudslides after heavy rains.

Thus new models for biodiversity conservation need to be developed, involving effective links among the fields of farmers, the pastures of ranchers, the managed forests of foresters, and the protected areas managed especially for wild biodiversity. Conservation options are available besides just "locking away" resources on which the poor depend for their survival and assets that low-income countries could use to promote development and national food security. Agricultural landscapes can be designed more creatively to take the needs of local people into account while pursuing biodiversity objectives.

Ecoagriculture

A central challenge of the twenty-first century, then, is to achieve biodiversity conservation and agricultural production goals at the same time—and, in many cases, in the same space. In this book the management of landscapes for both the production of food and the conservation of ecosystem services, in particular wild biodiversity, is referred to as *ecoagriculture*. For a start, improved natural resource management and technological breakthroughs in agriculture and resource use is essential to enhance our ability to manage biodiversity well. Genetic improvements in the major agricultural crops that feed the world will continue to be essential for maintaining and increasing productivity. But a much wider range of genetic, technological, environmental management, and policy innovations must be developed to support wild biodiversity in the world's bread baskets and rice bowls as well as in the extensive areas where food production is more difficult.

Diverse approaches to make agriculture more sustainable, while also more productive, are flowering around the world; many of these reduce the negative effects of farming on wild species and habitats. Such approaches need to be integrated more intentionally with conservation objectives, particularly in biodiversity "hotspots" (Myers 1988) and areas where the livelihood of the poor depends on ecosystem rehabilitation. New approaches to agricultural production must be developed that complement natural environments, enhance ecosystem functions, and improve rural livelihoods. While trade-offs between agricultural productivity and biodiversity conservation often seem stark, some surprising and exciting opportunities exist for complementarity. Local farmers and institutions, such as universities and agricultural research centers, are leading the way through active experimentation and adaptation of existing knowledge. But more targeted research on ecoagriculture is needed, and such research must be considered a global priority if major advances are to occur. Environmental and agricultural researchers must learn to work closely together to resolve existing conflicts between natural biodiversity and agricultural production in different ecoregions and under different management systems.

This book examines some of the current linkages between wild biodiversity and agriculture. It suggests strategies for improving agriculture while maintaining or enhancing wild biodiversity, assesses dozens of systems where this is already being done, and describes how research and policy action can contribute to conserving wild biodiversity. The book is structured in three parts. The first part describes the challenge of reconciling conservation and agricultural goals in areas important for both. The second part discusses the ecoagriculture approach and presents diverse case studies illustrating key strategies. The third part explores how policies, markets, and institutions can be re-shaped to support ecoagriculture in areas that are hotspots for both bio-diversity and food security.

The emphasis here is on tropical regions of the developing world, where increased agricultural productivity is most vital for food security, poverty reduction, and sustainable development, and where so much of the world's wild biodiversity is threatened. But the book also highlights lessons learned in developed countries (for example, California Wildlife Coalition 2002) where these are of wider relevance. While profitable ecoagriculture systems can and must be developed for large-scale commercial farming enterprises that are operating in areas of threatened biodiversity, most examples in this book emphasize strategies for small-scale, low-income farmers involved in commer-cial or subsistence production.

The biodiversity of domesticated crop and livestock species, and the com-plex of wild species that directly support agriculture (such as wild pollinators), is also critically important to future prosperity and is also suffering from numerous threats. This book will address how increased agricultural diversity can enhance habitat for wild species, and how strategies to enhance wild bio-diversity can build on the beneficial effects of many wild species for agricul-tural production and sustainability. However, it will not address the topic of genetic diversity of domesticated agricultural species, which has recently begun to receive wide attention from ministries of agriculture and the many agencies that support them (Gemmill 2002).

As the distinguished British ecologist Norman Myers pointed out, "It is in the common interest of both agriculture and the natural world that a mutu-ally supportive relationship be developed between them. Production of food need not destroy the wild ecosystems of the world and their wealth of bio-logical diversity. And preservation of wild ecosystems does not pose a threat to humanity feeding itself. In fact, just the opposite is true. Sensible use of nature, which includes substantially increased nature conservation efforts, is essential to feed the planet. . . . Nature equals food. Without wild places, we cannot hope to have food on our tables" (Myers 1987). And without healthy agricul-ture, we cannot expect nature to prosper.

Chapter 2

Wild Biodiversity under Threat

The variety of life on earth includes the millions of animals, plants, and microorganisms, the genes they contain, and the complex ecosystems they help form. These plants, animals, and microorganisms, evolving over hundreds of millions of years, have made our planet fit for the life we know today. This chapter provides a brief discussion of the value of wild biodiversity, its geography in relation to human populations, and the trends that reveal globally significant threats.

Definitions of Biodiversity

In the United Nations Convention on Biological Diversity (Box 2.1), governments agreed on an "official" definition of biological diversity (sometimes shortened to "biodiversity"). It is "the variability among living organisms from all sources including, inter alia, terrestrial, marine, and other aquatic ecosystems and the ecological complexes of which they are part; this includes diversity within species, between species, and of ecosystems." But this simple definition hides a much more complex picture, including diversity of genes, populations, landscapes, and biomes (Table 2.1).

Diversity is a characteristic of all living organisms, and thus it is just as relevant to agricultural crops as to wildlife in remote wildernesses. "Wild" biodiversity does not mean "pristine" or untouched by humans, because virtually all ecosystems have been profoundly affected by people. For example, many of the tree species now dominant in the mature vegetation of tropical areas were, and still are, the same species that were protected, spared, or planted in land

9

Box 2.1. Major International Conventions Relevant to Biodiversity

Many international conventions have been negotiated under the auspices of the United Nations that are highly relevant to the subject of this book. The major relevant conventions are summarized below.

Convention on Biological Diversity (CBD). The CBD was agreed at the Earth Summit at Rio de Janeiro in 1992, entered into force in 1993, and now has over 180 parties. It has three interlinked objectives: to conserve biological diversity; to utilize biological resources sustainably; and to share equitably the benefits arising from the use of genetic resources. The CBD has included in its program of work some specific activities that relate to agriculture and that are supportive of ecoagriculture. (http://www.biodiv.org)

Framework Convention on Climate Change (FCCC). The FCCC was also agreed at the Earth Summit. It is designed to limit human-induced disturbances to the global climate system by seeking to achieve a stable level of greenhouse gases in the atmosphere. Under the Kyoto Protocol of the FCCC, negotiated in 1997, governments are expected to make major investments in sequestering carbon and carrying out other activities that will mediate the impacts of climate change and help adapt to it. Many of these activities involve agriculture. (http://www.unfccc.de)

The Convention on Wetlands of International Importance (Ramsar Convention). Agreed in 1971, the Ramsar Convention promotes the conservation and wise use of wetlands through national action and international cooperation. Each party is required to designate at least one wetland site, which is expected to contain populations of plants and animals important for maintaining biological diversity. These sites may include local and indigenous communities, and the Ramsar Convention has placed considerable emphasis on sustainable development. (http://www.ramsar.org)

cleared for crops (Gomez-Pompa and Kaus 1992). Furthermore, virtually all tropical forests have been cleared at least once and probably several times over the past 10,000 years (Spencer 1966), and the temperate forests are likely to have been similarly treated (at least in areas accessible to people). As a result, the current pattern of habitats reflects complex interactions among physical, biological, and social forces over time. The landscapes that we see today form an ever-changing mosaic of unmanaged and managed patches of habitat that vary in size, shape, content, and arrangement in accordance with the history of resource exploitation by people (Redman 1999).

Table 2.1. Components of Biological Diversity

Components	Structure (physical organization of elements)	Composition (identity and variety of elements)	Function (ecological, evolutionary processes among elements)
LANDSCAPE (regional mosaic of land uses, land forms, and ecosystem types)	Areas of different habitat patches, perimeter area relation of patches, inter-spatial linkages	Identity, distribution, and proportion of habitat types	Patch persistence, inter-patch flow of energy, species, resources
ECOSYSTEM (interactions between members of a biological community and their physical environment)	Vegetation biomass, soil structural properties	Biogeochemical standing stock	Biogeochemical and hydrological cycling
COMMUNITY (guilds, functional groups, and patch types occurring in the same area and strongly interacting through trophic and spatial biotic relationships)	Vegetation patterns and organization of food chains	Relative abundance of species and guilds	Flow between patch types, disturbance regimes (e.g., fire, flood), successional processes, species interaction
SPECIES/POPULATIONS (variety of living species and component populations)	Population age structure, species abundance, and distribution	Particular species present	Demographic processes
GENETIC VARIABILITY WITHIN SPECIES (variations in genes within particular species, subspecies, or populations)	Genetic distance between populations in different patches	Alleles present, in what proportion	Gene flow, genetic drift, or loss of allelic diversity in isolated populations

Source: Based on Chapter 2 of Putz et al. (2000).

Value of Wild Biodiversity

We humans find beauty and pleasure in the diversity of nature. This diversity is also a foundation for human creativity and a subject for study. Many argue that biological diversity should be conserved as a matter of principle, because all species deserve respect regardless of their use to humanity, and because they are all components of our life-support system. But others need to see economic benefit in order to appreciate such a resource. Biodiversity plays a critical utilitarian role as well, supporting rural livelihoods, agricultural production, and ecosystem functions. All societies, urban and rural, industrial and nonindustrial, draw on a wide array of ecosystems, species, and genetic variants to meet their ever-changing needs. Biodiversity is the source of all biological wealth, supplying all of our food, much of our raw materials, and a wide range of goods and services, plus genetic materials for agriculture, medicine, and industry. The combined commercial worth of these genetic materials is estimated at U.S.\$500–800 billion per year (ten Kate and Laird 1999). People spend additional billions of dollars to appreciate nature through recreation and tourism.

While the value of genetic resources—nature's goods—is substantial, the value of nature's services is far higher. Because natural ecosystems and associated wild biodiversity help maintain the chemical balance of the earth's atmosphere, protect watersheds, renew soils, cycle nutrients, and provide many other ecosystem functions essential to human welfare, they are in a sense priceless because they are essential to life and cannot be replaced. The relationship between species diversity and ecosystem functioning remains a central issue on the global biodiversity agenda. For example, it is not yet known to what extent ecosystem structure, including the network of species interactions, affects how the system will respond to the climate changes expected to result from increased levels of carbon dioxide produced from burning fossil fuels. This limits the ability to predict which species or functional groups will benefit or suffer from such changes and to understand the underlying mechanisms.

Value for Rural Peoples

Wild biodiversity plays an important direct role in rural livelihoods, through wild products and critical environmental services such as waste treatment, watershed protection, and topsoil formation and protection (Salafsky and Wollenberg 2000). Rural people may depend heavily on wild plants, animals, and aquatic resources as enhancements to their diets; as emergency provisions in famines or delayed harvests; as sources of income, fuel, and fodder for domestic livestock; and as raw materials to be used in medicines, tools, and construction (Scherr, White, and Kaimowitz 2002). Wild animals account for 75 percent of protein consumed in the Congo; in Botswana fifty different species account for 40 percent of protein consumption. Wild animals also play important symbolic, ritualistic, and religious roles in rural societies (McNeely and Wachtel 1988) and serve as sources of inspiration for dances, stories, and other art forms.

Wild species are the source of traditional medicines basic to the health care of about 80 percent of people in developing countries. Over 5,000 species of plants and animals are used for medicinal purposes in China alone. By maintaining stable pest-predator relationships, diversity can help control infectious diseases and parasites (Grifo and Rosenthal 1997). Protection of natural forests and other wild vegetation can secure water quality, greatly reduce soil erosion, and reduce health threats from forest fires and flooding (Box 2.2).

Box 2.2. Biodiversity and the Functioning of Ecosystems

Natural ecosystems provide numerous services, in addition to biodiversity, that are essential to human existence: cycling and movement of nutrients, air and water purification, decomposition and detoxification of organic wastes, mitigation of floods and droughts, renewal of soil fertility, pollination of crops and natural vegetation, dispersal of seeds, protection from the sun's harmful ultraviolet rays, stabilization of the climate, moderation of weather extremes and their impacts, and the aesthetic beauty and intellectual stimulation that lift the human spirit (Daily 1997).

A hot topic for debate among ecologists is the relationship between biodiversity and these other ecosystem functions, especially ecosystem productivity, stability, and adaptability (Loreau et al. 2001). What happens to these functions as ecosystems are simplified? Research indicates that increased species diversity generally provides more opportunities for species interactions, which in turn improves the rates of resource use that govern ecosystem efficiency and productivity. The implication is that reducing species diversity can have significant adverse impacts on the functioning of ecosystems. For example, research has found that higher-diversity grassland systems typically outperform the best monocultures, achieving both greater productivity and more storage of carbon (Tilman et al. 2001).

That said, ecosystems have a great capacity to respond to disturbance, and many disturbed ecosystems may have more species than older and less disturbed systems under the same general conditions. It appears that when many species are present, ecosystem processes may be relatively insensitive to considerable variation. Using analogies from reliability engineering, Naeem (1998) suggests that reliability always increases as redundant components are added to a system; it helps to have a safe car, but seat belts still save lives. Similarly, species redundancy,

(continues)

Box 2.2. *Continued*

where several species provide similar functions, helps the system to function reliably and provide goods and services to people in a predictable manner. Even if one or several species are hit by a disease, for example, others are available to fill in. In fluctuating environments, biodiversity may insure ecosystems against declines in their functioning because many species provide guarantees that some will keep functioning even if others fail (Yachi and Loreau 1999).

Much more long-term research is needed to understand the interactions among species that affect the structure and dynamics of ecosystems. Programs in the relatively simple desert shrubland and pinyon-juniper woodland in the United States have shown that some environmental disruptions can lead to wholesale reorganization of ecosystems because the disturbance exceeded the ecological tolerances of dominant or keystone species that did not have an ecologically similar species to take their place. Other changes may be buffered because other species perform complementary functions in the system (Brown et al. 2001). Research on lakes, coral reefs, forests, and arid lands indicates that even gradual changes in climate, the flow of nutrients, extraction of natural resources, and habitat fragmentation can lead to sudden drastic switches from one kind of ecosystem to one of quite a different character (from a forest to scrubland or from a productive lake to a depleted one), and often one that is far less useful to people and the rest of nature (Scheffer et al. 2001). These findings have profound implications for the way agricultural systems should be managed. While many different factors can lead to ecosystem shifts, one critical factor is the loss of resilience due to declining biodiversity. Thus biodiversity helps farmers to hedge bets against uncertainty and adapt to changing conditions.

Value for Agriculture

Wild biodiversity contributes significantly to the productivity and sustainability of agriculture, forestry, and fisheries. By feeding on leaves, fruits, and seeds, animals greatly influence the composition and structure of natural vegetation, affect the reproductive success of plants, make soils more fertile, and regulate populations of pests. Biodiversity creates the conditions that maintain healthy ecosystems. Various species pollinate plants (birds, bees, bats—see Box 2.3); decompose wastes (earthworms, microorganisms, dung beetles, vultures, crows); disperse seeds (birds, primates, squirrels, ants, fish); and maintain a sort of balance among species through predation and grazing. Microorganisms in the soil may have a total biomass that adds up to as much as 1.5 tons per hectare. Soil faunal groups, including arthropods, earthworms, nematodes, and mollusks, facilitate movement of air and water within the soil matrix, regulate nutrient cycling, and build soil (Pfiffner 2000). Wild relatives of domestic plants and animals may provide valuable genetic material for crop and livestock

---

Box 2.3. Pollination, Biodiversity, and Agriculture

Pollination and cross-pollination help flowering plants reproduce. While wind-blown pollen works for some, for many other plants an animal pollinator is required. Bats, bees, beetles, and other insects are the principal pollinators of fruit trees; most important oil crops; coffee; coconut; and major staple food crops, including potato, cassava, yams, sweet potato, taro, and beans (Prescott-Allen and Prescott-Allen 1990).

Worldwide declines in populations of pollinators for economic plants are threatening both the yields of major food crops and the biodiversity of wild plants. Entire groups of pollinators, including indigenous birds, bats, and insects, are disappearing from the Pacific islands (Cox and Elmqvist 2000). A quarter of North America's wild and domestic honeybees have disappeared since 1988, primarily due to an epidemic of mites that prey on the bees. The cost to American farmers of the declining honeybee population is U.S.$5.7 billion per year (Nabhan and Buchmann 1997). Declines in pollinators, including 1,200 wild vertebrate pollinators, are leading to depressed yields of blueberries and cherries in Canada, cashew nuts in Borneo, Brazil nuts in South America, and pumpkins in the United States. Animal pollinators themselves help to maintain biodiversity, allowing numerous species of flowering plants to coexist rather than letting a few species dominate a flora by outcompeting other plants. Many of the flowers that are most beloved by gardeners have evolved particular colorations and patterns that attract various kinds of pollinators.

---

improvement (see Box 2.4). Key sources of DNA are banked in the cells of wild ancestral species of crops and in their close relatives.

Distribution of Wild Species Biodiversity

Wild species are not evenly distributed across geographic areas. Biodiversity, however, is important wherever it is found. The relatively few species found in low biodiversity systems—such as arid lands—are each particularly important for the functioning of those ecosystems, and therefore for people who live there; with little redundancy in such systems, one could argue, each individual species is even more important than species in ecosystems with great species diversity. Relatively small percentages of species live in extreme environments such as sand dunes, hot springs, and deep oceans, but most of these species are found nowhere else, making them particularly valuable because of their distinctive features. The tundras and open seas also have relatively modest numbers of species, though the populations of these species may be vast. Higher concentrations of species are found in grasslands and coniferous forests in temperate latitudes; even more inhabit tropical savannas, marshes and swamps,

Box 2.4. Potential Contribution of Wild Relatives to Domestic Livestock Improvement

Asia is home to at least twenty-two wild species that are either close relatives to domestic livestock (in the same genus) or are raised in captivity in a domestic form of the same species (elephant, pig, camel, water buffalo, yak, Bali cattle). The main wild populations of most of these species are found within protected areas. The World Conservation Union (IUCN) considers fourteen of the twenty-two species to be threatened. But modern techniques of embryo trans- plants, embryo splitting, and artificial insemination are now sufficiently well developed to enable even threatened wild relatives to make a real contribution to the livestock industry. Because embryos are free of many diseases, they can be shipped between nations without expensive quarantine precautions, and cryopreservation methods enable tiny bundles of cells to be sent by air in small insulated containers at little expense (National Research Council 1983). Genetic engineering has made it possible to introduce new genes into the germ line of an animal, and thereby produce proteins outside their normal environ- ment, significantly improving both daily weight gain and feed efficiency. But these have also led to a higher incidence of gastric ulcers, arthritis, heart enlargement, dermatitis, and renal disease (Pursel et al. 1989). Genes from wild relatives—which are often hardier than domestic breeds—may provide the raw material for dealing more effectively with such challenges, even using conven- tional breeding techniques.

rivers and lakes, coastal tidal zones, and nutrient-rich marine shoals. Some semiarid areas (such as western Australia and the miombo woodlands of Africa described in Example 11 in Chapter 6) and Mediterranean ecosystems (such as the South African Cape or the Balkans) have great species diversity. Most of the world's species, though, are found in the humid tropics, frequently in countries with the least financial, technical, and institutional means to con- serve biodiversity (Table 2.2).

Though comprising only 2.3 percent of the surface of the earth, lowland and montane tropical rainforests probably hold more than 65 percent of all ter- restrial species (Wilson 1992). Tropical coral reefs—sometimes called the "rain- forests of the oceans" for their species richness—come a close second. The 600,000 square kilometers of reefs comprise 0.1 percent of the earth's total surface, but they may hold as many as 950,000 species (though only 10 per- cent have been described) (Burke et al. 2000).

The tropical forests of Amazonia may be the habitats that receive the greatest share of attention from conservationists. But the vast majority—more than 70 percent—of South America's threatened species of birds are found in montane forests and high altitude wetlands in the northern and central Andes,

Table 2.2. The World's Most Species-rich Countries

Country	Mammal Species		Country	Mammal Species
1. Indonesia	515	6.	Peru	361
2. Mexico	449	7.	Colombia	359
3. Brazil	428	8.	India	350
4. Democratic		9.	Uganda	311
Rep. of Congo	409	10.	Tanzania	310
5. China	394			

Country	Bird Species		Country	Bird Species
1. Colombia	1,721	6.	Venezuela	1,275
2. Peru	1,701	7.	Bolivia	±1,250
3. Brazil	1,622	8.	India	1,200
4. Indonesia	1,519	9.	Malaysia	±1,200
5. Ecuador	1,447	10.	China	1,195

Country	Reptile Species		Country	Reptile Species
1. Mexico	717	6.	Colombia	383
2. Australia	686	7.	Ecuador	345
3. Indonesia	±600	8.	Peru	297
4. Brazil	467	9.	Malaysia	294
5. India	453	10.	Thailand/Papua New Guinea	282

Country	Amphibian Species		Country	Amphibian Species
1. Brazil	516	6.	China	265
2. Colombia	407	7.	Peru	251
3. Ecuador	358	8.	Democratic	
4. Mexico	282		Rep. of Congo	216
5. Indonesia	270	9.	United States	205
		10.	Venezuela/Australia	197

Country	Flowering Plant Species		Country	Flowering Plant Species
1. Brazil	55,000	6.	South Africa	21,000
2. Colombia	45,000	7.	Indonesia	20,000
3. China	27,000	8.	Venezuela	20,000
4. Mexico	25,000	9.	Peru	20,000
5. Australia	23,000	10.	Russian Federation (former USSR)	20,000

Source: McNeely et al. 1990.

deciduous and semiarid Pacific woodlands along the coast from western Colombia to northern Chile, and the native grasslands and riverine forests of southern and eastern Brazil. These areas are particularly threatened by expanding agriculture (Myers et al. 2000).

Public attention often focuses on charismatic species such as giant pandas, tigers, rhinos, the great whales, parrots, and cranes. But soil invertebrates and microorganisms invisible to the naked eye may be even more important for the functioning of ecosystems. Though the biodiversity of soils is remarkably rich, science has yet to describe an estimated 98 percent of the species living in soil. For six taxonomic groups (including pot worms, centipedes, and millipedes), 100 percent of known species are found in the soil; for another eight taxonomic groups (including bacteria, fungi, mites, termites, ants, sowbugs, earthworms, and flies), half or more of known species are found in the soil (Wall and Moore 1999). These tiny soil organisms provide critical links between the atmospheric, terrestrial, and aquatic realms. Unfortunately, current knowledge is insufficient to identify geographical patterns of diversity, ranges of species of microorganisms, the kinds of species that may be particularly important to ecosystem processes, and the major threats to these organisms.

A notable and disturbing characteristic of the geographic distribution of wild biodiversity is that it corresponds strongly with human population density. More than 1.1 billion people live within the twenty-five global biodiversity hotspots that ecologists describe as the most threatened species-rich regions on earth (Myers et al. 2000). Population density in seven of these hotspots reached more than 100 people per square kilometer in 1995 (see Table 2.3). On the other extreme, only about 75 million people, or 1.3 percent of the world's population, live within the three major tropical wilderness areas: the Upper Amazonia and Guyana Shield, the Congo River Basin, and the New Guinea–Melanesia complex of islands. Together these cover about 6 percent of the earth's surface (Cincotta and Engelman 2000), as shown in Map 2.1.

But wild biodiversity is not only under pressure in the species-rich hot spots or the tropical wilderness areas. Densely populated South and East Asia—dominated by India and China—have converted extensive areas of habitat to agriculture and thus have more limited areas devoted to wild biodiversity.

Protected Areas

Over the past 100 years or so, but especially over the past few decades, most countries have established protected areas designed to achieve various conservation objectives. As shown in Figure 2.1, some 8 percent of all land is now legally protected by sites meeting the criteria of the World Conservation Union (IUCN), although many declared protected areas are de facto occupied

Table 2.3. Population Density and Habitat Conservation in the 25 Global Hotspots

Hotspot	Hotspot Area (thousands of sq. km)	Human Population, 1995 (thousands)	Population Density, 1995 (per sq. km)	Population Growth Rate, 1995–2000 (percent per year)	Extent of Original Vegetation (thousands of sq. km)	Extent Remaining (percent of original)	Extent Protected (percent of original)
1 Tropical Andes	1,415	57,920	40	2.8	1,258	25	6.3
2 Mesoamerica	1,099	61,060	56	2.2	1,155	20	12.0
3 Caribbean	264	38,780	136	1.2	264	11	15.6
4 Atlantic Forest Region	824	65,050	79	1.7	1,228	8	2.7
5 Chocó-Darién-Western Ecuador	134	5,930	44	3.2	261	24	6.3
6 Brazilian Cerrado	2,160	14,370	7	2.4	1,783	20	1.2
7 Central Chile	320	9,710	29	1.4	300	30	3.1
8 California Floristic Province	236	25,360	108	1.2	324	25	9.7
9 Madagascar and Indian Ocean Islands	587	15,450	26	2.7	594	10	1.9
10 Eastern Arc Mts and Coastal Forests	142	7,070	50	2.2	30	7	16.9
11 Guinean Forests of West Africa	660	68,290	104	2.7	1,265	10	1.6
12 Cape Floristic Province	82	3,480	42	2.0	74	24	19.0
13 Succulent Karoo	193	460	3	1.9	112	27	2.1
14 Mediterranean Basin	1,556	174,460	111	1.3	2,362	5	1.8
15 Caucasus	184	13,940	76	−0.3	500	10	2.8
16 Sundaland	1,500	180,490	121	2.1	1,600	8	5.6
17 Wallacea	341	18,260	54	1.9	347	15	5.9
18 Philippines	293	61,790	198	2.1	301	8	1.3
19 Indo-Burma	2,313	224,920	98	1.5	2,060	5	7.8
20 Mountains of South-Central China	469	12,830	25	1.5	800	8	2.1
21 Western Ghats and Sri Lanka	136	46,810	341	1.4	183	7	10.4
22 Southwest Australia	107	1,440	13	1.7	310	11	10.8
23 New Caledonia	16	140	8	2.1	19	28	2.8
24 New Zealand	260	2,740	11	1.0	271	22	19.2
25 Polynesia/Micronesia	46	2,900	58	1.3	46	22	10.7

Sources: Cincotta and Engelman 2000; global biodiversity hotspots defined by Conservation International.

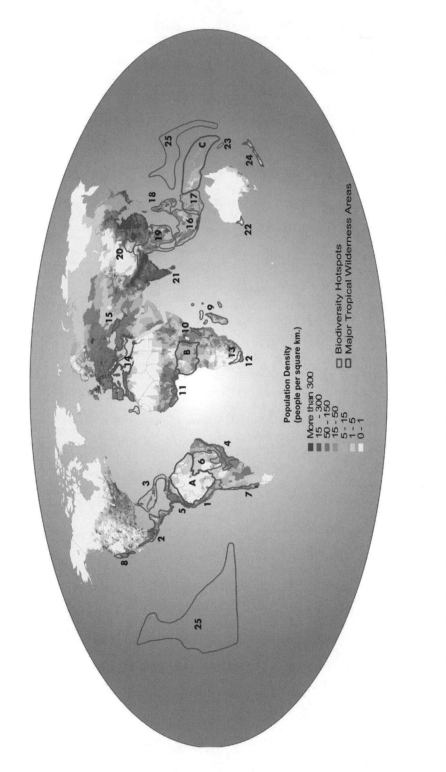

Map 2.1. Population in the Global Biodiversity Hotspots, 1995. *Source:* Cincotta and Engelman, 2000.

Square kilometers

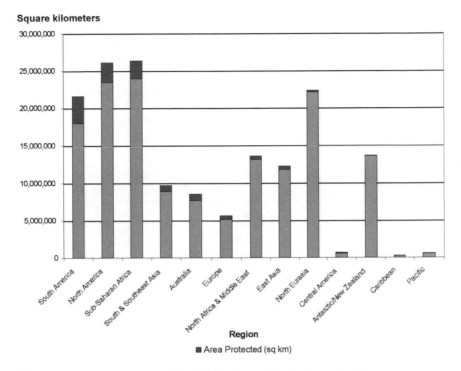

Figure 2.1. Protected Areas of the World. *Source:* McNeely et al. 1994.
Note: Includes sites meeting IUCN criteria only.

or used fairly intensively by local people. On the other hand, many more areas are too small to meet IUCN's size criterion of 1,000 hectares but are still valuable for biodiversity. The latest compilation of global statistics on protected areas lists some 44,197 sites covering 13,279,127 square kilometers, nearly 10 percent of the earth's land surface (World Conservation Monitoring Centre 2000).

The overlap of protected areas with agricultural areas (defined as more than 30 percent of land cover under crops or planted pastures) is notable, as shown in Map 2.2. Agriculture is largely absent in 45 percent of the world's major protected areas, which account for 71 percent of the total area protected. But agriculture is practiced in the other reserves, which account for nearly 29 percent of globally protected areas. In over 17 percent of the global area in protected reserves, agriculture occupies more than 30 percent of the land. Farming activity almost certainly has a marked influence on adjacent nonfarmed parts of those reserves (Sebastian 2001).

Map 2.2 also shows that many protected areas are islands of wild habitat within a sea of agricultural activity. In Central America many small protected

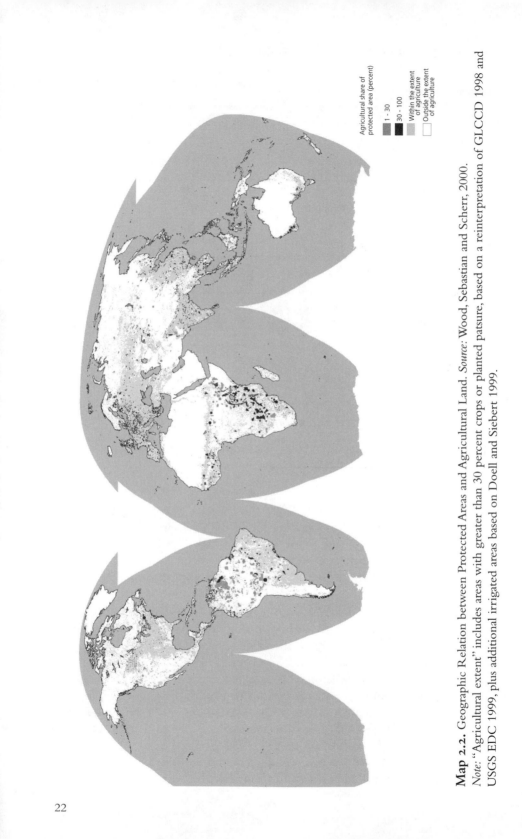

Map 2.2. Geographic Relation between Protected Areas and Agricultural Land. *Source:* Wood, Sebastian and Scherr, 2000.
Note: "Agricultural extent" includes areas with greater than 30 percent crops or planted pasture, based on a reinterpretation of GLCCD 1998 and USGS EDC 1999, plus additional irrigated areas based on Doell and Siebert 1999.

Agricultural share of
protected area (percent)

1 - 30

30 - 100

Within the extent
of agriculture
Outside the extent
of agriculture

areas are interspersed with agricultural lands. Because of limited total land area, agricultural pressures on the protected areas are intense and likely to increase. Many protected areas in South America are located along the agricultural frontier of the Amazon. Whether they can be effectively protected depends greatly on the dynamics of frontier settlement and employment opportunities in other areas. In Africa and in South, Southeast, and East Asia the overlap of protected areas with agricultural lands is striking. Protecting these areas effectively seems especially challenging.

On the other hand, Java—one of the most densely populated parts of the world, with 120 million people packed onto an island the size of New York State—still has an outstanding system of twenty-two national parks and nature reserves covering nearly 650,000 hectares. This demonstrates that even the densest human populations can conserve wild habitats, and can justify doing so even in times of economic hardship. Some Javanese farmlands also have great agrobiodiversity, with one survey of 351 Javanese home gardens recording 607 plant species (Dover and Talbot 1987).

Status of and Trends in Wild Species Populations

Trends in wild biodiversity suggest troubling patterns of species loss, population loss, and extinction. About 1.5 million species have been named, described, and maintained in museum collections (Wilson 1992). The most current scientific consensus of total living species ranges between 7 and 15 million, though some estimates approach 100 million. Mammals and all other vertebrates (animals with backbones) account for only around 40,000 species, or fewer than 1 percent of all species. Plant life, consisting of flowering plants, conifers, mosses, and ferns, probably makes up fewer than 5 percent of all species, or about 300,000 in total. Arthropods—a group that includes the insects, crustaceans, spiders, mites, centipedes, and their relatives—are by far the most diverse and widespread of all organisms, though they are not well studied. They are thought to make up at least 40 percent of living species. A world of discovery remains in invertebrates (mollusks, echinoderms, jellyfish, worms) and microorganisms (fungi, protozoa, algae, bacteria, viruses), likely to number in the millions of species (see Box 2.5).

Historical Species Extinction

Life on our planet has gone through a tumultuous history, experiencing both flowerings and mass extinctions. Scientists have identified at least six major extinction episodes, and it appears that we are in the midst of the sixth extinction at present. This last episode has involved humans in three major "waves" of species extinction (Martin and Klein 1994). The first was in prehistoric

---

Box 2.5. Knowledge Gaps on the Status and Distribution of Biodiversity: Example from Sri Lanka

Ignorance about the status and distribution of many species is a hallmark of our current understanding of biodiversity, because so little fieldwork has been done on the issue in most parts of the tropics. For example, an IUCN project team conducting surveys in Sri Lankan forests in 1993 made several remarkable discoveries. The team found a population of rock balsam (*Impatiens repens*), an endemic creeping herb that had not been seen for 136 years; *Semecarpus pseudoemarginata,* an endemic member of the mango family that was known previously from a single specimen collected in 1853; two species previously known only from India—*Carissa inermis* and *Eriolaena hookeriana*—and a new species of ebony (*Diospyros*), a tree of considerable economic importance. These discoveries, on an island that is already well studied, indicate that basic survey and inventory work is still needed (Jayasuriya 1995). Wild species that are still uncollected may be lost, along with their unknown potential for future food production and environmental services.

---

times and resulted primarily from human overhunting as people moved into new regions for the first time. Within a few thousand years after humans first arrived on the Australian continent some 50,000 years ago, the continent lost 86 percent of its large marsupial mammals, plus some egg-laying mammals and giant lizards. About 12,000 years ago, human hunters migrated from Asia to North America across the land bridge that then existed. Within a thousand years, North America lost at least fifty-seven species of large mammals—73 percent of all large mammals on the continent. These included horses and camels, giant sloths, glyptodonts (which resembled giant armadillos), mammoths, and mastodons. Their remains line the flooring and garbage pits of Pleistocene human settlements. Losses in Europe were roughly comparable, but Africa and Asia fared somewhat better.

The second wave of extinctions—notable for the loss of bird species—was associated with human settlements of oceanic islands within the past 1,000 years. Nearly all the diverse and often extraordinary bird species of New Zealand were lost by the mid-1700s, a result of overhunting and the introduction of pigs, dogs, and rats. More than half a million skeletons of the huge flightless birds known as moas have been found in ancient Maori settlements in New Zealand. Similar processes occurred in Madagascar, Cyprus, the Azores, the Caribbean islands, and Polynesia, where more than 1,000 bird species—over 10 percent of the birds then alive on earth—became extinct after people first arrived (Flannery 1995).

Current and Projected Species Extinctions

The third major extinction wave is taking place now. The basic facts are reasonably clear. About two-thirds of all terrestrial species are found in tropical forests, which covered 14 to 18 million square kilometers some 400 years ago but have now been reduced to half that area. At least 1 million square kilometers are cleared each decade, with several times that amount severely damaged by burning and selective logging. Biologists convert this habitat loss to species loss based on principles of island biogeography. They assume that terrestrial islands of habitat remaining in a sea of converted lands will behave like oceanic islands, where the relationship between island size and the number of species that can be supported in the long term is well known. Of course, this assumes that fragments are comparable to oceanic islands, that the surroundings will be incompatible to most species, and that conditions before fragmentation were stable, all of which have been questioned (Haila 2002). Though the rate of species extinction from habitat fragmentation is poorly known, Pimm and Raven (2000) project that current extinction rates are somewhere around 1,000 species per decade per million species. Because species loss accelerates as habitat area declines, forest-clearing to date has only eliminated an estimated 15 percent of the species contained in the lost forests. But if the rate of forest clearing remains constant, the extinction curve is predicted to accelerate rapidly to a peak by the middle of the twenty-first century.

The third wave of extinction has been building over the past 400 years. Unlike the early waves, it is affecting species of all evolutionary forms and sizes, from all regions and habitats. The consensus of scientists is that biodiversity today is being lost at a rate that is two to three orders of magnitude faster than is normal in geological history (Wilson 1985). IUCN has developed a new system of assessing threats to individual species, based primarily on extinction probabilities, which in turn are based on the latest thinking of conservation biology. The resulting Red Lists (IUCN 1996, 1997, 2000b) have stimulated the preparation of many species-specific action plans. While these action plans are extremely important, they address only the most obvious expression of the extinction crisis. Thousands, perhaps even millions, of lesser-known and unstudied organisms are moving toward extinction without anyone marking their passing. One leading expert in biodiversity estimates that due to deforestation alone, as many as 27,000 species are doomed to extinction each year (Wilson 2000). Worse, the more scientists know about a group of species, the farther along the road to extinction that group appears to be; the groups that are best known, such as primates and parrots, tend to have the highest percentage of threatened species. It is highly likely, for example, that additional research would reveal that the number of threatened invertebrates is several orders of magnitude higher than now recorded. Extinction risks may also be higher than assumed. For example, although the effects of overfishing on single fish species are generally reversible, most marine fisheries involving multiple species do not rapidly recover from prolonged declines (Hutchings 2000).

It is particularly regrettable, from an ecosystem perspective, that species targeted by human hunters often play important roles in maintaining habitats. Especially important for dispersing seeds are large birds such as hornbills and currassows; these are the very species most favored by human hunters, and hence often the first ones to disappear. The mammals most responsible for dispersing many species of fruit trees are the primates, and they too are widely admired as prey by hunters and therefore under great threat of extinction. No fewer than 116 primate species are considered threatened by IUCN (2000b).

Table 2.4. The Top 25 Countries in Terms of Threatened Species

Country	Mammals	Birds	Reptiles	Amphibians	Fishes	Total
Indonesia	128	104	19	0	60	311
United States	35	50	28	24	123	260
China	75	90	15	1	28	209
Mexico	64	36	18	3	86	207
Australia	58	45	37	25	37	202
India	75	73	16	3	4	171
Philippines	49	86	7	2	26	170
Peru	46	64	9	1	0	120
Colombia	35	64	15	0	5	118
Papua New Guinea	57	31	10	0	13	111
Thailand	34	45	16	0	14	109
Madagascar	46	28	17	2	13	106
Malaysia	42	34	14	0	14	104
South Africa	33	16	19	9	27	104
Vietnam	38	47	12	1	3	101
Myanmar	31	44	20	0	1	96
Ecuador	28	53	12	0	1	94
Kenya	43	24	5	0	20	92
Tanzania	33	30	4	0	19	86
Argentina	27	41	5	5	1	79
Laos	30	27	7	0	4	68
Venezuela	24	22	14	0	5	65
Nepal	28	27	5	0	0	60
New Zealand	3	44	11	1	8	59
Ethiopia	35	20	1	0	0	56

Source: IUCN 1997.

Note: "Threatened" refers to the following IUCN categories: critically endangered, endangered, and vulnerable.

While most of the recorded extinction in historical times has taken place in the United States, in Australia, and on Pacific islands, the threat of modern extinction is especially pronounced in the developing world (Table 2.4). The developing countries threatened with the most notable biodiversity losses are Indonesia, China, and Mexico—perhaps not surprising given their great species richness and high population pressure.

Third-wave extinctions are not yet catastrophic—just 1 percent of bird species (131 species) and 1.8 percent of mammal species (87 species) have been exterminated in the past 400 years. But the latest IUCN figures indicate that far higher numbers are poised at the precipice. Nearly 24 percent of all species of mammals, 12 percent of birds, and almost 14 percent of plants are currently threatened with extinction (Figure 2.2).

While scientists are still uncertain about how many species currently exist, some experts calculate that if present trends continue, at least 25 percent of the world's species could become extinct, or be reduced to tiny remnants, by the

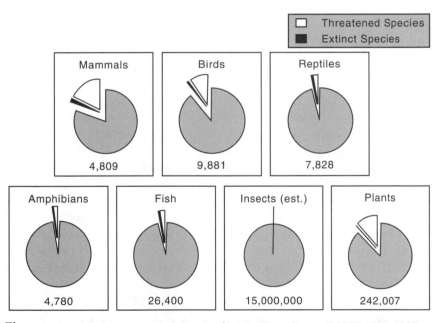

Figure 2.2. Global Conservation Status of Major Taxa. *Source:* IUCN 1996, 1997. *Notes:* Total species of plants are only those assessed (total number might reach 300,000); total species of insects is highly speculative, with estimates varying by an order of magnitude. "Threatened species" refers to the following: critically endangered, endangered, and vulnerable. "Extinct" includes extinct in the wild since 1600; for plants, it includes those suspected to have become extinct recently. The total number of each respective species is indicated below each pie.

middle of this century. Many more species are losing a considerable part of their genetic variation, making them increasingly vulnerable to pests, disease, and climatic change. As discussed in more detail in the next chapter, this rapid rate of loss is due especially to the transformation of natural habitats into agricultural lands. Other significant factors include excessive harvesting of particular species of economic value; the effects of invasive alien species, including diseases; the impacts of various environmental pollutants; and changes in climate. Virtually all of these are due ultimately to increasing demands of humans on the environment—in other words, the competition between humans and the other species with whom we share our planet.

Predicting which species are likely to be most at risk is an extremely difficult business (Lawton and May 1995). Pimm (1996), based on two decades of work, found no clear single cause for the decline of Hawaiian birds and concluded that extinctions are the result of interactions between multiple factors: "The consequence of multiple factors working together is that once the rot begins, extinctions will be fast, furious, multifactorial, and in greater numbers than predicted from habitat destruction alone."

Population Extinction

Biodiversity loss is not measured simply by the loss of species. At least as important is the loss of the geographically distinct populations that together form species. Each vertebrate species has on average about 220 genetically distinct populations, though this number varies widely between species. The planet supports about 11 million populations of vertebrates. Most of the benefits of biodiversity depend on large numbers of populations of each species, because each population ordinarily provides an incremental amount of an ecosystem good or service (Hughes et al. 1997). The existence of multiple populations, for example, provides the genetic diversity that is crucial for the development and improvement of pharmaceuticals and agricultural crops.

Species extinction often represents the end point of a process of progressive population extinctions. The deletion of populations over much of the range of a species is likely to be of at least as much concern as the final extinction of that species (Hobbs and Mooney 1998). Assuming that population extinction is a linear function of habitat loss, about 180 populations of vertebrates are being destroyed per hour in tropical forests alone. This is an absolute rate three orders of magnitude higher, and a percentage rate three to eight times higher, than conservative estimates of species extinction. Hughes and her coworkers conclude that "the consequences for human well-being of the rapid loss of populations will depend in part on the degree to which their functions can be replaced by populations of 'weedy' species, but they are likely to be severe" (Hughes et al. 1997).

Population extinctions often result from habitat destruction and modification, which push species out of most of their former range and into small refuge areas. Moreover, species additions, in the form of invasive species, can be instrumental in hastening extinction of native populations (see Box 2.6). Species additions are frequently more numerous than extinctions in any given area and often result in dramatic changes in ecosystem structure or function. Examples from California and Western Australia (Hobbs and Mooney 1998) show broadly similar trends in species extinction, range contractions, and invasions. Thus, by concentrating on species extinction alone, many of the important human effects on biodiversity may be overlooked.

Implications

Every species alive today has an unbroken line of ancestors back to the very beginnings of life. Knowingly allowing any such lineage to become extinct is both tragic and, many people believe, immoral. At a more practical level, scientists do not know how much population diversity and how much species diversity can be lost without severely impairing the ecosystem services upon which all life depends. Prudence suggests that we should not assume that populations and species are freely interchangeable (Ehrlich and Daily 1993). Indeed, it seems most prudent to assume that the health of the human economic system depends on the maintenance of both population diversity (that is, genetic

---

Box 2.6. Biodiversity Dynamics in a Forest Fragment in Singapore

An isolated 4-hectare fragment of lowland tropical rainforest has been preserved in the Singapore Botanic Garden since it was founded in 1859. A comparison of a recent inventory of woody plants with the historic record of the flora based on collections of herbarium specimens dating back to the 1880s shows that of the 448 historically recorded native species, only 220 are still present along with 80 introduced species that may have replaced some of the native species. On the other hand, 94 native species were found for which no historic records exist. The nearly 51 percent loss of plant species richness over the last 100 years or so has not been distributed uniformly across plant life-form groups. Tree species have been less likely to go extinct than shrubs, climbers, or epiphytes. But half of the tree species present in 1994 were represented by only one or two individuals. Individual longevity may be the major factor correlating with persistence of plant species in isolated forest fragments. But these species may not be reproducing and in fact may be among the "living dead" (Turner et al. 1996).

diversity within species) and species diversity. Any other assumption amounts to gambling with the future of civilization.

The path to an unsustainable future is visible and we are on it. Natural diversity is more threatened now than at any time since the extinction of the dinosaurs 65 million years ago. The trend is steadily downward, as more habitats are converted to exclusively human uses. Preserving the habitats that support species populations is clearly the critical task at hand for both conserving biodiversity and maintaining ecosystems. No habitat destruction can be taken lightly, and because agriculture accounts for the most extensive habitat changes, it deserves special attention.

Chapter 3

Agriculture and Human Welfare

In seeking to secure wild biodiversity it is critical to acknowledge the equally pressing need for lands to produce food, secure rural livelihoods, and contribute to economic development, especially in poor and rapidly changing countries of the developing world. This chapter presents an overview of global agricultural and other production systems and lays out the challenges of food supply, rural employment, and poverty reduction facing the world over the next few decades.

Agriculture: A Dominant Global Land Use

The term *agriculture* comes from the Latin words *agre* (field) and *cultivare* (to cultivate). It is defined by *Webster's Dictionary* as the "art of preparing soil, sowing and planting seeds, caring for plants, and harvesting crops." More broadly, agriculture is a process of modifying natural ecosystems to provide more goods and services for people, through the nurturing of domesticated species of plants and animals. In this book, agriculture is taken to include all forms of husbandry, including crop and livestock production, aquaculture, and tree plantations. Most nonagricultural lands also contribute to food production, both directly in the form of plants or animals harvested and indirectly in the form of grazing for livestock or building up of soils.

Throughout history, rural people have depended on nature to provide numerous goods and services, ranging from water to medicines to firewood to various kinds of food. Hunting, fishing, and gathering from forests, swamps, grasslands, freshwater systems, and the coastal zone have always played an important role in the human diet, especially in seasons when agricultural products have been in short supply. Human communities have used at least 3,000 plants as food, of the 75,000 or more plants that are potentially edible. In Southeast Asia local cultures have used at least 300 species of vegetables, including about 80 that grow only in wild habitats. Thus managing natural systems as well as converting them has always been part of the way that people relate to their environment. Further, the way that people relate to their land and resources—including wildlife, crops, medicinal plants, and livestock—helps to define their culture.

Agriculture has played a central role in the success of the human species, enabling the human population to expand far beyond what would have been possible with only hunting and gathering. Our population density as hunters and gatherers might only be between one and two individuals per square kilometer, with higher densities in some particularly favorable habitats. Today's global population density averages about forty-four people per square kilometer, about thirty times what would be expected for a mammalian species of our size and diet. This high population density is made possible by agriculture, as well as our success in controlling parasites, predators, and competitors (Cincotta and Engelman 2000).

Map 3.1 shows the global pattern of land use. The area with over 60 percent of land in "agriculture" (the definition used in this map) is only slightly greater worldwide than protected areas (about 10 percent). But another 17 percent of land is in mosaics with 30 to 60 percent of agricultural land intermixed with grassland or forest (Wood, Sebastian, and Scherr 2000). In tropical countries, 13 percent of land is dominated by agriculture and 19 percent is in land use mosaics mixing agriculture with forest or grassland (Sebastian 2002). About 35 percent of total area is in grasslands, of which roughly half is used for grazing of domestic livestock. A further 29.6 percent of land is forest and 1.4 percent is forest plantations (FAO 2001), of which about half is used for industrial forest production and the rest for subsistence and small-scale commercial uses. The remaining 7 percent is under other uses, such as urban areas and barren lands (World Resources Institute 2000).

Globally, agricultural production systems are highly diverse, varying according to the mix of annual crop, perennial crop, and livestock components; irrigated or nonirrigated water sources; and management intensity. The major types of agricultural systems include intensive irrigated cropping systems; intensive annual cropping systems in high-quality and lower-quality rain-fed lands; perennial tree crop systems; extensive fallow-based systems; pastoral and ranching systems; intensive livestock production; and forest and aquaculture production systems (Matthews et al. 2000; Ruthenberg 1980; Wood, Sebastian, and Scherr 2000). The scale and basic characteristics of these systems, and general features influencing interactions with wild biodiversity, are described in Table 3.1.

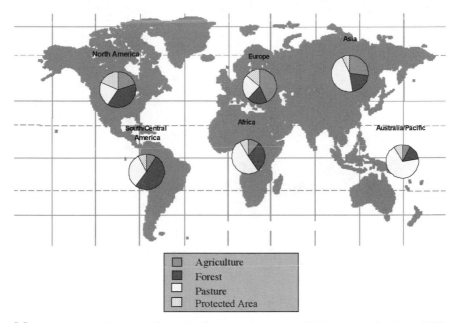

Map 3.1. Agricultural Land Use, by Region. *Sources:* World Resources Institute 1998 and IUCN.
Notes: South/Central America includes Mexico; Europe includes all of Russia, but its former southern republics are included in Asia, as is the Middle East; Pacific includes New Zealand and Papua New Guinea but not Irian Jaya (which is included in Asia). Protected areas are in IUCN Categories I–V (see Box 9.3). Many protected areas are included in forests. The agricultural area shown here includes only lands where over 60 percent of cover is in crops or planted pastures.

When intensively farmed, agricultural lands are subject to regular distur-bance of the soil and dominated by species of interest primarily to humans. Figure 3.1 illustrates how the diversity of agricultural species varies across sys-tems, from very low-diversity systems such as intensive cereal monocultures to high-diversity systems that include at least some fields with natural vegetation (e.g., fallow plots). Even some intensive systems using annual crop monocul-tures in small plots (such as many small-scale farms in China) may have inter-mediate to high inter-plot diversity. In general, greater wildlife diversity appears to be associated with a diverse assortment of agricultural systems. Indeed, considerable wild biodiversity still persists in many agricultural land-scapes, though this is not always recognized. Even in fairly intensive systems, farmlands may support many wild species of plants, many invertebrates (including those in the soil), reptiles, amphibians, small mammals, and birds (both resident and migratory).

Table 3.1. Overview of Agricultural Systems and Associated Threats to Wild Biodiversity

Type of System	Extent	Rural Population	Role in Food Supply	Ecosystem/Habitat Characteristics	Threats to Wild Biodiversity
Intensive irrigated cropping systems	7.5% of all arable, grazing lands; 17.5 % of cropland globally; 34% in East, South Asia	One third of rural population in developing countries	Up to 40% of world's crop output; 2/3 of rice and wheat production; major focus of Green Revolution	Secure water supplies through groundwater, rivers, lakes, artificial tanks. Extensive transformation of water flows, nutrient flows, soil quality and landforms; multiple-cropping; intensive use of inputs. Yields very high and more stable due to more controlled growing conditions.	Salinization of soil and water; agrochemical pollution; clearing of wild vegetation; drainage of wetlands; changed drainage patterns; monocultures common
Intensive annual cropping systems in high-quality rainfed lands	In developing countries, 23% of arable and pasture land		No good estimates, probably account for another 40% of crop output; major focus of Green Revolution	More fertile soils, relatively flat but well drained terrain; resilience under continuous cultivation; reliable rainfall; trend to monocultures and clearing of non-crop vegetation	Soil degradation, loss of ecosystem functions, agrochemical pollution, clearing of wild vegetation
Intensive annual crop systems in lower-quality rainfed lands	No good estimates, but likely over half of total	Over half rural population in developing countries	No good estimates, but probably around 20% of crop output	Lands steeper or less fertile; subject to drought or intensive rains. Reduced use of fallow due to high population density. Mosaic land use pattern due to varying topography/quality.	Soil degradation, agrochemical pollution, clearing of wild vegetation
Perennial crop systems (trees, shrubs, palms)	Plantations cover 10% of cropland; widespread in agroforestry	No estimates	Supply most fruits, beverages, nuts, some vegetable oils, spices, fibers	Pure plantations, mixed stands, inter-cropped in annual crops, border plantings; some wild stands managed for commercial harvest. Prominent in areas too steep, dry, or wet for arable crops	May convert natural forest; agrochemical pollution; where converted from annual crops, wild species increase

Extensive agricultural systems (most use longer fallows)	No good estimates, mostly in areas	Possibly half billion people	Small share of crop output, most for subsistence use.	Farmers use longer fallows to renew field productivity, thus cultivated "swiddens" are a small proportion of total land cover; fallows are often modified. Land often cleared using fire. Often found where biophysical conditions make continuous annual crop production uneconomic, risky or unsustainable at low input levels	Uncontrolled fires destroy seeds, wildlife, vegetative capacity to propagate; high erosion after clearing, extensive areas cleared for low crop yields
Pastoral and ranching systems	26% of world's land area; 23% land in developing countries	Est. 25 million people rely on pastoralism; no estimates on ranching	Animals provide food, income, savings, crop nutrients, draft power; Major source of beef, goat, mutton	Important in parts of world where slow plant growth and high risks to crops due to drought, frost, poor drainage. Low human population densities. Animals use diverse landscape niches. Natural rangelands have high species diversity.	Damage to vegetation, soil compaction; shift of grass/range species; destruction of predators; fencing prevents animal movement, disease transmission
Intensive livestock production systems	Small, esp. found in peri-urban areas	No estimates	A third of livestock supply; most urban milk supply	Small areas of cattle feedlots, intensive dairy, poultry and pig farms. Use feed and concentrates produced elsewhere.	Consume 39% of world's grain; waste management; control of disease
Forest production systems	29.6% global land in forest or woodlands; 1.4% in forest plantations; agroforestry widespread in regions cleared from forest	A quarter of world's poor and 90% of poorest depend significantly on forests	Forest foods important seasonal supplement for rural poor; fuel used for most cooking; Forest income from timber and non-timber forest products used to purchase food	About half of forests supply industrial products, half subsistence and small-scale commercial uses. A third of industrial wood is from plantations. Almost a quarter of forests in developing world owned or controlled by indigenous or local communities. Diverse agroforestry systems mixing crop, livestock, fish production with trees and farm forests.	Deforestation, forest degradation, pests and diseases; hunting around logging operations

(continues)

Table 3.1. Continued

Type of System	Extent	Rural Population	Role in Food Supply	Ecosystem/Habitat Characteristics	Threats to Wild Biodiversity
Natural fisheries	Coastal marine and freshwater fisheries (three quarters of fish consumed)	30 million fishers (84% in Asia); number has doubled since 1970	Fish and shellfish account for 16.5% of total animal protein consumed worldwide; 21.8% in low-income, food-deficit countries; 26.2% in Asia	Artisanal fishers in small boats for most lake and coastal fisheries; large-scale commercial fishing-processing operations in marine fisheries	Depletion of fishery stocks; 70% of major fisheries are fully fished or over-fished already (while demand is rising rapidly)
Aquaculture	Freshwater aquaculture, mariculture (a quarter of fish consumed)	30 million fishers (84% in Asia); number has doubled since 1970	Fish and shellfish account for 16.5% of total animal protein consumed worldwide; 21.8% in low-income, food-deficit countries; 26.2% in Asia	Commercial farms with intensive methods; family and cooperative farms relying on extensive and semi-intensive methods. Fish enclosed in secure tanks/ponds, competitors and predators controlled; food supply enhanced. 220+ species of finfish and shellfish produced.	Over-use of inputs, high wastes, spread of pathogens

Notes: Estimates of area extent and population in agricultural systems are drawn from Nelson, et al. (1997). Estimates of importance for food supply and ecosystem/habitat statistics are taken from Wood, Sebastian and Scherr (2000); for forestry from Scherr, White and Kaimowitz (2002) and FAO (2001); and for fisheries from Burke, et al. (2000), Revenga, et al. (2000), and FAO fisheries statistics on-line (www.fao.org).

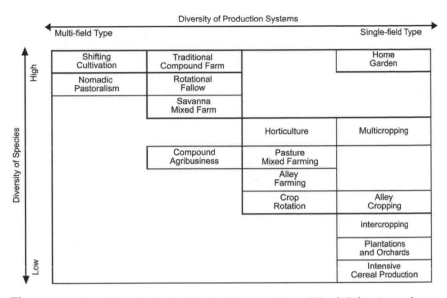

Figure 3.1. Agro-Diversity in Production Systems. *Source:* Wood, Sebastian, and Scherr 2000.

The Challenge of Ensuring Future Food Supply

High levels of global food production and historically low food prices have lulled many observers into a sense of complacency about food security. Yet increasing the world's food supply to meet demand remains an acute challenge because of population growth, changes in patterns of food demand, and emerging constraints to continued productivity growth on many existing farmlands. Because food accounts for such a high proportion of the budget of the poor (even for many farmers), major price increases for food—which might induce higher production, but also additional land clearing—are not necessarily desirable. Neither are inexpensive, "dumped" agricultural surpluses from the highly subsidized agriculture of industrialized countries, because these surpluses can dampen prices paid to poor farmers, thereby decreasing their capacity to make long-term investments in the productivity of their land. In any case, the needs of the rural poor, especially in regions with limited non-farm employment opportunities, are best met by increasing the productivity on their own farmlands. The fundamental interlinkages among population growth, agricultural demand and supply, rural poverty, and economic development are briefly described below.

Recent Trends in Global Food Demand and Supply

In 1900, 160 million people, or 10 percent of the world's 1.6 billion people, lived in cities. One hundred years later, fully 50 percent of the world's 6 billion people live in urban areas. These urban populations rely on surplus food and fiber from the agricultural sector (World Bank 2000).

During the past few decades, consumption of agricultural products has grown more rapidly than population, as incomes also rose. Between 1950 and 1984 per capita grain production increased by 40 percent, and between 1950 and 1990 the per capita supply of beef and mutton increased by 26 percent. For all developing countries combined, per capita consumption of beef, mutton, goat, pork, poultry, eggs, and milk rose by an average of 50 percent between 1973 and 1996 (Fritschel and Mohan 1999). In addition, world fish catches underwent a 4.6-fold increase between 1950 and 1989, doubling the per capita production of seafood. Aquaculture production tripled between 1984 and 1995, from 7 million to 21 million tons per year; in 1995 it accounted for 19 percent of the global fish harvest (Postel 1998). World consumption of wood also increased 2.5-fold between 1950 and 1991, with per capita consumption increasing by a third during this period (Durning 1994). As with food consumption, most of the growth in total and per capita wood consumption has occurred in the developing world. Cereals account for nearly 60 percent of total consumption in developing countries, in terms of

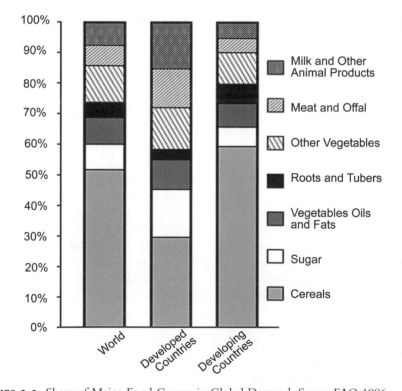

Figure 3.2. Share of Major Food Groups in Global Demand. *Source:* FAO 1996.

energy supply, while in developed countries meat and animal products and sugar consumption are almost as important as cereals (Figure 3.2).

Overall, food prices have declined since 1960, enabling higher consumption (Figure 3.3). At least part of the price decline is due to agricultural subsidies. In countries belonging to the Organization for Economic Cooperation and Development (OECD), these subsidies amounted to U.S.$297 million in 1997, more than $14,000 per full-time farmer (Myers and Kent 2001).

Agricultural production has risen to meet growing demand. Yield increase was the key factor in this production growth. Between 1950 and 1995, human population increased by 122 percent while area planted with grain expanded by only 17 percent. But grain productivity increased by 141 percent, largely due to the use of improved varieties and application of chemical fertilizers. Average cereal yields have increased from just over one metric ton per hectare to nearly three metric tons per hectare, and the use of nitrogen fertilizer has increased dramatically, from about 5 million to 80 million metric tons between 1950 and 1995 (Cohen 1998). Irrigation has also contributed greatly to higher productivity, with irrigated cropland area increasing from about 140 to 270 million hectares between 1950 and 1995.

Agriculture has become much more energy-intensive, typically consuming far more energy than it produces. And with the development of extensive trading and transportation networks, virtually all agricultural systems are integrated to some extent into commercial product and labor markets. The sustainability of energy-intensive production and transportation systems in the long term may be threatened as fossil fuels eventually become more expensive.

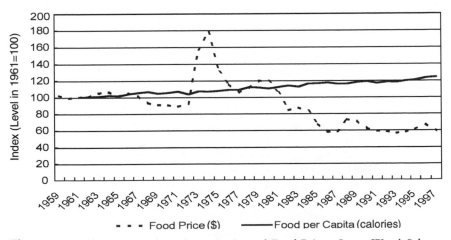

Figure 3.3. Global Index of Food per Capita and Food Prices. *Source:* Wood, Sebastian, and Scherr 2000.

The Impact of Population Growth on Future Food Demand

The central factors influencing land use and agriculture in the next few decades will be population growth and income changes that modify demand. By the year 2020, global population is projected by demographers to reach around 7.7 billion, of which well over 80 percent will be in developing countries. The population of Africa, for example, could increase 70 percent by 2020. According to long-range population projections prepared by the United Nations (1998), by the end of the twenty-first century global human population could reach as high as 17 billion. Or it could peak at less than 7.5 billion around 2040 and return again to below 5.5 billion by the century's end. Table 3.2 presents recent shorter-term estimates between these two extremes.

Most demographers believe that the actual trajectory will depend on what happens to human fertility rates. In most of the fifteen developing countries that have the greatest diversity of wild species, fertility dropped by more than half between 1960 and 2000. Important exceptions were the Democratic Republic of Congo, India, Madagascar, and Papua New Guinea (Cincotta and Engelman 2000), where fertility remained high. Fertility rates are heavily influenced by location and income. Typically, rates are lower among urban women and higher-income women (Cohen 1995a).

By far the greatest population increase is expected to take place in the biodiversity-rich countries of the tropics. In nineteen of the world's twenty-five biodiversity hotspots, population is growing more rapidly than in the world as a whole. Population is decreasing, moderately, in only one hotspot (the Caucasus). Because of rapid migration and high rural fertility rates, population in the tropical wilderness areas is, on average, growing at an annual

Table 3.2. Global Population Projections by Region, 2000–2025 (in millions)

	2000	*2025*	*% Increase*
Africa	793	1,358	71
North America	314	383	22
Latin America and the Caribbean	518	694	34
Asia	3,672	4,776	30
Oceania	30	40	33
Europe	727	683	−6
World Total	6,056	7,936	31

Sources: United Nations 2002. Available at: www.un/unpa/.

Note: These numbers represent mid-level estimates, neither high nor low (medium variant).

rate of 3.1 percent, more than twice the world's average rate of growth. The most serious population pressures on biodiversity hotspots are in the Caribbean, the Philippines, Sri Lanka, and the Western Ghats of India. The hotspots are also rapidly urbanizing. Currently 146 major cities are located in or directly adjacent to a hotspot; 62 of these cities have over 1 million inhabitants (Cincotta and Engelman 2000; also see Map 3.2), and many have high levels of urban and peri-urban (areas around cities) agricultural production (Cheema et al. 1996).

Changing Patterns of Food Demand

History has shown that increasing wealth raises per capita consumption of agricultural products, especially meat, fruits, vegetables, spices, and fish. Meat demand is projected by IFPRI to increase by 63 percent between 1993 and 2020. Demand for the products of aquaculture will also increase, despite climbing prices. Global demand for cereals is expected to increase by 41 percent and roots and tubers by 40 percent, with 80 percent and 90 percent of increased demand for these crops, respectively, coming from developing countries (Pinstrup-Andersen, Pandya-Lorch, and Rosegrant 1997).

As demand for animals higher on the food chain increases, so too does the demand for animal feed inputs, including grains, cassava, fish meal, and other commodities. Cultural changes could reduce consumption of grain-fed meat for religious, ethical, human health, or environmental reasons. While such changes are not yet evident, the spread of foot-and-mouth disease across Europe in 2001 indicates the potential of such changes. It certainly is possible for people to have perfectly adequate diets with relatively small amounts of meat, or even with no meat at all. In fact, feeding lower on the food chain—eating more vegetable matter—would enable more people to be supported on our planet (Wilson 2002). However, it is unlikely that our species will ever become totally vegetarian, because ranching appears to be the most productive land use for our species in many arid and semi-arid parts of the planet.

Global production and consumption of industrial forest products are projected to rise by 25 percent between 1996 and 2010, with Asia and Oceania likely to have the highest rates of expansion (Bazett 2000). Demand for other nonfood products, such as fiber from cotton or hemp, will also rise. Farm-produced timber and pulp are likely to replace natural forest sources in many parts of the world, given rapid depletion and higher productivity of industrial and farm plantations (Dewees and Scherr 1996).

Although food production has grown faster than population at a global level, most countries consume more grain than they produce and thus import grain, mainly from the OECD countries. Some observers believe that these importing countries, especially China, will greatly expand their impact on

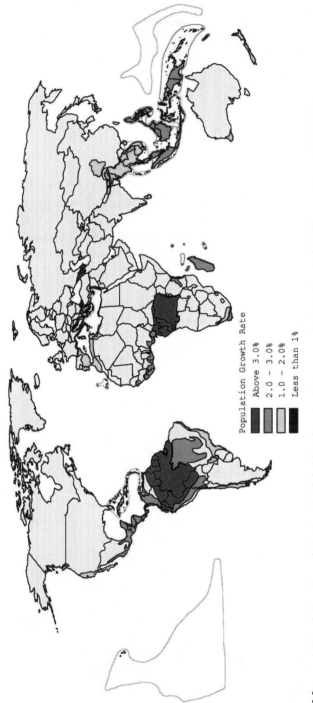

Map 3.2. Population Growth in the 25 Global Biodiversity Hotspots and Major Tropical Wilderness Areas, 1995–2000.
Source: Cincotta and Engelman 2000.

Population Growth Rate

Above 3.0%

2.0 – 3.0%

1.0 – 2.0%

Less than 1%

world food demand in the years ahead (Brown 1995, Paarlberg 1996). In any event, developing-country imports of temperate-zone agricultural products often will be purchased with income earned from exports of agricultural products from tropical environments, further increasing the integration of the global agriculture system.

Food Production Constraints

As indicated above, the world has enjoyed historically unprecedented increases in agricultural production and productivity growth over the past few decades. A critical question is how long that growth can continue. In the past four decades, cultivated land in developing countries increased by 40 percent (Fischer et al. 2001). Even with this extensive clearing, arable land per capita declined from just under 0.5 hectare in 1950 to just under 0.3 hectare in 1990 as populations increased faster than lands devoted to crops. Per capita landholdings in developing countries are expected to decline to 0.1–0.2 hectare by 2050 (FAO 1993). Asia and North Africa will reach this low level by 2025. China, a country with significant scarcity of arable land, now has a "no-net-loss" land policy. Thus, while large areas of environmentally sensitive land are being taken out of production in some parts of China, land conversion and intensification are accelerating in other parts.

Although an additional 1.5 billion hectares theoretically could be converted to cropland, mainly in South America and sub-Saharan Africa, the areas with the best cropland are already being fully farmed. Only areas with greater production constraints, such as steep hillsides, tropical forests on infertile soils, semiarid regions, and forest areas that support a great deal of the world's biodiversity, remain for conversion (Fischer et al. 2001). Most are inherently unsuitable for continuous annual crop production, although research innovations may make them more attractive for other forms of agriculture.

Some indicators suggest that ecosystem and resource limits for traditional agriculture are already being reached. On a per capita basis, irrigated lands have shrunk by 6 percent between 1978 and 1990 and are expected to contract by a further 12 percent per capita by the year 2010 as population growth outruns new irrigation and as water availability becomes a limiting factor. Even worse, nearly 10 percent of irrigated lands have become so saline as to reduce crop yields, and another 20 percent are suffering from a buildup of salts in the soil caused by poorly managed irrigation systems. Costs to farmers in terms of reduced income amounts to U.S.$11 billion per year (Postel 1999).

While irrigation expanded by an average of 2.8 percent per year between 1950 and 1980, subsequent expansion has been much slower, falling to only 1.2 percent per year in the 1990s. Falling production levels also suggest ecosystem and resource limits. World fish harvests peaked at 100 million tons in

1989; by 1993 they had declined 7 percent from 1989 levels. Growth in grain production has slowed since 1984, with per capita output falling 11 percent by 1993. The Worldwatch Institute, extrapolating from historical data, forecasts that "if current trends in resource use continue and if world population grows as projected, by 2010 per capita availability of rangeland will drop by 22 percent and the fish catch by 10 percent. And cropland area and forestland per person will shrink by 21 and 30 percent respectively" (Postel 1994). Adding to the challenge of resource limits is global climate change. It is projected to exacerbate problems such as unpredictable rainfall, pests, and diseases, especially in the tropical zone, although where local impacts will be positive or negative is highly uncertain (Watson et al. 2000).

While feeding growing populations will be a serious challenge in developing countries, some parts of the world where agriculture is heavily subsidized—notably the United States, Canada, and western Europe—are removing land from agriculture and taking steps to reduce what is seen as overproduction. It is conceivable that the world may be capable of producing sufficient food to support a much greater population in the year 2025, and that the real constraint will be the price of food. In that case, poor countries may be unable to purchase developed-country surpluses to feed their populations. Most governments are seeking to implement a policy of cheap food, in part to assist the urban poor and industrial workers. But this also means that poor farmers are unlikely to earn a sufficiently high price for their products to enable them to escape the poverty trap, unless their own farm productivity can be increased.

Rural Poverty Reduction and Economic Development

Increasing agricultural production by low-income producers in the developing countries is critical not only for meeting market demands, but also for reducing rural poverty and raising living standards. The impressive increases in agricultural production described earlier could, theoretically, ensure food security to the world's entire population. However, food access is determined not only by the available supply, but also by having the arable land and inputs needed to produce it or the cash income to purchase it. The world's poorest have none of these. Some 1.2 billion people worldwide earn less than one dollar a day (a poverty line that adjusts for differences across countries and times in purchasing power). Forty-four percent of these poor live in South Asia, about 24 percent each in sub-Saharan Africa and East Asia, and 6.5 percent in Latin America and the Caribbean (IFAD 2001).

Thus, even in the agricultural breadbaskets of the tropics, where food is amply available, poor people with low cash incomes are unable to buy enough. Even farmers cannot ensure their families' food security, if they are small-scale farmers facing significant production constraints and are too poor to invest in farm improvements or inputs. Poor transportation, food storage, and marketing systems throughout large areas of the developing world mean that even food that can be imported cheaply to port cities becomes too expensive for people in the hinterland to buy, if it even makes it to such remote areas.

This poverty is concentrated in many of the areas where wild biodiversity is richest or most threatened (Nelson et al. 1997). Seventy-five percent of those earning a dollar a day or less work and live in rural areas. Projections indicate that more than 60 percent of the poor will continue to be rural people in 2025 (IFAD 2000). An estimated 325 million poor people live on favored agricultural lands in developing countries, while 630 million live on marginal agricultural, forested, and arid lands (Nelson et al. 1997). Some 300 million people, most of them poor, live in forested areas and another 200 million live around them (Panayotou and Ashton 1992). Many indigenous ethnic groups, among the most impoverished and marginalized peoples, live in lands where extensive wild biodiversity remains. In Mexico, for example, most of the remaining forests, natural habitats, and protected areas are in lands occupied by indigenous peoples. These areas were often deprived of infrastructure and other investment that would have induced major land use and landscape changes. On the other hand, local people also often were, or had to be, more ecologically sensitive because they depended solely on local agriculture. The long-term land security they enjoyed also encouraged resource husbandry.

A major consequence of poverty is hunger and malnutrition. In 2000, 182 million preschool children—33 percent of all children under five in the developing world—were stunted or chronically undernourished; 27 percent were underweight. While the percentages appear to be dropping in Asia, they are escalating in Africa (McClafferty 2000). Fourteen million children, most of them in developing countries, die every year from hunger-related disease—a number equivalent to three jumbo jets crashing every hour, every day of the year (World Food Programme 2000).

In 1995–97, 18 percent of the total population of the developing world was undernourished. Countries with biodiversity hotspots that also have more than a fifth of their population undernourished include Bolivia, Cameroon, Cambodia, Democratic Republic of Congo, Dominican Republic, Haiti, Honduras, India, Kenya, Laos, Madagascar, Namibia, Nepal, Nicaragua, Papua New Guinea, the Philippines, and Thailand. Several relatively large countries have undernutrition (or lack of sufficient food) rates that are much higher in the vicinity of biodiversity hot spots than for the country as a whole; these include Brazil, China, Ecuador, Guatemala, Indonesia, Mexico, Peru, and Vietnam. Many dryland areas, though not rich in biodiversity, have rare habitats and highly endangered wild species. Dryland countries that have both highly endangered species and especially high rates of undernutrition (more than 35 percent of the entire population) include Afghanistan, Chad, Ethiopia, Niger, Tanzania, and Yemen (World Food Programme 2000).

In most developing countries, the rural poor depend more on agriculture than the rural nonpoor and derive more of their income from common property than the nonpoor (Hopkins et al. 1994; Malik 1998). Because the rural poor have limited access to external or industrial agricultural inputs, natural capital—the inherent productivity of their natural resource base, including soils, forests, and water—is of particular importance to their livelihood security. The term *ecological poverty* has recently come into use to describe the type of widespread poverty that arises from degradation or loss of such natural capital (Coward et al. 1999). But ecological poverty both leads to poverty and results from it. When poor people have trouble finding food because of insufficient agricultural production or income, they may become even more dependent on gleaning the products of wild biodiversity, clearing new fields from natural habitat, and poaching and encroaching on protected areas. Such measures may provide emergency relief, but they are not sustainable and may reduce the area's natural capital or even destroy natural food supplies.

Rural Populations Are Still Growing

While from 1960 to 1995 rural populations declined in most developed countries and a few developing countries (such as Brazil), the absolute number of rural dwellers in the developing countries rose by almost 40 percent, from 2.0 to 2.8 billion. The total rural population in Africa grew by 68 percent. The industrialized countries reached their peak rural population of 366 million in 1950 and have shown a decline since then to 301 million in 1990, a number that is projected to decline further to 198 million by 2025 (Lutz, Prinz, and Langgassmer 1993). This trajectory represents a 40 percent decline in 75 years. Where economic development provides real alternatives to agriculture on marginal land, declining rural populations may expand the area available for conservation purposes (Sankaram 1993). Alternatively, a declining rural population may simply mean increasingly mechanized and extensive resource extraction and agricultural production to support growing urban populations (Shiva 1991).

In the biodiversity-rich developing countries, rural populations will probably continue to remain large. In recent decades, rural populations have grown along with urban populations, though at slower rates because of rural-to-urban migration. Rural populations in developing countries are projected to peak at 3.09 billion in 2015 (accounting for 94 percent of the world's total rural population), then decline over the next ten years to 3.03 billion. The additional people will require additional land not only for food and income, but also for settlement and infrastructure. They will also require natural resources to meet subsistence food, fuel, water, and raw material needs. Most will continue to rely on agriculture as their livelihood. Population growth affects poverty, because increasing family size makes economic opportunities harder to find. This in turn leads to exploitation of lower quality agricultural lands and to the breakdown of traditional mechanisms for sustainable resource management.

Agriculture and Economic Development

Unquestionably, the path of economic development involves growth in non-farm economic activity. Indeed, as the labor force shifts to nonfarm activities, some types of environmental pressures on biodiversity may be eased so long as local food production and availability do not decline. But this does not contradict the central importance of agricultural development not only for poor farmers, but also for the national economies in low-income developing countries. In most of the regions where biodiversity loss is most severe, the economy is highly dependent upon agriculture for income generation, employment, and development (Table 3.3). The share of agriculture in the gross domestic product (GDP) is only 2 percent in high-income countries and 11 percent in middle-income countries, but in low-income countries agriculture accounts for 28 percent of GDP. In five Asian and eleven African countries, more than 40 percent of the GDP comes from the agricultural sector; in sub-Saharan Africa the share of the GDP in agriculture rose between 1980 and 1997 (World Resources Institute 2001). In the poorest countries, agriculture provides livelihoods for 69 percent of the workforce and 76 percent of economically active women, and it accounts on average for half of the GDP (World Bank 2000).

Moreover, in poor countries, agricultural prosperity is central to the growth of nonfarm sectors. In West Africa, for example, because of the pattern of consumption expenditures by farm households, adding one dollar of new farm income to the economy resulted in a total increase of household income ranging from U.S.$1.96 in Niger to U.S.$2.88 in Burkina Faso (Delgado et al. 1998). These impacts are higher than from any other sector.

Despite the dominance of agriculture in the economies of most developing countries, the tropics currently are clearly at a disadvantage in international agricultural trade because of the geographic constraints of poorer soils, greater pest problems, and more difficult climates (Gallup, Sachs, and Mellinger 1999). Because of these disadvantages and the structure of subsidies and trade policies in the developed countries, some development analysts have suggested that tropical countries cannot use agriculture as their engine of growth and instead should rely on food imports and investment in other sectors. Even some environmentalists have argued that because tropical countries have an absolute advantage in biodiversity conservation and ecotourism, they should give priority to protecting biodiversity and not aggressively pursue agricultural development.

These arguments may seem sensible, but they are based on several flawed assumptions. First, it is indeed true that with current agricultural practices and trade agreements, tropical producing areas are at a relative disadvantage. Only a few tropical farmlands can compete effectively in international markets with an intensive rain-fed monoculture of cereal grains. But the potential for primary biological production in the tropics is high. Food production using crop plants better adapted to tropical conditions and in production systems designed explicitly to fit tropical ecosystems could boost yields significantly. At the moment many tropical production systems use technologies originally developed for temperate climates. Agricultural innovation and research in the tropics have enormous scope for establishing a strong comparative advantage for production of many commodities. Meanwhile, global demand for higher-value tropical crops (mainly tree and palm crops such as coffee, tea, cocoa, and palm oil, which have the potential to be grown sustainably) is increasing.

Second, the importance of international trade to global food security has been somewhat exaggerated. While food trade has grown dramatically in the

Table 3.3. Economic Role of Agriculture in Developing Regions, 1995–97 Average

	% GDP in agriculture 1997	Total population (millions)	% Rural population	% Agri-cultural population	Labor force (millions)	% Labor force in agriculture
North America	2	299.5	23.6	2.5	153.2	2.4
Europe	11	517.3	25.8	8.1	245.1	8.2
Oceania	NA	28.9	29.8	18.3	13.6	18.4
Russian Federation (former USSR)	NA	291.8	31.8	17.0	146.8	16.0
Latin America and the Caribbean	10	487.8	26.2	22.9	203.0	21.9
West Asia/ North Africa	NA	350.8	39.1	29.9	126.1	33.2
Southeast Asia	19	488.2	65.7	51.7	238.8	55.5
South Asia	27	1,273.0	73.3	57.3	551.3	60.6
East Asia	19	1,435.4	62.4	60.7	839.2	61.7
Sub-Saharan Africa	25	581.1	68.4	63.0	258.1	64.7
World	4	5,753.7	54.2	44.2	2,775.1	46.4

Source: GDP from World Bank 2001; other data adapted from Wood, Sebastian, and Scherr (2000), Table 14; original data from FAO 1999a.

Note: "NA" means not available.

past few decades, the share of food that is traded—10 percent—has remained relatively constant since 1960. Thus most food is grown and consumed within national borders, and this is likely to remain the case for most developing countries (McCalla 2000). Food imports are indeed likely to continue growing as a proportion of total food supply in many countries, particularly as population growth rates peak. But for most tropical countries, food security will depend mainly on heightened investment in domestic agricultural production. Countries with poor transportation and food marketing infrastructure will depend even more on locally produced food. Moreover, while increasing food imports may slow rates of agricultural extensification (opening of new farming lands) and intensification, they will rarely reverse them in countries with large rural populations, unless patterns of economic development enable very high rates of nonfarm employment and foreign exchange earnings.

Third, it must be remembered that comparative advantage does not require an absolute advantage in a given line of production. Rather, it refers to the choices within a country between the production options available to it. And for many of the poorest, biodiversity-rich countries, nonagricultural economic options do not appear to be nearly as promising engines of growth as agriculture. Nor are nonagricultural options likely to have the employment multipliers (causing further expansion in employment) needed to encourage broad-based development. Exceptions might be countries that have reached middle-income status, such as Mexico. But even there economic development is likely to bypass large parts of the population unless accompanied by dynamic agricultural growth. Moreover, many governments will consider a strong agricultural sector essential to their national security and independence.

Implications

These factors—the need for continued growth in food supply and the importance of agricultural development in reducing rural poverty—demonstrate that accelerated agricultural development is essential in most developing countries. Because of the impact of agricultural development on poverty and on farmer incentives for good land management, it is also a necessary part of any effort to conserve wild biodiversity.

Indeed, the past few years have seen a resurgence of interest among the international community in poverty reduction in the developing world. The World Food Summit of 1997 called for a halving of malnutrition rates worldwide by 2015 (FAO 1997). The sixteen Future Harvest agricultural and natural resource management research centers supported by the Consultative Group on International Agricultural Research (CGIAR) are committed to focus their research to support poverty reduction (Serageldin 1996). The World Bank has renewed its commitment to poverty reduction as its core mission. Many bilateral aid donors have done the same. And after a long period of reduced investment in agriculture, some international and national aid agencies have begun again to focus on opportunities for poverty reduction through agricultural investments, especially in Africa. What happens in the agricultural sector will have profound implications for human welfare and economic development. For both pragmatic and ethical reasons, poverty-reducing agricultural innovation must be an integral part of global and national strategies to conserve biodiversity. The challenge to achieve such integration will be especially great in tropical developing countries.

Chapter 4

Agriculture and Wild Biodiversity

Historically, agricultural expansion and intensification have had a largely negative effect on wild biodiversity. But until this century, agriculture in most parts of the world represented only islands within a large sea of wild lands, so that much biodiversity could survive away from farming areas. Other factors—urban development, hunting, mining, infrastructure construction, invasive alien species, industrial pollution, energy development—have often posed greater threats to wild biodiversity than farming. But today the scale and technology of agricultural land use have made it the leading threat.

This chapter presents an overview of the various ways agricultural change has impinged on wild biodiversity, through land use conversion, monocultures of domesticated species, modification of water resources, pollution from agricultural chemicals and livestock operations, and fragmentation and degradation of habitats and wild gene pools. Analysis is handicapped by the paucity of data on the relationship between specific types of agricultural practices and wild biodiversity. To the extent studies are undertaken, they mainly look at birds and some mammals. While the challenge for biodiversity conservation is to manage resources at a landscape scale, most case studies focus on individual farm fields. Only by understanding the large-scale effects of agriculture on biodiversity can we reasonably expect to find effective ways to integrate farming with the conservation of wild biodiversity, or at least find accommodations between the two.

Historical Relation of Agriculture and Wild Biodiversity

Historically, many subsistence-oriented farming communities or nomadic pastoral groups depended on a complex patchwork of field, pastures, and forest areas that provided different goods and services throughout the year. These farming cultures may have increased or at least sustained wild biodiversity by encouraging a variety of habitats, only a small portion of which were intensively cultivated. Lacking modern conveniences such as supermarkets, freezers, microwave ovens, or global transportation, these peoples instead managed their environments to provide a somewhat predictable flow of the products they needed. Such approaches to habitat management often enabled wildlife to prosper, along with wild relatives of domesticated plants.

Many of these traditions remain alive today among indigenous populations, and some modern societies are seeking to re-create them. What we call forest reserves, for example, were maintained as sacred groves in some parts of Manipur in eastern India. According to Gadgil, Hemam, and Reddy (1996): "While these refugia in Manipur are no longer considered to be inviolable as abodes of spiritual beings, the system of community-based vigilance and protection is identical to that prevailing with the sacred groves. Notably, taboos against extraction of any plant material from these refugia are often total and may extend even to the extraction of rattan, which is in much commercial demand." They conclude that human societies at different stages of development employ different mechanisms to promote ecological resilience and sustainable use of biological resources. Small-scale horticulturally based societies depend on a respect for the sacred or on social conventions, but the large agrarian societies require state-sponsored regulation, often at the cost of respect for the sacred.

Elsewhere, however, even before modern inputs such as chemical fertilizers, pesticides, and machinery became available, the ancient Mesopotamian, Javanese, Chinese, Aztec, Incan, Indian, and Cambodian civilizations thrived. They were able to produce a sufficient agricultural surplus to support high population levels, priesthoods, armies, vocational specialization, and impressive aesthetic development. Considerable evidence indicates that during prehistoric and early historic times, these intensive systems, as well as many of the more extensive forms of agriculture that were basically unmanaged extraction, resulted in widespread land degradation. Some researchers attribute the decline of some civilizations at least in part to this degradation, including in Mesopotamia, Central America, and Cambodia (Hillel 1991; Redman 1999).

More productive land may have been irreversibly degraded in the past 10,000 years than is currently under agricultural production, with climatic factors probably playing a role. Humus (the living part of the soil) is roughly estimated to have been lost at a rate of around 25 million metric tons per year, on average, since agriculture began 10,000 years ago. That loss rate has accelerated to 300 million tons per year in the past 300 years and 760 million tons per year in the past fifty years. Nearly 16 percent of the original stock of organic soil carbon may have been lost. Within the past 300 years, 100 million hectares of irrigated land alone apparently have been irreversibly degraded, while another 110 million hectares have suffered diminished productivity due to human-induced salinization (Rozanov, Targulian, and Orlov 1990). Millions of hectares of forests and natural vegetation have been cleared for agricultural use and for harvesting timber and fuelwood (see Figure 4.1). Such large-scale cutting and land conversion often degrade the soil, and they cause radical changes to the biodiversity of an area.

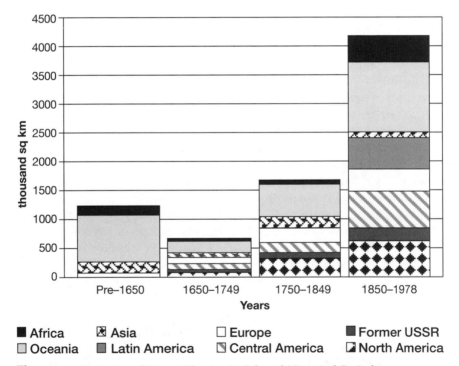

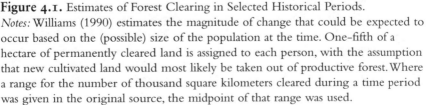

Figure 4.1. Estimates of Forest Clearing in Selected Historical Periods.
Notes: Williams (1990) estimates the magnitude of change that could be expected to occur based on the (possible) size of the population at the time. One-fifth of a hectare of permanently cleared land is assigned to each person, with the assumption that new cultivated land would most likely be taken out of productive forest. Where a range for the number of thousand square kilometers cleared during a time period was given in the original source, the midpoint of that range was used.

Why Farmers Feel Threatened by Wild Biodiversity

Ever since crops were first domesticated over 10,000 years ago, farmers have had reason to be concerned about nearby wildlife populations. Farmers' resistance to increasing wildlife populations can be considerable, even among individuals with a strong philosophical commitment to environmental values. Wild predators kill domestic livestock. Rodents and other mammals feed on crops in the field and on stored grain. Wild birds damage fruit and unprotected grains such as millet, while underground creatures munch on roots and vegetables. Snakes hiding in undergrowth are feared by both livestock and people. Weeds compete with crops and pastures for water and nutrients, reduce crop yields, and increase production costs. Bacteria, fungi, and viruses cause diseases, especially in the tropics where no winter breaks their life cycle. Today, farmers living near protected areas must sometimes contend with hippos trampling their crop fields, elephants uprooting their trees, birds devastating their rice fields, fruit bats harvesting their mangoes, or lions threatening their cattle. It is not surprising that wariness of the wild is shared by farming cultures around the world, even as they are typically more aware and appreciative of environmental values than their urban counterparts.

The economic cost of pests and diseases is high. It is estimated that one-fifth to one-fourth of global cereal output is lost to pests, diseases, and weeds before harvest; another large share is lost postharvest to rodents, fungi, and other pests. Detailed data on wildlife damage globally are difficult to collect and not widely available. In the United States, Pennsylvania suffers annual losses of $74 million statewide from wildlife damage to corn, alfalfa, oats, and cabbage (Brittingham and Tzilkowski 1995). Wildlife damage cost farmers in Ontario, Canada, an estimated U.S. $41 million in 1998. More than 50 percent of the field-crop, fruit, and vegetable farmers in Ontario reported some losses to wildlife that year. Coyotes, wolves, and dogs caused losses for 10 percent of Ontario's beef producers and 29 percent of sheep producers. More than 15 percent of fruit producers on average have to replant and/or reprune annually as a result of wildlife losses. Four percent of Ontario's vegetable producers and 2 percent of field-crop producers replant some crop as a result of damage caused by wildlife (Ontario Soil and Crop Improvement Association 2001).

Losses of crops to wildlife are even more significant in many developing countries. In Africa, the red-billed quelea (*Quelea quelea*) and several species of weaverbirds (*Ploceus* spp.) can lead to losses of cereals amounting to tens of millions of dollars (Ntiamoa-Baidu 1997). Various species of rodents also feed on cereals, peanuts, rice, corn, sugar cane, and cassava. Elephants, baboons, bush pigs, and various antelope can also be highly destructive of field crops. In Sri Lanka, elephants are notorious crop raiders and even attack villagers; in 1997 nearly thirty houses were destroyed by marauding elephants (Corea 2001). Rodents can cause considerable damage to stored products; in Indonesia, rats alone eat enough rice each year to feed 20 million people (Ylonen 2001). For all these reasons, efforts to protect wildlife have been minimal in most efforts to increase agricultural production.

Land Conversion

The loss and fragmentation of native habitats caused by agricultural development and conversion of agricultural lands into urban sprawl are widely recognized as the most serious modern threats to the conservation of biodiversity (Main, Roka, and Noss 1999; Sala et al. 2000; Heywood and Watson 1995). They are estimated to affect 89 percent of all threatened birds, 83 percent of threatened mammals, and 91 percent of threatened plants (IUCN 2000b). While some land has reverted to wilderness following agriculture-related degradation, in many cases fundamental ecological changes have prevented reestablishment of the original plants and animals.

Conversion of Natural Habitat to Agricultural Use

By far the most significant impact of agriculture on biodiversity is the conversion of highly diverse forests and other natural habitats into much simpler pastures or agricultural systems. An ecological rule of thumb is that a 30 percent or greater loss of natural vegetation leads to major shifts in wild biodiversity (Carl Binning, CSIRO, personal communication, 2001). Map 4.1 illustrates the global extent of agriculture by habitat type, with "agriculture" defined to mean at least 30 percent agricultural cover (crops, perennial crops distinguishable from forests, and cultivated pastures) as revealed by remote-sensing images. Converting native forests into grasslands, or forests or grasslands into croplands, leads to the loss of most native plant species and the animals that depend on them. Populations of below-ground organisms are also affected when the main vegetation cover is radically altered. While a majority of food production comes from intensive crop and livestock systems, these account for only a quarter and perhaps as little as a sixth of the total land area that is being utilized for—and thus ecologically impacted by—agriculture. Estimates from satellite imagery indicate that agriculture is the dominant land use in 10 percent of land area and part of the landscape mosaic in another 17 percent of land area, while 10 to 20 percent of land is under extensive grazing systems. About 1 to 5 percent of food is produced in natural forests.

Figure 4.2 summarizes the proportion of each major global habitat type that has been converted into agriculture, weighted according to the proportion of land cover. The two habitats most radically affected are temperate broadleaf and mixed forests, and tropical and subtropical dry and monsoon broadleaf forests. Nearly half the global area of these habitats has been converted to agricultural use. While the rates of tropical-forest clearing are very

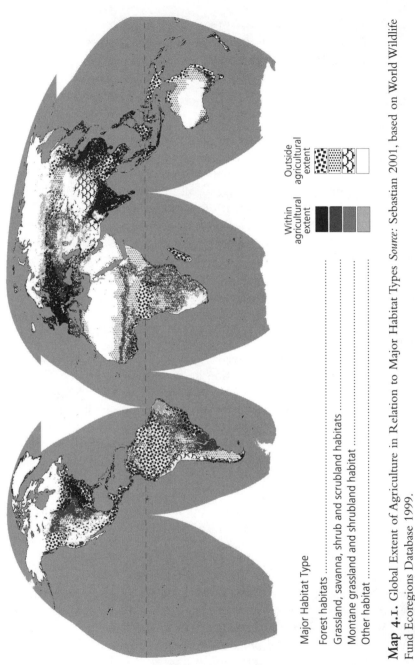

Major Habitat Type

Forest habitats ....
Grassland, savanna, shrub and scrubland habitats ....
Montane grassland and shrubland habitat ....
Other habitat ....

Within
agricultural
extent

Outside
agricultural
extent

Map 4.1. Global Extent of Agriculture in Relation to Major Habitat Types *Source:* Sebastian 2001, based on World Wildlife Fund Ecoregions Database 1999.

Note: The PAGE "agricultural extent" includes areas with greater than 30 percent crops or planted pasture, based on a reinter-pretation of GLCCD 1998 and USGS EDC 1999, plus additional irrigated areas based on Doell and Siebert 1999.

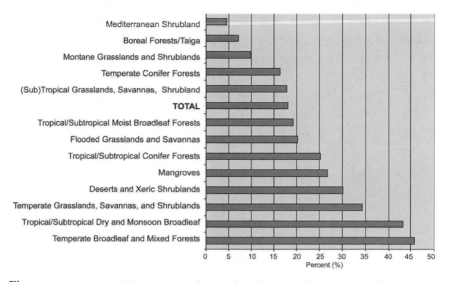

Figure 4.2. Estimated Percentage of Agricultural Land within Major Habitats
Source: WWF 1999 (Ecoregions Database).
Note: The agricultural area within a habitat type was determined by applying a weighted percentage to each PAGE agricultural land cover class: 80% for PAGE class >60% agricultural; 50% for class 40–60% agricultural; and 5% for class 0–30% agricultural (see Wood, Sebastian, and Scherr 2000).

high, only a fifth of all tropical and subtropical moist broadleaf forests ("rainforests") appear to have been converted to permanent agricultural use. This finding may reflect measurement errors, because of the difficulty in interpreting satellite imagery to detect agricultural land uses in and around abundant natural tree cover. Also, in long-rotation fallow systems the forest-fallow phase shows as "forest" in satellite images (Wood, Sebastian, and Scherr 2000). The greatest impact of agriculture on vegetation classes has been on grassland, which is not surprising given that the most extensively grown agricultural crops are cereal grasses (corn, rice, and wheat).

Caution is required in interpreting data about land conversion. Most statistics report only the net results of change, when in fact land use change is highly dynamic, with particular parcels of land moving in and out of different uses. For example, China in 1994 had a total stock of 131.5 million hectares of cultivated land. In 1995, the net change of increases and declines, some 409,000 hectares, reduced the stock of cultivated land only slightly. But this apparent stability masked major changes. In that year, China lost some 798,000 hectares of land for food production, most of it converted to horticulture, used for reforestation, or lost in natural disasters. At the same time, China's farmers expanded cultivated land by nearly 389,000 hectares, mainly by reclaiming previously unused areas, but also by converting areas previously used for other purposes (Wood, Sebastian, and Scherr 2000).Thus the area of land whose wild biodiversity was affected by agricultural use is much larger than is implied by net figures.

Critical Areas of Biodiversity Loss Due to Conversion

The global trend shows agricultural lands increasing by only 1.4 percent from 1987 to 1997 (World Resources Institute 2000), but some parts of the world are suffering particularly egregious losses of biodiversity due to the conversion of habitats to agricultural uses. Widespread burning of forests for commercial logging and plantation clearing in the late 1990s in Indonesia accelerated the losses in that megadiversity country. In South America land-use changes are rapid and widespread. Deforestation in the Brazilian Amazon increased by 27 percent in 1998 when 1.68 million hectares were cleared. Since 1972, more than 50 million hectares, some 13 percent of the entire Amazon region, has been converted to crops and pastures. A similar amount has been severely damaged by logging crews and fire. In Central America, pasture areas increased from 3.5 million hectares in 1950 to 9.5 million hectares in 1992 (Blackburn and de Haan 1999).

Within the Southeast Asian biodiversity hotspots alone—the Philippines, Wallacea (the islands of Borneo, Bali, Sumatra, and Java), Sundaland (the Indonesian islands from Sulawesi and Lombok east to Halmahera and Ceram), and Indo-Burma—cropland grew by some 11 million hectares in the decade prior to forest surveys conducted in 1992–94. Nearly all of this new cropland was cleared from forests, and since then rates of deforestation appear to have accelerated, assisted by burning in 1997, 1998, and 1999, and by commercial logging that was stepped up during the recent Asian economic crisis (Matthews 2000). Tree-felling, a very broad term most often associated with land-clearing for agriculture, is considered to have the most important impact on tree diversity, and IUCN now lists more than 8,700 tree species as threatened (Oldfield, Lusty, and MacKinven 1998).

Recovery of Agricultural Land Back to Natural Vegetation

Land conversion has not been unidirectional. An estimated 5 to 7 million hectares of land formerly in cropland (about 0.3 to 0.4 percent of the world's arable land) are abandoned annually due to severe soil degradation (Scherr 2001). Economic changes have also led to abandonment of cropland, particularly in land-abundant countries where market forces have encouraged regional economic specialization.

This cyclical process is nothing new; many of the areas that currently are considered "natural habitat" of greatest importance for biodiversity are abandoned fields or pastures. For example, the forests of Southeast Asia have been cleared repeatedly by shifting cultivators over the past several thousand years (Spencer 1966; Bellwood 1985). Many of the forested areas of Borneo and Sumatra supported ancient civilizations (see, for example, Schnitger 1964, Coedes 1968, and McNeely and Wachtel 1988). If a scientist had assessed the fate of Sumatran forests in, say, 1000 A.D., he or she quite likely would have concluded that the habitat had been ruined and that any remaining forest would soon disappear. But the forests of Southeast Asia recovered from past abuses, though many species were lost, especially in Java (McNeely 1978). Similar stories can be told about many other forested regions (McNeely 1994) and arid lands (Redman 1999). Few developing countries have yet experienced the trend in the United States and Europe of taking large areas out of crop production for conservation purposes. History suggests, however, that areas inappropriate for agriculture that are forced into annual crop production will be abandoned eventually and revert to more natural habitats.

Where even relatively small areas of habitat refuge have survived, they can provide the source of plants and animals that can repopulate abandoned fields. Nature has demonstrated remarkable resilience in many such situations. But even nature has her limits. In some cases, the original biota cannot reestablish on their own, due to fundamental changes in soil chemistry and structure, hydrological systems, seed sources, breeding populations, and crucial pollinators. The conversion of tropical forests to monocultural plantation cash crops, for example, is hard to reverse. When the price of coffee, cocoa, or palm oil drops so that the plantation is no longer economically viable, it cannot quickly revert to the biologically diverse forest that preceded it (WRI, IUCN, and UNEP 1992). Regeneration of biodiversity is problematic if important functional groups of organisms are lost, such as symbiotic mycorrhizal fungi. Thus degraded semiarid areas are often difficult to revegetate, if, as is usual, soil inoculum of these species has been lost (Roger Leakey, James Cook University, personal communication, 2000).

Modification of Landscapes

Landscape characteristics in and around agricultural areas also may have significant effects on wild biodiversity. Landscape structure, such as the size of patches of land under different uses, the mosaic of different land uses, and the distribution of disturbance patterns such as road networks, can affect the mix and numbers of wild biodiversity. For example, landscape structures may favor species that inhabit the edges of habitats (e.g., where forest meets grassland or cropland) rather than the interior, or they may interfere with the mobility wild species require to find food or mates (Forman 1995). Regional or global evidence is available for only two variables: the degree of landscape fragmentation and the amount of perennial vegetation retained in agricultural landscapes.

Fragmentation

Natural forests, woodlands, and savannas in all parts of the world are becoming increasingly fragmented by conversion to agricultural uses, as Map 4.2 illustrates for the Lake Victoria region in East Africa. These small, isolated forest fragments will lose many of their original species as a result, and the abun-

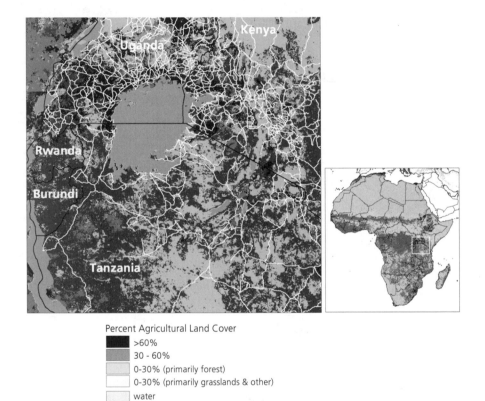

Percent Agricultural Land Cover
- >60%
- 30 - 60%
- 0-30% (primarily forest)
- 0-30% (primarily grasslands & other)
- water

Country boundaries

Highways & partly surfaced roads   Highlighted area

Map 4.2. Land Use Fragmentation around Lake Victoria, East Africa.
Source: Sebastian 2001.
Note: The PAGE "agricultural extent" includes areas with greater than 30 percent crops or planted pasture, based on a reinterpretation of GLCCD 1998 and USGS EDC 1999, plus additional irrigated areas based on Doell and Siebert 1999.

dance of particular species will change as will interactions among remaining species. It appears that in the first fifty years after emerging from isolation, tropical forest fragments of roughly 1,000 hectares lose half the total number of species they will eventually lose (see Box 4.1). This rate sets the time frame during which humanity must act to conserve wildlife in fragmented tropical forests and indicates the considerable extent of future global extinction. A key question is whether the fragmentation is permanent or reversible. Historically it has been reversible in Southeast Asia (Spencer 1966), Central America (Gomez-Pompa and Kaus 1992), West Africa (Fairhead and Leach 1998), and Amazonia (Roosevelt 1994), though undoubtedly some species were lost.

In tropical forests, large-canopy trees are critically important sources of fruits, flowers, and shelter for animal populations; they strongly influence forest structure, composition, hydrology, and carbon storage. Forest fragmentation in central Amazonia is having a devastating effect on these large trees, which die at a rate nearly three times faster when they are within 300 meters of edges than they do in forest interiors (Laurance et al. 2000). The rapid rate of mortality of large trees may diminish forest volume and structural complexity, promote the

---

Box 4.1. The Forest Fragments Project

A biological experiment planned to last a century was begun in Brazilian Amazonia in the late 1970s. The Forest Fragments Project was designed to determine how much rainforest is needed to maintain 99 percent of native species for a century. Brazilian ranchers and large-scale farmers agreed to leave their remaining rainforest in square tracts of varying sizes, from 1 to 1,000 hectares, so that scientists could monitor the effects of fragmentation on biological diversity. Changes were observed quickly, especially in the smaller forest tracts where numerous midsize plant and animal species dwindled in a matter of years, then disappeared. Many of the missing species were casualties of biodiversity "chain reactions." For example, colonies of army ants were lost from woodlands of 10 hectares or smaller. Then came the loss of ant-birds, which prey on insects taking flight to escape marauding ants. Vegetation along forest borders dried and was soon supplanted by nonrainforest plants. Bee populations plummeted even in the 100-hectare plots, placing at risk many species of orchids and other flowering plants that depend on them for pollination. The large mammals with extensive home ranges and big appetites—jaguars, pumas, and peccaries among them—simply left the area, abandoning even the larger tracts. Without peccaries digging wallows, which fill with rainwater, three species of frogs failed to breed and thus disappeared. Further chain reactions are expected, but ironically the project is threatened by re-encroachment of forest because the "isolated" patches have been difficult to maintain as isolated (Wilson 1992).

proliferation of short-lived pioneer species, and alter natural cycles affecting evapotranspiration, carbon cycling, and greenhouse-gas emissions.

Research has shown that animal movements and species richness are positively affected by the existence of corridors (strips of vegetation serving as habitat) and landscape connectivity (the extent to which different patches of habitat are linked together) (Debinski and Holt 2000). For some tree species, high tree density in fragments can partially offset the loss of genetic diversity that may result from fragmentation. But studies also show a wide range of species-specific responses to fragmentation, demonstrating the need for a better understanding of the ways species respond to fragmentation over time (Boffa, Petri, and do Amaral 2000). On the other hand, a substantial element of the rainforest plant community may be relatively resistant to fragmentation, so the dire predictions of plant extinction through loss of pollinating and seed-dispersing vertebrates may be overly pessimistic. But while such fragments may maintain a relatively high diversity of tree species, the vertebrate fauna, especially mammals, may be greatly reduced. Where hunting pressures are high, even structurally intact forests may be empty of large vertebrates, as is now the case in much of mainland Southeast Asia and parts of Amazonia (Robinson and Redford 1991).

Tree Cover in Agricultural Lands

Perennial vegetation in and around farm fields, especially native perennials and those found in ecologically strategic niches (for example, along stream banks or on steep slopes), form a landscape mosaic that can expand critical habitats for wildlife. Contrary to popular perception, many farming systems retain a high proportion of tree cover in agricultural land. Map 4.3 shows major regional differences in tree cover on lands classified as "agricultural extent," that is, having at least 30 percent agricultural land cover. This visual depiction is based on 1-kilometer resolution advanced very high resolution radiometer (AVHRR) satellite data. The proportion of tree cover is lowest in West Asia and North Africa, which had limited forest cover historically and is long-settled, dry, and dominated by livestock and cereals. Cover is also relatively low in Oceania, probably due to drier conditions, lower initial forest cover, and lack of agricultural land. In North America, Europe, the former USSR, and East Asia, more than half of all agricultural land has little or no tree cover, despite high original extent of natural forests in many agricultural regions. This is due in some cases to patterns of agricultural land-clearing and mechanized operations, in others to land scarcity, and elsewhere to habitat conditions.

In Latin America, sub-Saharan Africa, and South and Southeast Asia, by contrast, a high proportion of agricultural land (46, 36, and 25 percent, respectively) has more than 30 percent tree cover, except in the intensely cultivated irrigated

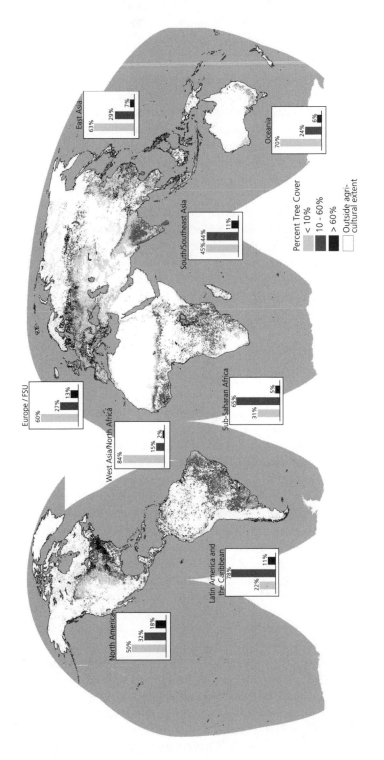

Map 4.3. Tree Cover in Agricultural Lands. *Source:* Sebastian 2001, adapted from Wood, Sebastian and Scherr 2000, based on DeFries, et al. 2000.

Note: The PAGE "agricultural extent" includes areas with greater than 30 percent crops or planted pasture, based on a reinterpretation of GLCCD 1998 and USGS EDC 1999, plus additional irrigated areas based on Doell and Siebert 1999.

lands in some of the drylands of South Asia. Several factors explain this. Tree and shrub fallows are still widely used in less densely populated areas; tree crop plantations (tea, coffee, oil palm, rubber, and so on) are a more important component of the agricultural economy in the tropics; and agricultural landscapes vary more in the wetter tropics, with many parts of farms that are unsuitable for crops. Farmers historically have protected or established trees on farms for economic reasons—to supply subsistence needs for fuelwood, building materials, fencing, medicines, fruits, nuts, and other products, or to trade. Indeed, in many humid and subhumid tropical areas, a pattern is observed whereby tree cover first declines as the population grows, but then reaches a point where local people begin to conserve and then reforest parts of the landscape so that tree cover increases (Scherr and Templeton 2000). Where appropriate incentives are available, farmers will continue this rational response of maintaining a wide diversity of crops, including trees; but where powerful economic incentives discourage such diversity by favoring monocropping, tree cover—along with the multiplicity of species that accompanies trees—will be lost.

Modifications in Hydrological Systems

To achieve higher and more stable agricultural yields, human societies have made dramatic changes in patterns of water flow and storage by building dams, controlling floods, draining wetlands, and pumping groundwater. In addition, humans use rivers, lakes, and swamps for transportation, sewage disposal, and daily water needs, so that much of the earth's accessible fresh waters are already co-opted by humans (Sala et al. 2000). Water diversions for agriculture are as serious a problem for many aquatic species as land conversion is for terrestrial species. Given the myriad ways humans use water, the biodiversity of freshwater ecosystems is far more threatened than that of terrestrial ecosystems. Of the estimated 10,000 freshwater fish species, about 20 percent are threatened, while nearly 35 percent of turtle species are also in danger of extinction. Freshwater snakes and crustaceans are even worse off, with population declines of over 80 percent in some major river systems. Most of these threats to aquatic habitats are related to agriculture (World Conservation Monitoring Centre 2000).

Irrigation

As noted above, more than 250 million hectares globally are under irrigation for agriculture, of which three-quarters is in developing countries, particularly India, China, and Pakistan (Map 4.4). Of the 9,000 to 12,500 cubic kilometers of water estimated to be available globally for use each year (Postel, Dailey, and Erhlich 1996), between 3,500 and 3,700 cubic kilometers were extracted in 1995 (Shiklomanov 1996). Of that total, around 70 percent—

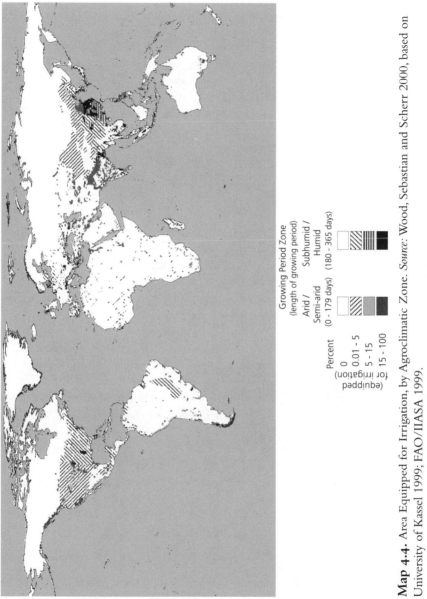

Map 4.4. Area Equipped for Irrigation, by Agroclimatic Zone. *Source:* Wood, Sebastian and Scherr 2000, based on University of Kassel 1999; FAO/IIASA 1999.

Note: The percent equipped for irrigation is within a 50 by 50 kilometer area.

some 2,450 to 2,700 cubic kilometers—was used for irrigation (Postel 1999; World Conservation Monitoring Centre 2000). In low-income countries, 87 percent of water extraction was used for agriculture, compared to 74 percent in middle-income countries and 30 percent in high-income countries (World Bank 2000). Most experts believe that countries should withdraw no more than 20 percent of their fresh water supplies annually for all purposes, in order to maintain healthy hydrological systems (Postel 1999; Abramovitz 1996).

Decreased river flows and falling groundwater levels are pervasive in irrigated areas, as there are usually few incentives not to overuse water. As a result, most irrigation systems are highly inefficient in their use of water. In the United States, farmers are extracting groundwater at greater than the recharge rate for roughly one-fifth of the irrigated area (Postel 1999). Water tables in the North China Plain have been falling by up to 1 meter per year, and by 25 to 30 meters in a decade in parts of Tamil Nadu, India. These massive transfers of water from rivers, artificial reservoirs, and underground aquifers, with associated flood control, canal, and drainage infrastructure, have reshaped hydrological regimes, created barriers to annual fish migrations, and altered water dynamics. Disrupting the natural flow of rivers can lead to disruption of the complex biological communities that depend on the natural variability of the river, change water temperature and chemistry, affect sedimentation rates, and cause numerous other impacts on wild biodiversity (World Commission on Dams 2000). Dam building is considered one of the top two causes (along with biological invasion) of freshwater species extinction. Almost 60 percent of the world's largest 237 rivers are greatly or moderately fragmented by dams, diversions, or canals, often leading to overextraction of water from rivers. This not only reduces water levels, but also changes water chemistry in both the river and the bodies of water into which it drains. A notorious case is the remarkable shrinkage and salinization of the Aral Sea and the consequent loss of fish species and fishing livelihoods (Gleick 1993; Postel 1999). The White River National Wildlife Refuge in Arkansas, USA, is currently threatened by an irrigation project whose design will disrupt the natural river flow and destroy rare forested wetlands (National Audubon Society 2000).

Drainage of Wetlands

Wetlands (which include swamps, marshes, lakes, rivers, estuaries, and peat lands) are among the most species-rich natural habitats. Half the world's wetlands are estimated to have been lost in the twentieth century. While urban and infrastructure development, recreational use, and control of diseases such as malaria have been important rationales for wetland conversion, land development for agricultural use has long been the leading factor. In the United States, up until the 1950s approximately 87 percent of all wetland conversion was attributable to agriculture. Recent legislation restricting wetland conversion changed that share; between 1982 and 1992 only 20 percent of wetland losses were due to agriculture, while 57 percent were due to urban expansion (Lacher et al. 1999). Wetlands drainage, canal construction, and other changes in the water regime resulting from agricultural use have radical impacts on the numerous species that require particular characteristics of water chemistry, temperature, and flow patterns (Revenga et al. 2000). For example, when the floodplain of Denmark's Skjern River was converted to agriculture, otters disappeared, nesting waterfowl declined, and the country's last wild population of salmon was reduced to a small fraction of its peak. A major restoration project is now under way to correct these past abuses.

While most developed countries have established controls restricting further wetland drainage, and even initiated some wetland habitat restoration, in many developing countries conversion of wetlands is still seen as a crucial mechanism to relieve agricultural land scarcity. Even officially protected areas remain under threat. In more than half of the nearly 1,000 Wetlands of International Importance listed under the Ramsar Convention, agriculture is considered to be a major cause of change to wetlands. Of 957 designated Ramsar sites, 250 are currently being used for agriculture and 95 for aquaculture (Frazier 1999). Negotiations are under way among national and regional stakeholders regarding the extent to which the Mekong wetlands should be converted to agriculture (Ahmed and Hirsch 2000), though the information needed to make an informed decision is only partially available.

Changing Vegetation in Watersheds

Less visible and sometimes more pernicious hydrological impacts have resulted from converting natural vegetation in major watersheds to agricultural use. A global digital map of watersheds shows that a high and growing proportion of land in the world's river basins is now under agricultural use (Map 4.5), most notably in Europe and South Asia. The construction of roads and footpaths and increases in rapid runoff from new construction and land-clearing associated with farming often affect hydrological patterns as much as the farming itself. Conversion reduces rainfall infiltration into the soil and thus water storage in the soil; it increases the speed of water flow over the land surface, which in turn accelerates soil erosion and increases flooding risks downstream. In turn, these changes modify aquatic, soil, and riverine habitats and their associated biodiversity. Natural vegetation usually does the best job of intercepting rainfall and slowing surface water flow, so rain has time to filter through the soil and recharge local water supplies.

New research shows that plant monocultures near fresh waters may result in dramatic declines in the biodiversity of aquatic sediments, because diverse

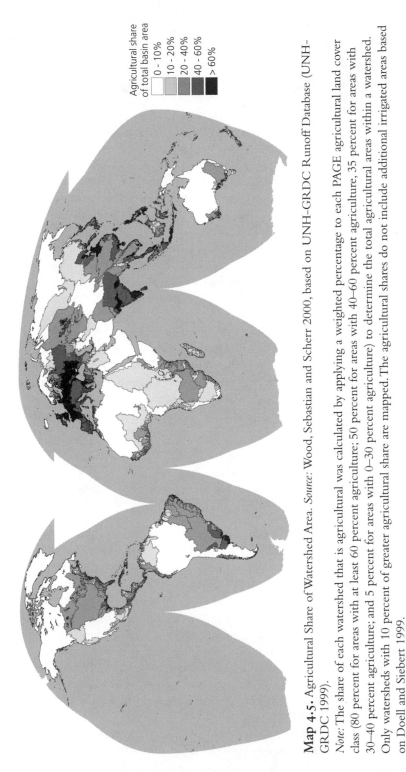

Map 4.5. Agricultural Share of Watershed Area. *Source:* Wood, Sebastian and Scherr 2000, based on UNH–GRDC Runoff Database (UNH–GRDC 1999).

Note: The share of each watershed that is agricultural was calculated by applying a weighted percentage to each PAGE agricultural land cover class (80 percent for areas with at least 60 percent agriculture; 50 percent for areas with 40–60 percent agriculture, 35 percent for areas with 30–40 percent agriculture; and 5 percent for areas with 0–30 percent agriculture) to determine the total agricultural areas within a watershed. Only watersheds with 10 percent of greater agricultural share are mapped. The agricultural shares do not include additional irrigated areas based on Doell and Siebert 1999.

Agricultural share
of total basin area

☐ 0 - 10%
10 - 20%
20 - 40%
40 - 60%
> 60%

plant communities provide diverse food, habitat, and shade for sediment biota. Thus land-use and management practices may need to be revised to focus on the maintenance of plant diversity adjacent to freshwater bodies, not only on the maintenance of a sizable plant buffer zone (Palmer et al. 2000).

Habitat Pollution and Degradation

Habitat pollution and degradation are considered a critical threat to most wild species. They are a major threat to 85 percent of threatened species in the United States (Wilcove et al. 1998). Negative impacts from agricultural production include agrochemical pollution, soil and rangeland degradation, and loss of critical habitats and associated wild biodiversity. Agricultural intensification in much of northern, western, and central Europe is leading to alarming declines in the wildlife inhabiting farmland. Birds, butterflies, beneficial insects, wildflowers, and game species are disappearing (Sotherton 1998).

Pesticide Pollution

Pests and diseases that compete with crops, or feed on crops and livestock, have been the perennial scourge of farmers around the world. It should come as no surprise that farmers take strong measures to fight these pests. With technological breakthroughs beginning in the mid-1800s, but greatly accelerating after World War II, farmers have been using chemicals to fight crop-eating insects, competing weeds, and fungal diseases. As shown in Table 4.1, pesticide use has expanded dramatically, from around U.S.$20.5 billion in 1983 to U.S.$34.1 billion in 1998 (these figures include nonagricultural uses). Globally, about 45 percent of the value of pesticides used is herbicides, 29 percent insecticides, 19

Table 4.1. Global Pesticide Consumption, 1983–98

Region	Value in 1983[a] (in U.S. $)	Value in 1998[a] (in U.S. $)	Percentage increase 1983–1998	1998 value per hectare of cropland[b]
North America	3,991	8,980	125	40
Latin America	1,258	3,000	138	19
Western Europe	5,847	9,000	54	102
Eastern Europe	2,898	3,190	10	14
Africa/Mideast	942	1,610	71	5
Asia/Oceania	5,572	8,370	50	16
Total	20,507	34,150	67	23

Source: (a) Yudelman, Ratta, and Nygaard 1998; (b) IFPRI calculation based on Yudelman, Ratta, and Nygaard 1998 and FAOSTAT 1999, reported in Wood, Sebastian, and Scherr 2000, Table 9.

percent fungicides, and 6 percent other chemicals (Yudelman, Ratta, and Nygaard 1998). Developed countries spend roughly two-thirds of the total amount spent on pesticides. Use intensity is highest in Western Europe, where the value of pesticides applied per hectare is four times the global average, and twenty times higher than in Africa. Many pesticides have made a significant contribution to the growth in crop yields, though development of pesticide resistance has eroded some of those gains (Wood, Sebastian, and Scherr 2000).

Unfortunately, many pesticides have had a disastrous impact on biodiversity, through direct ingestion by wild animals, poisoning of wild plants, and pollution of freshwater and coastal habitats. Pesticide residues enter waterways via runoff, affecting water quality, entering the food chain, and disrupting freshwater and coastal ecosystems, including coral reefs, mangrove forests, and sea-grass beds.

The insects that are the direct target of insecticides are obvious victims, but other species are also affected. A particular problem for biodiversity is the impact of persistent organic pollutants (POPs) such as DDT, dieldrin, and heptachlor. Rachel Carson's classic book, *Silent Spring* (1962), described the impact of these pesticides on wildlife populations. POPs accumulate in fatty tissues and are found in increasing concentrations higher up in food chains. By disrupting their endocrine systems (which produce hormones), these chemicals disrupt reproduction and lead to loss of fertility, birth defects, and other problems (which also affect humans) (Colborn, Dumanoski, and Myers 1996). Top carnivores such as hawks and eagles, for example, can have concentrations of POPs that are 10 million times higher than those of grazing species. Otters and other fish-eating species have also been affected by direct poisoning, disruption of reproduction, and increased susceptibility to disease. The impacts on nontarget invertebrates and decomposer microorganisms are poorly known but could be disastrous if the experience with larger animals is an indicator of potential effects.

New generations of pesticides have been produced since the alarms about pesticide use were first raised in the 1970s. Many of these new pesticides target pests with greater specificity and/or have lower toxicity levels. Efforts are expanding to develop and promote integrated pest management systems that reduce or avoid use of pesticides toxic to wildlife, as well as to educate farmers to use economically optimal levels of pesticides, rather than levels that eradicate all pests as well as beneficial insects. Nonetheless, pesticide pollution remains a critical threat to biodiversity in many parts of the developing world, where regulations are poorly enforced, farmers are poorly informed, and high-value crops, such as cotton, are grown using high levels of pesticides.

Fertilizer and Organic Nutrient Pollution

Maintaining high levels of agricultural productivity depends on continual renewal and enrichment of a soil's organic matter to replace nutrients "exported" through crop harvest. These nutrients may be obtained through organic sources such as animal manure, sludge, compost, and mulch (all of which also replenish organic matter) or through inorganic fertilizers. For farmers with ready market access, inorganic fertilizers developed since the mid-nineteenth century have been a relatively inexpensive source of nutrients. European farmers used them widely by 1900. Globally, application of inorganic fertilizers increased from 14 million metric tons (5.5 kilograms per capita) in 1950 to 137 million tons (23.1 kilograms per capita) in 1998. See Map 4.6 for an overview of the current extent of fertilizer application. Availability of this low-cost nutrient source, and the breeding of crop varieties that can take advantage of nutrient-rich fields, is one of the key factors behind historic increases in crop yields in recent decades (Pinstrup-Andersen, Pandya-Lorch, and Rosegrant 1997). The importance of maintaining a soil's organic matter for long-term sustainability of production—and even for optimal performance of inorganic fertilizers—has become evident in recent years, especially in many types of tropical soils (Palm, Myers, and Nandwa 1997). But reliance on chemical fertilizers, at least to supplement organic inputs, is likely to remain essential to maintaining high yields in many farming systems. Where synthetic fertilizers are not used, high levels of organic nutrient sources such as animal manures are required, as in some systems of intensive, organic agricultural production. The explosive growth in intensive livestock operations has led to large accumulations of organic waste materials (bedding and manure) that potentially can be used as fertilizers.

Fertilizers are boons to farmers, but they also have ecological costs. Some—especially concentrated inorganic products but also some organic fertilizers—are easily washed away by heavy rains. Nutrients from these sources are deposited directly in waterways and water supplies, or indirectly through soil infiltration into groundwater or surface runoff. The nutrients nitrogen and phosphorous cause the most problems, for drinking water supplies as well as ponds and lakes.

Effluent discharge from large-scale feedlot operations is a major source of water pollution. Even small farms can pollute nearby streams and other water bodies. For example, in 1,785 bodies of water in thirty-nine states of the United States, livestock waste has been identified as the principal pollutant (Cincotta and Engelman 2000). While in some places waste is applied to crop fields as fertilizer, currently much livestock waste in the United States is not suited for such use for complex reasons, including geography, lack of sufficient local carbon source for composting, high energy requirements for drying slurry, and so forth.

Effluent discharge associated with aquaculture also contributes to nutrient pollution and poor water quality. Particularly near intensive aquaculture systems (for which the fish farmer supplies nearly all nutrients), untreated wastewater and sedimentation from uneaten feed often collect because farmers

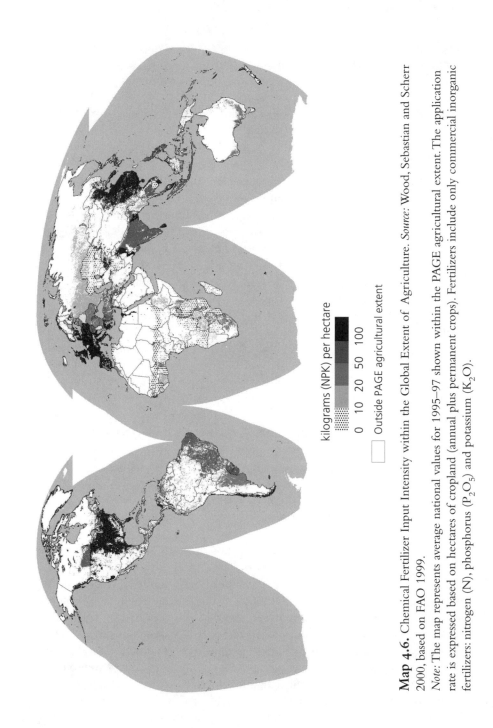

kilograms (NPK) per hectare

0   10   20   50   100

Outside PAGE agricultural extent

Map 4.6. Chemical Fertilizer Input Intensity within the Global Extent of Agriculture. *Source:* Wood, Sebastian and Scherr 2000, based on FAO 1999.

Note: The map represents average national values for 1995–97 shown within the PAGE agricultural extent. The application rate is expressed based on hectares of cropland (annual plus permanent crops). Fertilizers include only commercial inorganic fertilizers: nitrogen (N), phosphorus (P_2O_5) and potassium (K_2O).

lack sufficient knowledge of dietary requirements for different species. Inefficiencies in the production process lead to overexploitation of capture fisheries, discharge of effluents (often into fragile marine ecosystems such as mangrove swamps), and the discard of large amounts of by-catch (nontarget fish inadvertently harvested along with target fish) (Naylor et al. 2000).

Excessive nutrients poured into lakes, rivers, and the coastal zone through runoff and leaching (whether from organic or inorganic fertilizers or animal wastes) can cause serious harm to wild biodiversity. Excessive growth of aquatic plant life resulting from overabundant nutrients (known as eutrophication) can turn wetlands into wastelands. The resulting long-term increase of aquatic plant life depletes oxygen over large areas, killing fish and dramatically altering ecosystems. One oxygen-depleted "dead zone" near the outlet of the Mississippi River in the Gulf of Mexico covers 18,000 square kilometers (an area larger than Kuwait). The loss of oxygen has dramatically reduced populations and species diversity, increased mortality among bottom-dwelling communities, and put fishery resources under stress, thereby threatening the livelihoods of local fishermen (Alexander, Smith, and Schwarz 2000). Even-larger dead zones are reported in the Baltic and Black Seas (Cincotta and Engelman 2000). Eutrophication of surface waters and oxygen depletion in bottom waters, caused by runoff of nutrients from land, have also been increasing along the west coast of India, leading to serious damage to marine life.

Soil Erosion and Degradation

All agricultural production depends upon the thin layer of biologically active soil that lies above bedrock; this layer provides nutrients, circulates and stores water, and hosts micro-flora and -fauna that are beneficial for plant growth. Annual crops typically rely only on the top meter of soil, though perennial grasses and trees have roots that can seek sustenance far deeper. Certain types of soils are especially desirable for crop production, and regions where such soils are abundant are the "bread baskets" of the world. But only 16 percent of soils in the world's agricultural areas fit this profile, and about 60 percent of these are in temperate areas. Only 15 percent lie within the tropics (Wood, Sebastian, and Scherr 2000). The 84 percent that does not fit the ideal profile has significant inherent constraints that limit crop production and require special management to maintain the properties needed for productive agriculture. These include characteristics such as erodability, high acidity or alkalinity, or poor capacity to hold nutrients.

Maintaining adequate soil quality and depth for agriculture has always been a challenge for farmers, particularly on the more constrained soils. The challenge is greatest in the tropics, where warmer temperatures break down organic matter rapidly and high-intensity rainfall and winds present more

powerful erosive forces. These pressures, combined with inadequate land husbandry, have led to widespread declines in soil quality and loss of the most biologically active topsoil. Degradation variously affects soil depth, organic
matter content, water-holding capacity, physical structure, and chemical characteristics such as acidity and salinity. As a result of degradation, it is roughly
estimated that potential productivity has declined substantially on 16 percent
of land in developing countries and half of intensively managed cropland. The
most extensive degradation is found on cropland in Africa and Central America, pasture in Africa, and forests in Central America. Global data from quali-
tative expert assessments suggest that almost 75 percent of Central America's
agricultural land has been seriously degraded, as has 20 percent of Africa's and
11 percent of Asia's. More detailed data for Southeast Asia shown on Map 4.7
suggest even greater degradation there. Each year some 5 to 7 million hectares
of cropland are abandoned due to soil erosion, nutrient depletion, salinization,
and waterlogging (Scherr 1999). These changes damage the habitat for wildlife
directly dependent on soil resources and for wildlife all along the food chain.
They also require the conversion of more land to agriculture to replace the
abandoned lands.

Frequent cultivation to create a high-quality seedbed for annual crops, and
to control weeds on otherwise bare soil during seed establishment, contributes
greatly to the degradation of habitats of soil microorganisms by breaking down
soil structure, which makes soil prone to compaction or erosion by wind and
water. Erosion removes the most productive topsoil and leads to sedimentation
of waterways, making the water murky, smothering organisms on the bottoms
of streams and other water bodies, and in severe cases clogging small streams.
Cultivation causes changes in the chemical and physical environment of soil
habitats that directly threaten the survival and reproduction of flora and fauna
sensitive to soil conditions, including microorganisms, thereby disrupting the
natural processes that maintain soil productivity. Once soils are degraded to the
point that agriculture is no longer profitable, some pioneer species will take
advantage of existing conditions, but those are often weedy species; the species
characteristic of more mature ecosystems are unlikely to find these habitats
suitable for many decades.

Rangeland Degradation

Far more of the earth's land area is used as rangeland, to feed domestic livestock,
than as cropland (see Map 3.1). These same lands are commonly shared with
wildlife—and have considerable potential for ecologically stable coexistence
when properly managed. However, in many parts of the developed and developing world, overgrazing by livestock, coupled with episodic droughts, has
caused widespread rangeland degradation and loss of plant and animal diversity,

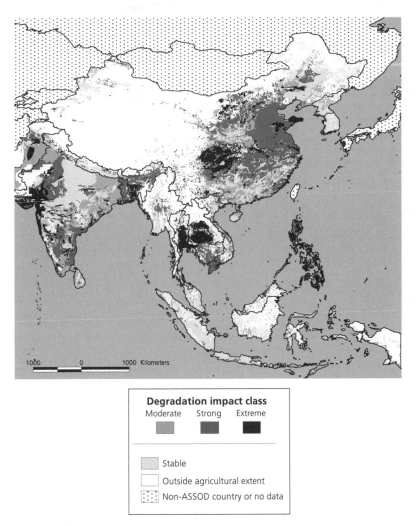

Degradation impact class

Moderate   Strong   Extreme

Stable

Outside agricultural extent

Non-ASSOD country or no data

Map 4.7. Soil Degradation in South and Southeast Asia within the Extent of Agriculture. *Source:* Wood, Sebastian and Scherr 2000, based on Assessment of the Status of Human-Induced Soil Degradation in South and Southeast Asia (ASSOD) (van Lynden and Oldeman 1997).

Note: The PAGE "agricultural extent" includes areas with greater than 30 percent agriculture, based on a reinterpretation of GLCCD 1998, plus additional irrigated areas based on Doell and Seibert 1999. Wind erosion includes: loss of topsoil by wind action, terrain deformation and overblowing; water erosion includes: loss of topsoil and terrain deformation; chemical degradation includes: fertility decline, reduced organic matter content, salinization/alkalization, dystrification/acidification, eutrophication and pollution; physical degradation includes: compaction, crusting, sealing, waterlogging, lowering of the soil's surface, loss of productive function, and aridification; stable lands include land under human influence with no or little degradation impact; and wasteland is land with minimal vegetation and with little human influence on soil stability.

despite high natural resilience. For example, the world's greatest diversity of grazing animals is found on the African savannas. The exceptional diversity of these species, and their size and numbers, is directly linked to the great heterogeneity of African savanna ecosystems. Herbivores of different types and sizes feed in different habitats and have different geographical ranges. Intact communities of these grazing animals in the savannas, where species are distributed across body size classes and a variety of feeding habits (grazing, browsing, or a combination of the two), help to regulate the structure of the savanna ecosystem and the way that it functions. In many parts of Africa, however, these highly diverse systems have been replaced with livestock systems of low diversity and high biomass density within a narrow body size range (primarily cattle and sheep), leading to the simplification and, hence, vulnerability of ecosystems. The removal of competitors, pathogens, and predators and the widespread provisioning of water have led to ecological dominance of these livestock. Even if the populations of the domesticated animals are reduced, the original biodiversity is unlikely to recover very quickly (du Toit and Cumming 1999).

Forest Degradation

Natural forest systems that have avoided complete deforestation have often been subject to significant degradation. Factors include overexploitation of selected timber and nontimber species; disease conditions exacerbated by human use patterns, such as the spread of malaria-carrying mosquitoes; trampling and browsing by domestic livestock; compaction of soils and plant damage by logging; and damage from uncontrolled fires. Putz et al. (2000) compare the expected effects of a range of forest uses on the components of biodiversity, as shown in Figure 4.3. They conclude that nontimber forest products, reduced-impact logging, reserves, and forestry techniques such as selective thinning mostly conserve biodiversity components at landscape, ecosystem, and community scales, but they still affect species and genetic diversity. Genetic, species, and community biodiversity are mostly reduced as a result of conventional logging and enrichment planting (establishment of higher-value species or improved varieties in the regrowing forest).

Reduced Pollination and Seed Dispersal

A major negative outcome of habitat pollution and degradation has been the decline in species responsible for the pollination and seed dispersal of wild plant species. The vast majority of species found in many tropical forests come from animal-dispersed seeds (Howe and Smallwood 1982), so the loss of animals

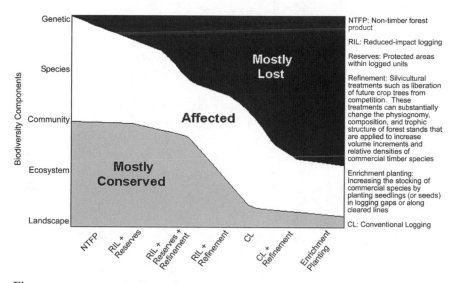

Figure 4.3. Expected Effects of Management on a Range of Forest Components of Biodiversity. *Source:* Putz et al. 2000.

inevitably affects the recovery rates of forests. While insects are the most important animal pollinators, vertebrates including birds, bats, and squirrels pollinate about 5 percent of rainforest canopy trees and 20 to 25 percent of the sub-canopy and understory plants (Buchmann and Nabhan 1996). Important issues to study include interactions between plants and their pollinators across space; the effects of fragmentation of nesting sites as well as floral resources; the level of redundancy within groups of pollinators; the fragility of pollinators and their ability to move through unfavorable habitat types; and the identification of essential resources provided not only by native habitats but also by crops and other domesticated plants. Conserving or restoring pollination will require a far greater understanding of ecological interactions than we have now (Kremen and Ricketts 2000).

Threats to Wildlife from Loss of Farmland Habitats

Some types of wild biodiversity depend upon human agricultural management practices. A study of two oases in the Sonoran Desert on the Mexico–U.S. border indicated that the customary land-use practices of Papago farmers on the Mexican side contributed to biodiversity of both oases (Nabhan et al. 1982). Shifting cultivation communities have played a custodial role in preserving forest ecosystems and natural species (De Foresta and

Michón 1997). Indeed, many of today's "natural" ecosystems are anthro-pogenic—that is, developed through human interventions (Hecht 1993). In tropical Asia, for example, the welfare of the large mammals of the region, such as kouprey (*Bos sauveli*), banteng (*Bos javanicus*), gaur (*Bos gaurus*), ele-phant (*Elephas maximus*), and tiger (*Panthera tigris*), may depend on the con-tinuation of a certain amount of shifting cultivation (Wharton 1968). Wild pollinators or fauna that provide seed dispersal for wild plants may also rely on farmland habitats. In some habitats, wildlife can be harmed both when farming is abandoned and when farming systems are simplified (as is common when small, diverse farms are converted to large monoculture plantings), which removes key crop species or microhabitat conditions required by wild species.

The best-documented evidence of the role of agricultural lands in main-taining wild biodiversity is from Europe, which has a long history of agricul-ture encouraging many wild species adapted to open habitats and selecting against those that could not adapt to such conditions. One example from Spain is described in Box 4.2. Agricultural intensification in much of north-ern, western, and central Europe is now leading to sharp declines in birds, but-terflies, beneficial insects, wildflowers, and game species inhabiting farmland (Sotherton 1998). According to BirdLife International, agricultural activities take place in 65 percent of the Important Bird Areas (IBAs) in Europe and are predominant in 17 percent of them. Agricultural intensification is identified as the most serious threat affecting IBAs in Europe. It is cited as a threat at 1,300 sites (36 percent) and has a high impact at nearly 400 of these. Ironically, land abandonment is also a significant threat to more than 600 IBAs in parts of cen-tral and eastern Europe (Heath and Evans 2000). Without improved manage-ment, the 40 or so species of birds that are dependent on arable lands, such as the great bustard (*Otis tarda*), skylark (*Aluda arvensis*), and corn bunting (*Mil-iaria calandra*), are likely to continue their precipitous decline (Tucker and Heath 1994). Cereal-producing agricultural lands have historically contained up to 700 species of plants in Europe and some 2,000 species of insects and spiders in the United Kingdom (Potts 1991). Therefore, appropriate manage-ment of agricultural lands could enable the many species dependent on arable land to survive, if not prosper.

Overexploitation and Eradication of Wild Species

Much of the overharvesting of wild species around the world is due to the activ-ities of specialized extractive industries, commercial traders, and poachers who have little relation to agriculture. However, overexploitation by local people for their own use—and their willingness to participate in unsustainable trading—is significant in many places. Overexploitation of wild species is often intimately related to the state of agricultural systems and their capacity to provide food security. Poverty and low agricultural productivity heighten dependence, for example, on supplemental wild foods, especially as famine foods. Wild plants in forests, wetlands, or riverine areas of rangelands may be important sources of fodder for domestic livestock, especially when pastures or rangelands are over-grazed or diminished by drought. The popular novel *The Poisonwood Bible* by Barbara Kingsolver (1999) highlights the central role of bush meat in providing protein for West African farmers, with their notoriously protein-scarce, cassava-based farming systems. Introduction of more protein-rich plants and domestic livestock could potentially reduce their dependence on bush meat.

Human population growth also results in overexploitation of forest species

---

Box 4.2. Biodiversity in the Dehesa Agricultural Systems of Spain

Dehesas are silvipastoral systems in the Iberian Peninsula that cover nearly 3.5 million hectares. The trees found in these grazing lands include holm oak (*Quercus ilex*) and, to a lesser extent, olive (*Olea europeaea*), stone pine (*Pinus pinea*), and wild pear (*Pyrus bourgaeana*). Traditionally, agriculturalists grazed or herded sheep, livestock, and pigs on these lands and cultivated small areas using a short fallow system with oats, wheat, barley, rye, and hay. Dehesas have been managed for centuries in a way that supports levels of biodiversity that are very high for the region. High within-habitat heterogeneity is caused by the presence of ten to eighty scattered, isolated trees and their canopy effects on herbaceous plants. High between-habitat biodiversity results from a variety of grazed, shrubby, and cultivated types of dehesas and from differences in stand composition, density, and structure. Thirty percent of the vascular plant species of the Iberian Peninsula are found in the dehesas. The population viability of one endangered species (Spanish imperial eagle, *Aquila adalberti*), nine vulnerable ones (for example, black vulture, *Aegypius monachus*), and three rare ones (for example, black stork, *Ciconia nigra*) depends completely or partially on the structural integrity of the dehesas. Six to seven million woodpigeons (*Columba palumbus*), 60,000 to 70,000 common cranes (*Grus grus*), and a large number of songbirds depend on the dehesas as winter habitat. Typical management practices like tree pruning have a strong positive influence on wintering birds because they enhance food abundance. The productivity of acorns (the most important food resource for wintering birds) was shown to be ten times higher in a managed dehesa compared to a dense native *Quercus ilex* forest. Since the 1950s, however, the economy in many dehesa areas has been shifting from a pastoral to a ranching system. Shepherding is being replaced by large-scale free-range grazing, and product diversity has been reduced drastically—mainly to two products, lamb and beef. Without greater attention to management of the dehesas, their bio-diversity services are at enormous risk (Plieninger and Wilbrand 2001).

for fuelwood or organic materials for cropland nutrients. Many nontimber forest products, such as rattan and some species of orchids, have been over-collected in unregulated forests, to the point of local extinction. Economic hardship may make illegal hunting a more appealing livelihood, encouraging poor farmers to hunt and gather protected species, particularly primates, mollusks and corals, seahorses, and tropical birds and plants (see Box 4.3). Even with widespread acknowledgment that most capture fisheries are currently over-fished, increased demand for wild fish for poultry and swine feed and aquaculture (to feed carnivorous fish species) increases pressure to continue overfishing. Many biologists are skeptical that any significant commercial use of wildlife can be sustainable over the long term (Reynolds et al. 2001).

Many indigenous peoples have developed relationships with their natural resources that were sustainable over many generations. But modern technology, ranging from shotguns to wire snares to flashlights, has enabled hunters to harvest more meat, leading to the rapid decline in the population of many wildlife species. Similarly, the chainsaw has increased local people's capacity to cut trees. Furthermore, new immigrants into frontier areas typically ignore local hunting customs such as hunting seasons, restrictions on certain species, and other practices that local populations had developed through experience to maintain a balance between people and resources. It will be extremely difficult for wildlife to support rural populations in the future in the same way as it has in the past, because far more people are competing for wildlife resources. This implies even greater dependence on agriculture, thus continuing the practices that have created the current vicious circle.

Meanwhile, farmers' organizations and government agriculture departments in many parts of the world have organized systematic campaigns to

---

Box 4.3. Overexploitation of Turtles in Southeast Asia

As China enters the global economy and Chinese consumers have more money available to purchase the things they value, the turtle and tortoise populations of Southeast Asia have been devastated. Poverty-stricken rural people have been gathering every turtle in sight for sale as food and medicine in the turtle markets of China. More than 240 metric tons of turtles, representing more than 200,000 individual animals, were taken from Vietnam for sale in China in 1994. In Laos, Vietnam, and Myanmar today, it is difficult to find a single turtle, even in the protected areas of those countries. This represents a tragic lost opportunity. If means had been found to help local people manage and market turtles as a complement to agricultural production, the poor could have had a sustainable income supplement, local turtle habitats could have been protected, and turtle populations could have been maintained (Li and Wang 1999).

eradicate certain wild species from agricultural regions, in particular animal predators and major pests. Bounties have been placed on wolves, coyotes, lions, rats, moles, and some species of birds.

Impacts of Agricultural Species on Associated Wild Species

Agricultural species themselves may interfere with the wild species in and around agricultural areas. The principal negative impacts are habitat invasion by alien species and disease transmission to wild animals from domestic livestock. Unknown effects from genetically modified organisms are emerging as a new concern.

Invasive Alien Species

In most parts of the world, the great bulk of human dietary needs is met by species that have been introduced from elsewhere (Hoyt 1992). It is difficult, for example, to imagine Africa without cattle, goats, corn, and cassava, but all these species were introduced to the region. Species introductions, in this sense, are an essential part of human welfare in virtually all parts of the world. And maintaining the health of these introduced species may require introduction of additional species for use in biological control programs that protect against agricultural pests (Waage 1991; Thomas and Willis 1998).

However, some introduced species become invasive, taking over natural ecosystems and threatening native species (Mooney and Drake 1987; Carlton and Geller 1993). Such invasive alien species have major economic costs, as described in Chapter 3 for weeds and crop pests. Even fast-growing tree species introduced for agroforestry, such as species of *Leucaena,* have sometimes spread to become troublesome weeds. Fisheries have been profoundly affected by invasive species introduced for commercial or sport fishing or for contained use in aquaculture and mariculture facilities (which are seldom escape-proof). The introduction of Nile perch into African lakes, for example, has increased profits from commercial fishing and contributed to foreign exchange gains, but at the expense of more than 100 endemic fish species that are now extinct (the losses are greatest in Lake Victoria, as explained under Example 23 in Chapter 7). In Dianchi Lake in China, more than thirty alien species of fish were found in the 1970s, reducing the number of native species from twenty-five to eight over a period of twenty years (Xie 1999). The introduction of the African tilapia (*Oreochromis* spp.) into Lake Nicaragua in the 1980s resulted in the decline of native populations of fish and the imminent collapse of one of the world's most distinctive freshwater ecosystems (McKaye et al. 1995). The Atlantic salmon was eliminated from many rivers in Norway after the introduction of the Baltic salmon for aquaculture. Studies of the introduction of nonnative fish in Europe, North America, Australia, and New Zealand reveal that 77 percent of these introductions resulted in the drastic reduction or elimination of native fish species. In North America alone, twenty-seven species and thirteen subspecies of native fish became extinct in the last century due largely to introduced species (Stein et al. 2000).

The impact of alien plant pathogens, such as the fungus that causes Dutch elm disease, is even greater, with economic costs amounting to over U.S.$35 billion per year in India, $17 billion in Brazil, $1.8 billion in South Africa, $3 billion in Australia, and $23 billion in the United States (Pimental et al. 2001). All told, introduced or invasive alien species of animals, plants, and microbes may cause as much as U.S.$250 billion per year in losses to world agriculture (Bright 1999). The losses to global biodiversity are so serious that a major international effort under the United Nations Convention on Biological Diversity has begun to address the issue (McNeely et al. 2001).

Wildlife Diseases from Domestic Livestock

In many parts of the world, domestic livestock and wildlife have coexisted for centuries, especially in areas of low human population density. Unfortunately, along with sharing feed and water resources, they also share parasites and associated diseases. Indeed, wildlife diseases are becoming a more serious concern with growing human encroachment into wildlife habitats. Human population expansion has stimulated emerging infectious diseases (EIDs) by increasing population density, especially in urban areas (where dengue and cholera are becoming more common) and by encroaching into wildlife habitats (leading to, for example, Ross River Virus). Human encroachment on shrinking wildlife habitats can also increase wildlife population densities in remaining habitats, which in turn makes those populations more susceptible to diseases. Increased contact with wildlife and associated diseases, combined with international trade in livestock, has led to outbreaks of diseases such as rinderpest in Africa and foot-and-mouth disease in Europe (Daszak, Cunningham, and Hyatt 2000).

Disease transmission also occurs in the other direction, as livestock diseases can devastate wildlife. Bovine tuberculosis originating from domestic cattle has spread rapidly in recent decades among buffalo and lions, and among smaller numbers of cheetah and baboons, in parts of East Africa. This disease is also seriously compromising the rapidly expanding deer-farming industry of China and Southeast Asia. An active search is under way for a bovine tuberculosis vaccine that can be used to treat both cattle and wild animals. Many infectious diseases such as rabies, hog cholera, African swine fever, and screwworm are potentially dangerous to humans as well as wild and domestic animals (Woodford 2000). Introduced infectious diseases may have been responsible for many of the Pleistocene extinctions of large animals after humans colonized the continental land masses and large islands, leading to speculation that many wildlife diseases currently considered native may actually have originated from those early introductions (Daszak, Cunningham, and Hyatt 2000).

Emerging Concerns about Genetically Modified Organisms

Genetic improvements through crop and livestock breeding have played a major role in increasing production and may in the future contribute to conserving endangered species. A newly developed set of tools, generally referred to as genetic engineering, now enables specific traits to be directly inserted into the genetic material of a crop or animal, even traits from a very different organism. While farmers and scientists have been using selective breeding to modify crops and animals for centuries, genetic engineering allows much greater control over the modification process as well as new forms of genetic combination. In general, a single gene from an outside source (either from the same species or from an entirely different organism) that contains coding for desired characteristics—such as herbicide resistance or an antibacterial compound—is inserted into the recipient organism (Persley and Lantin 2000). For example, frost resistance in tomatoes has been enhanced using fish genes. Agribusiness is poised to introduce an array of "prescription" foods bioengineered to provide nutrients for those suffering from deficiencies, such as rice engineered to produce vitamin A. Bio-engineering may raise the potential yield of some crops by altering plant stomata and reengineering the photosynthesis process (Mann 1999), reducing the need for pesticides, reducing consumption of water, and adapting to saline soils.

However, economic, social, health, and ethical concerns have been raised by scientists and the general public about the rapid spread of genetically modified (GM) crops. Because GM crops are developed largely in the laboratory, key experiments on both the environmental risks and the benefits are lacking (Wolfenbarger and Phifer 2000). One concern, not yet much documented in the field, is the potential for genes to migrate from domesticated GM crops into wild plants. Genes have been moving for many years from conventionally bred crops to wild relatives. In the United Kingdom, for example, oilseed rape (*Brassica napus,* also known as canola) and native species such as wild turnip (*B. rapa*) do occasionally combine to form hybrids. Gene transfer is almost inevitable with crops that have close relatives in adjacent natural ecosystems.

Hybrid crops can be affected by GM crops if they accidentally receive "foreign" GM genes. The foreign genes may change the fitness (ability to reproduce) and population dynamics of hybrids, eventually back-crossing into the native species and becoming established (though this apparently has not yet happened in practice). Some foreign genes introduced into hybrids between cultivated crops and native species may decrease fitness in the wild, leading to rapid selection of these genes out of the population, thus avoiding ecological damage. But the transfer of genes for resistance to insects, fungi, and viruses could also increase the fitness of resulting hybrids, possibly forming aggressive weeds or plants that could swamp wild populations. Weeds having tolerance to a range of herbicides could emerge, making them difficult to control not only in agriculture, but also in natural ecosystems such as grasslands. If nontarget plants acquired insect resistance from GM crops, they could dominate ecosystems because they could no longer be controlled by those insects (and those insect populations could decline or die out). Development of crop varieties genetically modified to thrive in certain marginal environments could threaten remaining habitats whose only protection to date is their low productivity for agriculture, such as saltwater marshes. Further research and monitoring are needed to identify such potential impacts (Johnson 2000) as a means of helping inform decisions about whether the potential benefits of GM crops outweigh the costs.

Climate Change Induced by Agricultural Land Use

Scientific evidence indicates that the earth's climate is changing because of human economic activities that produce so-called greenhouse gases, the most important being carbon dioxide. The climate is projected to warm during the next 100 years at a rate more rapid than anything experienced during the last 10,000 years (Houghton et al. 1995). The biodiversity impacts are expected to be profound. The composition and geographic distribution of many biomes, including forests, rangelands, deserts, mountain systems, lakes, wetlands, and marine systems, will shift as individual species respond to changes in climate (Furtado and Kishor 1999). Climate change has already begun to damage coral reefs, the "rainforests of the sea," which are proving acutely sensitive to its effects (Pockley 2000).

Agroecosystems play an important role in stabilizing global climate by storing carbon. Carbon storage in soils and vegetation in farmed agroecosystems is estimated to be 18 to 24 percent of the global total, while forest ecosystems store about 40 percent (Wood, Sebastian, and Scherr 2000). On the other hand, while the primary cause of global warming is fossil fuel combustion, land-use practices account for 20 to 25 percent of greenhouse gas emissions. The primary sources of agriculture-based carbon dioxide emissions are deforestation (which inevitably involves burning and other releases of carbon dioxide), burning of trees and crop residues, and conversion of land from higher to lower carbon storage uses, including loss of soil organic matter associated with cultivation. The primary sources of emissions of methane, another greenhouse gas, are livestock and paddy rice production (Wood, Sebastian, and Scherr 2000; R. Watson et al. 2000).

Implications

The evidence is now overwhelming that many components of current agricultural production systems pose a globally significant threat to wild biodiversity. A recent global model projects that over the next fifty years—without changes in agricultural systems—another 1 billion hectares of natural ecosystems could be converted to agriculture, accompanied by a 2.4- to 2.7-fold increase in nitrogen- and phosphorous-driven eutrophication and pesticide use. The scientists conclude that ecosystem and biodiversity impacts would be unprecedented (Tilman et al. 2001). Thus, individuals and organizations concerned with biodiversity conservation must begin to focus far more of their resources and initiatives in agricultural areas. Agricultural systems must be transformed to be more "friendly" to wildlife, particularly in and around global and local biodiversity hotspots.

Part II

The Opportunity: Integrating Biodiversity Conservation in Agricultural Development

The challenge seems clear. Agricultural systems must be transformed to support wild species while simultaneously maintaining or improving productivity and reducing poverty. But how realistic is it to achieve biodiversity conservation in agricultural regions? Reintroducing large wild carnivores into densely settled agricultural communities is hardly realistic; the ecological roles they play in ecosystems will have to be met in other ways. Large mammals requiring large territories can only be managed in more extensive agricultural systems, and large tropical trees require large expanses of natural habitat. But the vast majority of species, from mammals and trees to soil and water microorganisms, can probably maintain reasonably healthy populations in habitats that are considerably modified by agriculture, so long as agricultural systems are designed to be compatible with their most critical needs. We have given this integrated strategy a name: *ecoagriculture,* which broadly defined is an approach that brings together agricultural development and conservation of wild biodiversity as explicit objectives in the same landscapes.

Ecoagriculture builds on the concept of ecosystem management that is being adopted by many conservation organizations around the world. It recognizes that good farm and landscape design, while crucial, is just a start; long-term success will also require improvements in the productivity and functioning of agricultural components and management systems.

Chapter 5 describes the evolution of the concept of ecoagriculture and introduces the major strategies for designing such systems. Chapters 6 and 7 describe six ecoagriculture strategies to make space for wildlife in noncultivated, nongrazed parts of the agricultural landscape, and to enhance the habitat value of the productive areas themselves. Thirty-six case studies are presented from diverse farming systems around the world. These examples show that such systems can be practical on a wider scale, socially and economically acceptable to farmers, and relevant in the dynamic economic environments within which farmers must operate. In all of the cases, wild biodiversity increased and farmer incomes and livelihoods improved; in most cases (and in all of those in developing countries) agricultural product supply also increased.

If ecoagriculture is successful, then populations of wild species will increase in and around agricultural communities. This will pose its own challenges. While many of these wild species will have a positive impact on farming and on farmers, and most will have a neutral impact, others may become pests or nuisances and must be actively managed. Chapter 8 discusses these challenges, and some of the methods that have been developed to facilitate the coexistence of wildlife and farmers.

While the long-term viability of the particular win-win cases documented here has yet to be confirmed, and the geographic spread of many is still limited, the cases demonstrate real potential for developing a biodiversity-friendly agriculture that also reduces poverty and meets the challenge of expanding demand for food. The potential for applying advanced tools and methods of agricultural and environmental science to the challenges of ecoagriculture is huge, and largely unexploited. The learning curve for this process is just beginning, and much more research and field experimentation is needed to understand how to rebuild biodiversity in ways compatible with food security and rural development.

Chapter 5

Ecoagriculture: Genesis of the Approach

It is easy to illustrate the danger of expecting isolated protected areas to carry the full responsibility for conserving wild biodiversity. To take an extreme example, if biodiversity is conserved on *only* the 10 percent of land currently under legal protection, 30 to 50 percent of existing species would be lost, according to projections based on island biogeography theory (MacArthur and Wilson 1967). If habitat in the remaining 90 percent of land were left in small isolated patches, then the number of species lost could be much higher, especially among large animals and plants that tend to have relatively small populations. Thus while protected areas are an essential element in plans to conserve biodiversity, they must be augmented by managing ecosystems as a whole to conserve biodiversity and to provide food and income. Because food and fiber production—produced from agriculture and harvested from natural systems—is such a dominant land use, and has such extensive influence on wild biodiversity, it needs to play a much larger role in biodiversity conservation.

At present, wild biodiversity and farming activities coexist to some extent in landscape mosaics where many niches are noneconomic for agriculture (for example, due to soil, moisture, or slope conditions) and thus function as semi-natural ecosystems. But farmers in more heavily populated and intensively managed areas often face a sharp trade-off between agricultural production and biodiversity. If farmers want a little more biodiversity, they must sacrifice a lot of production; if they want a little more production, they must sacrifice a lot of biodiversity. Under current technological, economic, and policy conditions, agricultural systems characterized by high plant diversity typically provide lower returns to farmers' labor than do most of the low biodiversity systems. Figure 5.1 shows an example from Rondônia, Brazil. Evidence from many parts of the world highlights trade-offs between agriculture, economic development, and environmental concerns that are difficult to reconcile under present conditions (Lee and Barrett 2001).

More environment-friendly practices developed in the past half-century, usually adapted from traditional farming systems, sometimes had higher costs or lower yields, leading to accusations that these indirectly harmed biodiversity by making farmers clear new lands to meet growing demand for food (Trewavas 2001). This view has persisted despite extensive evidence today of high productivity in many organic and other alternative farming systems. For example, a twenty-one-year scientific study of potatoes, barley, winter wheat, beets, and grass clover production in Switzerland found that organic fields receive between 34 and 53 percent less fertilizer and energy and 97 percent fewer pesticides. Per unit of energy, the organic systems produced more food, and the organic soils housed a larger and more diverse community of organisms. Organic fields received less than half the nutrients applied to conventional fields, yet averaged 80 percent of the yields; in the case of winter wheat, yields averaged 90 percent of those of conventional practice (Mader et al. 2002). In the United States, studies in three midwestern states found organic corn systems to be more profitable than continuous corn, even without price premiums, due to lower production costs and greater drought hardiness (Welsch 1999). But even

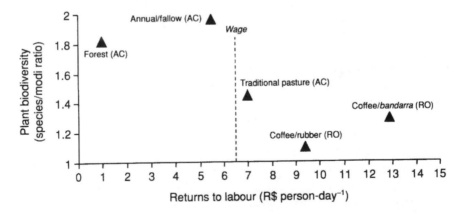

Figure 5.1. Trade-offs between Farm Income and Plant Diversity in the Brazilian Amazon. *Source:* Vosti et al. 2001.
Notes: The vertical "Wage" line equals the average daily wage for hired labor during the study period. All prices are in December 1996 Brazilian reales (U.S.$1 = R$1.04) Species/ modi ratio measurements are for the stable system land cover; for the annual/fallow system, the measurements are for the fallow phase. AC data from Pedro Peixoto, Acre; RO data from Theobroma, Rondônia.

organic farmers who carefully avoid negative impacts on wildlife from agricultural pollution typically remain cautious about increasing wildlife in or around their farms beyond those species that support sustainable production.

Fortunately—largely below the "radar screen" of most observers—a growing number of innovative agriculturalists and environmentalists have begun to tackle the agriculture-income-wild biodiversity challenge. Researchers, farmers, and community planners with diverse perspectives have begun working together to develop land-use systems managed for *both* agricultural production and the conservation of wild biodiversity and other ecosystem services. In this book, such systems are called *ecoagriculture,* and the component elements that contribute to that dual function are thus ecoagricultural practices.

What Do Wild Species Need?

Ecoagriculture systems not only produce food, but also accommodate the needs of wild native species. In some cases they may focus on key species of high biodiversity value (however that may be defined) or on species that can serve as "indicators" of habitat quality for a wide range of species. To design such systems, it is essential to determine what habitat features are critical for conserving or increasing populations of target species. Key features are summarized in Box 5.1.

Many wild species are flexible and hardy and have characteristics that enable their persistence even with substantial habitat modification. Other species are highly sensitive to habitat quality, and these are the ones that tend to be most threatened by agricultural expansion and intensification. In fact, biodiversity is probably the ecosystem service most sensitive to land use. While a wide range of landscape patterns and plant species can maintain watershed function or store carbon, some wild species may be so totally

---

Box 5.1. Habitat Features Critical for Wild Biodiversity

- Nesting sites
- Protective cover against predators
- Clean water
- Access to breeding territory (migratory routes unimpeded)
- Access to food sources in all seasons (varied habitats)
- Predator-prey balance
- Presence of other beneficial species, pollinators, and other interdependent species

dependent upon one particular plant species for food that they disappear when that species disappears. This habitat diversity needed by wild species conflicts directly with humans' historical tendency, particularly in the past century, to radically simplify agricultural systems.

A critical element in designing appropriate conservation programs is determining the minimum viable population (MVP) of a species, defined as the smallest number of individuals necessary to give a population a high probability of surviving over a significant period of time (Primack 1993). If a population of a species declines below the MVP, it probably is on the road to extinction, due to loss of genetic variability, inbreeding, random demographic fluctuations, and random environmental fluctuations (such as disease outbreaks, fires, droughts, or harvesting by humans). For vertebrates like birds and mammals, a common rule of thumb, based on population genetics and other variables (Frankel and Soulé 1991), is that 500 to 1,000 individuals are needed to ensure long-term survival, while 10,000 individuals may be required for species with extremely variable populations, such as some annual plants (like some grasses) or insects (like some butterflies) (Lande 1988). On the other hand, some species of plants, such as bristlecone pines, may be able to survive for long periods of time with very small populations.

Of course, the success of some invasive alien species demonstrates that, given favorable conditions, very small populations can soon expand. Also, some threatened species have been brought back from the edge of extinction, for example, the Arabian oryx (*Oryx leucoryx*). But it appears far preferable to prevent a species from reaching such a low population that it needs an expensive recovery effort (in drastic cases, through captive breeding or protection in nurseries for subsequent reintroduction), and this requires that an area of habitat be provided that will ensure survival of a minimum viable population. This area can be determined by studying the range size and habitat requirements of the species concerned. In the case of wide-ranging carnivores such as tigers or large birds of prey, or many species of uncommon tropical forest trees, this area can be very extensive.

The Ecosystem Approach

The conceptual framework from which ecoagriculture arises is based on the ecosystem approach as developed under the Convention on Biological Diversity (CBD). The ecosystem approach recognizes that ecosystems must be managed as a whole. Protected areas function as reservoirs of wild biodiversity within a matrix of land management that enhances habitat value and provides a range of benefits to people, from food supply and income to environmental services. Biodiversity protection in an ecosystem management framework calls for a coordinated strategy that clarifies goals and investment plans for land uses that influence wildlife protection. It encourages protected areas to be integrated fully within key planning frameworks, including land-use and development plans, national biodiversity strategies and action plans, and strategic plans for relevant economic sectors (agriculture, forestry, fisheries, tourism, energy, transportation, and even the military). Within this integrated strategy, agricultural lands need to be managed as part of the matrix surrounding protected reserves, while the protected areas are managed as part of the matrix surrounding agricultural lands (Pirot, Meynell, and Elder 2000).

Related concepts used by some environmental planners include bioregional planning, ecoregion-based conservation, ecosystem management, an ecosystem-based approach, integrated conservation and development projects (ICDP), biosphere reserves, landscape ecology, and integrated coastal zone management. All are based on more comprehensive approaches to resource management that protect and restore whole ecosystems or bioregions, to simultaneously conserve biodiversity and sustain farming, forestry, and other human uses. This idea that conservation problems should be addressed in whole ecological or landscape units based on integrated biological, physical, and socioeconomic assessments stretches back at least into the 1960s, but it could be argued that this has been the de facto approach of stable rural communities throughout history.

Although using a small area of land—such as an individual farm—for diverse purposes is challenging, implementing multiple uses over larger areas can be much easier. Over a large bioregion various parts of the landscape can be under different dominant land uses, with all uses contributing to the overall objectives of ecosystem management. Ecosystem management provides a comprehensive framework for bringing together different approaches to conservation, helping to integrate or coordinate the various sectors with an interest in biodiversity. The scope of ecosystem management efforts may include activities across the entire land and waterscape, crossing ownership, political, and even international boundaries. Conserving a species of rare or threatened plant involves conserving all parts of its ecosystem, including pollinators, seed dispersers, and other organisms that play significant roles in its life cycle. Ecosystem analysis can help decision-makers consider options for landscape-scale developments, as illustrated in Figure 5.2. The ecosystem approach ideally encourages decentralization of management to the lowest appropriate level. This approach uses adaptive management policies that can deal with uncertainties and can be modified in the light of experience and changing conditions; it draws on multidisciplinary inputs to address scientific, social, and economic issues (Slocombe 1991; Grumbine 1994; Miller 1996). Large-scale examples are now under way, such as the Mesoamerican Biological Corridor in Central America described in Box 5.2 (Miller 1996). In such initiatives, agricultural areas assume considerable importance; in this case,

Figure 5.2. A Governor's Dilemma—Planning Jointly for Production and Biodiversity in Tropical Hillsides

Suppose you are governor of a hilly rainforest province in the tropics. If your planners focus on production and erosion control, their plan might have most of the area in agriculture using contour planting and perhaps hedgerows, and leaving small scattered nature reserves on the steepest slopes. If your planners instead focus on protecting biodiversity, especially interior species, most of the area would be in large nature reserves, separated by small farmland areas where species richness (and perhaps soil quality) is low.

However, suppose . . . you seriously value both erosion control and biodiversity. Ask your planners to focus concurrently on both major objectives, and your plan will be unlike either of the preceding plans. A few large nature reserves will encompass steep slopes and many moderate slopes, farmland will cover the remaining land, and hedgerows will be arranged to connect nature reserves and minimize erosion. Not only will the third design optimize and accomplish both objectives, it should gain a wider range of political support under your leadership as governor. And . . . it is more sustainable, since both biodiversity and soil are less likely to degrade over human generations.

Source: Quoted from Forman 1995.

corridor development offers opportunities to include densely settled agricultural areas.

One limitation is that understanding of ecosystem functioning remains very incomplete (Loreau et al. 2001). Some ecologists argue that ecosystem properties are determined by the functional traits of dominant species, or by the composition of functional groups, implying that at least some species may be redundant. Others contend that any reduction in diversity may compromise the integrity of the system, and that some redundancy reinforces ecosystem reliability (Naeem 1998). While some ecosystem processes may reflect the activities of a few dominant species, ecosystems with greater diversity are more likely to contain the most productive species and the multiple interactions and feedback loops that characterize more complex food webs (Levin 1999). Sensible ecosystem management, therefore, calls for conserving all of the elements of the system and recognizing that the incompleteness of our knowledge makes it risky to lose any of the pieces (see Box 2.1). Managing ecosystems and landscapes with a unified strategy to save all of their inhabitants can be a cost-effective approach to biodiversity conservation. While species-based approaches may still be required in some cases, models based on new understandings of ecological relationships (such as the one described in Box 5.3) can help to inform ecosystem management and to support ecoagriculture.

Box 5.2. The Mesoamerican Biological Corridor

To develop the Mesoamerican Biological Corridor, the Central American Commission on Environment and Development coordinates regional policy of eight countries. The project began in 1994 as the "Panther Trail" (Paseo Pantera), defined in purely biological terms. In response to resistance by local people who feared their land would be expropriated, the project evolved by 1999 to the present corridor, whose objectives include achieving broad socioeconomic goals. The concept of the corridor is to maintain connectivity between biological populations, so these can find food, mates, water, and refuge from predators, and reduce the risk of local extinctions due to natural disasters. In this case, population biologists determined that to maintain the region's biota, habitats needed to be large enough to retain a minimum population of 500 to 5,000 individuals of each species. Various means are being used to foster biological connectivity between the main habitat areas: maintaining corridors under contiguous natural cover, providing "stepping stones" of small habitat areas, encouraging diverse cropping patterns in the intervening croplands, and retaining large live and dead trees in surrounding forest clearings (Miller, Chang, and Johnson 2001).

Box 5.3. Ecological Model of Wildlife, Livestock, and Human Populations in Conservation Areas

A computer-based ecological model called SAVANNA was developed jointly by Colorado State University and the International Livestock Research Institute to help land-use planners create long-term plans for savanna ecosystems where wildlife, human, and domestic livestock populations coexist. The original model focused on the Masai Mara ecosystem in Kenya, where wildlife populations declined by one-third between 1977 and 1996, largely due to land-use changes. The model forecasts wildlife populations, the health of ecosystems, and human conditions 5 to 100 years after human and natural activity have changed the landscape. It is comprehensive enough to take into account the constant change of the natural world across large regions, while at the same time forecasting the future of an area as small as a 50-meter-wide watering hole. The model is now being used by conservationists, development planners, and local people for land-use planning in the Masai Mara National Reserve and Amboseli National Park. Using such landscape models, and the visual display of various scenarios, local stakeholders can evaluate together the implications of different land-use strategies. The model has also been adapted for use in other ecosystems, including the western United States.

Evolution of Approaches to Wild Biodiversity in Agricultural Systems

A thumbnail sketch of the evolution of approaches to wild biodiversity in agricultural systems is shown in Figure 5.3. For the past 10,000 to 12,000 years, most farmers practiced subsistence agriculture, obtaining yields adequate for household consumption while generally tolerating and adjusting to surrounding wildlife, where these could not be driven away or hunted out. They produced their food by managing the natural environment through site modifications and selection of superior seed and planting materials. But in places where nonfarming populations or external trade increased significantly, producers often depleted the soil and other nutrients, obtaining high yields over the short term but then abandoning the land when yields decline. Where high levels of labor or capital were available, farmers made large landscape modifications to create irrigation or drainage for areas with highly fertile soils or to create terraces on sloping lands, thus radically changing habitat conditions for wildlife.

In the middle of the nineteenth century the era of modern agriculture began, with the development of synthetic fertilizers, large-scale production of scientifically improved seeds, and mechanization. Technology further advanced in the twentieth century with the development of inorganic pesticides, livestock vaccines, and improved transportation and storage systems. Since the mid-1900s,

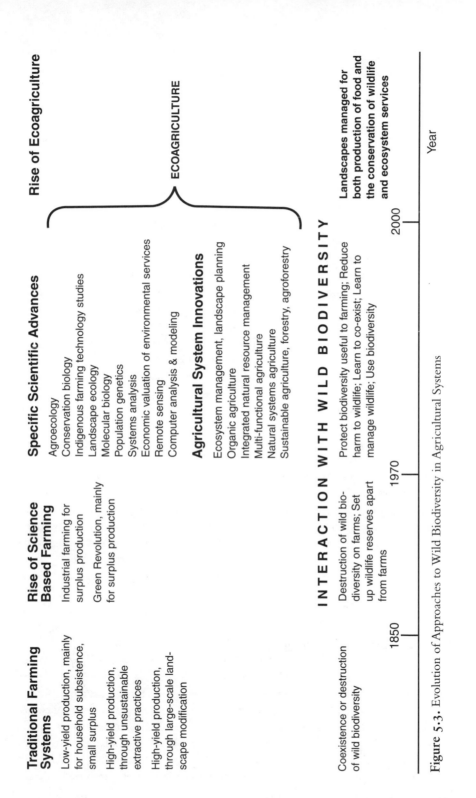

Figure 5·3. Evolution of Approaches to Wild Biodiversity in Agricultural Systems

97

science-based production systems have dominated developed-country commercial and industrial agriculture and high-value export crops in many developing countries. In a major international research effort to extend the benefits of these technologies to staple food production in developing countries, Green Revolution technologies began to be developed in the late 1960s. These latter technologies were enormously successful in raising agricultural production between the mid-nineteenth and late-twentieth centuries, especially in irrigated and high-quality farmlands, greatly increasing the human population carrying capacity of global agroecosystems (Wood, Sebastian, and Scherr 2000).

By the late 1960s and 1970s, however, the environmental side effects and inadequacies of new production technologies were becoming quite evident, including reductions in wild species populations. Conventional industrial agriculture and some Green Revolution technologies were critiqued as promoting yield growth for only a narrow range of crops grown in monocultures and eradicating other species. Supported by high external inputs, in particular agrochemicals, systems were criticized as having little regard for long-term sustainability of the resource base or for the supply or quality of other important ecosystem products and services (e.g., minor crops or biodiversity more generally).

Many poor farmers in developing countries, especially those farming in lower productivity, higher-risk soils and environments, benefited little from these technologies, either because they could not afford to use them or because they were not adapted to their conditions; thus they continued to use low-external-input systems. Under pressures of population growth, declining farm size spurred more intensive farming (for example, more frequent cropping). In some places, this intensification was sustainable, but elsewhere it led to significant soil erosion, excessive forest clearance, and ultimately reduced food security. Together with environmental concerns, such as increasing salinity in irrigated farming areas, these led to calls for more attention to agricultural sustainability (Conway and Barbier 1990). Meanwhile, accelerated farmland degradation in developed countries, such as the wind erosion suffered by dryland areas of Australia and the western United States, spurred similar concerns and the development of more environmentally sensitive approaches to farm and agricultural landscape planning (e.g., Campbell 1991).

In response to these symptoms of nonsustainability, a small segment of commercial producers—mainly in more developed countries—instead pursued organic farming, building on longstanding principles of organic gardening. Certification and marketing programs for organic products soon followed. Formal research on alternative systems began with the establishment in the United States of the Rodale Institute and the Henry A. Wallace Institute for Alternative Agriculture in the early 1970s.

The mainstream farming and agribusiness communities continued to resist more ecologically oriented production systems. This was due to a combination of public farm subsidies favoring conventional systens, the influence of agro-chemical, machinery, and agroprocessing companies, the lack of time or skills for intensive management on a large scale, and the short-term yield declines during system transition, which sometimes could be as much as 30 to 50 percent in cereals (Pretty and Hine 2000). But even large-scale commercial farming began to evolve, with support from public soil and water conservation and farmland reserve programs, the spread of productivity-increasing technologies such as minimum tillage, and the rise of consumer demand in developed countries for organic food. Still, the principal strategy during this period for addressing biodiversity threats arising from agriculture was to establish protected areas and land-use zones where farming activities were prohibited.

Yet scientific advances in a variety of fields were leading many researchers and practitioners to new strategies and principles for managing agricultural systems. The integration of ecological concerns and principles into modern agricultural research and technology development had become a mainstream enterprise by the 1990s. The fields of agroecology (Altieri 1990; Gliessman 2001), conservation biology (Soulé and Wilcox 1984), and landscape ecology (Forman 1995) were established, and new remote-sensing technology made rigorous landscape-scale monitoring and analysis possible for the first time. Many researchers took a new look at indigenous agricultural technologies in order to adapt their lessons for sustainable production. Advances in molecular biology and population genetics began to unveil the secrets of genetic diversity. Systems analysis informed both agricultural and ecological thinking. Economic valuation revealed the economic importance of biodiversity and other environmental services for agricultural production, sustainability, and other, broader concerns.

Responding to insights emerging from these new fields, agriculturalists began to experiment with a variety of models for sustaining and raising agricultural productivity that in many cases had positive impacts on wild biodiversity. The various approaches share key elements but put primary emphasis on different variables. These approaches are very briefly described in Table 5.1. While this table does not do justice to the richness of concepts, contributions, or dynamic evolution of these approaches, references are provided to enable the reader to examine them further. All of the case studies described later in this book had their origin in at least one of these models.

Some models embraced those elements of wild biodiversity that were recognized to contribute to agricultural sustainability and productivity (such as soil microorganisms or pollinators), or that could reduce the need for purchased inputs (such as cover crops). Some of the landscape-scale, agriculture and natural resource management models explicitly sought to increase wild biodiversity while maintaining production levels (Jackson and Jackson 2002).

Table 5.1. Models of Agriculture and Natural Resource Management

Approach	Concept	Relative Emphasis				Scale of Intervention
		Yield Growth	Sustainability of Production	Ecosystem Function (beyond agriculture)	Wild Biodiversity (beyond agriculture)	
Industrial Agriculture	Increase yields through improved varieties/breeds, agrochemical inputs, feed concentrates, mechanization, and system simplification.	XX				Large commercial farms, forests
Green Revolution	Increase yields through improved varieties/breeds, agrochemical inputs, and irrigation	XX				Small commercial farms in tropics
Soil and Water Conservation (Conservation Farming)	Undertake conservation practices that reduce the negative environmental impacts of agriculture, especially erosion and agrochemical pollution		X	XX		Mostly farm-scale; some landscape-level
Sustainable Agricultural Intensification	Improve natural resource use efficiency and sustainability through a science-based "doubly Green Revolution" (Conway 1997)	XX	XX			Small-scale farms
Agroecology	Develop agroecosystems that maintain the resource base, use minimum artificial inputs from outside the farm system, and manage pests and diseases through internal regulating mechanisms (Altieri 1995; Gliessman 2001; Uphoff 2002)	XX	XX	X		Small-scale, semi-commercial farms; communities

Approach	Goal				Scale of focus
Low-External-Input Agriculture	Develop agricultural systems for low-income farmers that can raise yields with minimal use of purchased inputs (ILEIA 1989; Reijntjes et al. 1999)	XX	XX	X	Semicommercial farms; communities
Organic Agriculture	Use natural ecological systems and processes to sustain agricultural production (Stolton, Geier, and McNeely 2000)	XX	XX		Small farms, large farms
Regenerative Agriculture	Make agriculture more productive and sustainable by investing in soil quality improvement (Rodale Institute 1999)	XX	XX		Individual fields; community focus
"Permanent" Agriculture	Develop perennial tree crops to substitute for products of annual crops grown where production is highly erosive and unsustainable (Jackson and Berry 1985; Smith 1958)	XX	XX	X	Farms, landscapes
Agroforestry	Integrate woody perennial plants (trees, palms, shrubs) into crop and livestock production systems to increase and diversify farm production, and to improve resource use efficiency and sustainability and ecosystem services (Steppler and Nair 1987)	XX	XX	X	Farms, landscapes
Permaculture	Use field and landscape planning to make diverse components of the landscape work together to create more productive and sustainable environments (Mollison 1990)	X	XX	X	Nonindustrial producers; communities
Natural Systems Agriculture	Draw lessons from the structure of local native biotic communities to design more sustainable agroecosystems (Soulé and Piper 1992; Lefroy et al. 1999)	X	XX	X	Farms, landscapes

(continues)

Table 5.1. Continued

Approach	Concept	Relative Emphasis				Scale of Intervention
		Yield Growth	Sustainability of Production	Ecosystem Function (beyond agriculture)	Wild Biodiversity (beyond agriculture)	
Holistic Resource Management	Promote integrated land management, particularly of rangeland and livestock, in line with natural ecological processes and local development priorities (Savory 1989)	X	XX	X		Community, landscape scales
Integrated Natural Resource Management (INRM)	Incorporate multiple aspects of natural resource use into a system of sustainable management to meet explicit production goals of users, as well as goals of the wider community (CIFOR 2000)	XX	XX	X	X	Diverse farm and landscape scales
Multifunctional Agriculture	Develop agricultural landscapes that are also compatible with provision of eco-system services such as water quality, landscape beauty, biodiversity, generally by compensating farmers for using lower-impact systems (Maltby et al. 1999)		X	XX	X	Community focus; various

Strategy	Description					Scale
Integrated Conservation and Development (ICDP)	Improve conservation of protected areas and their buffer zones by integrating components to promote socioeconomic development for local people compatible with park management objectives (Wells, Brandon, and Hannah 1992)	X	X	X	XX	Protected area and buffer zones
Integrated Farm and Catchment Planning	Reintegrate ecological, economic, and social values into land management, creating seminatural ecosystems with networks of perennial vegetation (Lefroy, Salerian, and Hobbs 1992)	X	XX	XX	XX	Diverse farm and landscape scales
Sustainable Forest Management	Develop management and harvest systems for natural forests that produce timber and nontimber forest products while protecting the ecosystem functions (Sharma 1992)	X	XX	XX	XX	Commercial and community forests
Ecoagriculture	Increase agricultural production and simultaneously restore biodiversity and other ecosystem functions, in a landscape or ecosystem management context (McNeely and Scherr 2001)	XX	XX	XX	XX	Landscape or ecosystem; small or large farms

Note: "XX" indicates principal impact expected from intervention; "X" indicates secondary impact expected from intervention.

Areas managed under more ecologically compatible production systems have grown rapidly, though they still represent a small share of total production. A study in seventeen African countries estimated that 730,000 households participating in forty-five projects or initiatives were practicing "sustainable agriculture," defined to include intensification of land use, diversification of crops and animals raised, better use of both renewable and nonrenewable resources, and other social criteria (Pretty 1999). In eight Asian countries, some 2.86 million households have increased food production on 4.93 million hectares using sustainable agriculture approaches. Approximately 1.13 million hectares of cropland worldwide are estimated to be under organic production systems (Clay 2002). As of the year 2000, approximately 82 million hectares of forest had been certified as being managed sustainably (Rametsteiner and Simula 2001). Management of buffer zones around protected areas now typically includes components to enhance farm productivity and conservation. In developed parts of the world such as Australia and Europe with food surpluses and economically valued ecosystem services, such as water quality and landscape beauty, management of farmlands for such ecosystem services has been officially encouraged through public policy.

All of the models described in Table 5.1 are still evolving, and in many cases that evolution is in the direction of integrating the objectives of yield growth, agricultural sustainability, ecosystem function, and wild biodiversity. Even approaches that began by working at field or farm level have begun to move their analysis and action to larger areas that may include protected areas and other land uses beyond agriculture. Those that began with a limited view of the role of wild biodiversity have begun to expand that view. For example, in May 1999 the International Federation of Organic Agriculture Movements, IUCN, and the environmental NGO World Wide Fund for Nature jointly produced the "Vignola Declaration" encouraging organic farmers to embrace objectives of biodiversity conservation in agroecosystems as well as other ecosystems and encouraging conservationists to recognize the contribution of organic agriculture to biodiversity (Stolton, Geier, and McNeely 2000). Thus ecoagriculture is not such a radical approach as it may first appear but rather is the logical outcome—and a useful conceptual umbrella—for many of the diverse strands of current thinking about agriculture and natural ecosystems.

Recent Scientific and Institutional Advances

Of critical importance to the development and feasibility of ecoagriculture are recent advances in basic science and tools from the agricultural, ecological, and land-use planning fields. These advances are transforming our capacity to investigate multiobjective, multicomponent, multiscale interactions between agriculture and biodiversity. They are thus revolutionizing our capacity to design and manage agricultural ecosystems (Scherr 2000a). Some of the most significant of these are summarized below.

- *New understanding of the role of agricultural lands in providing ecosystem functions.* Recent geographic and ecological studies have demonstrated that agricultural land-use and production systems have important impacts on ecosystem function, such as carbon sequestration, not just locally, but at sub-regional and global scales (Wood, Sebastian, and Scherr 2000), as described in Chapter 4.

- *New concepts and approaches for the study of complex agroecological systems.* Conceptual developments in the fields of agroecology, population biology, soil microbiology, biotechnology, and landscape ecology have opened up unforeseen avenues for agricultural system development. For example, tissue culture techniques make possible accelerated genetic improvement in tree crops. Recently developed approaches from complex systems science (using techniques from systems analysis, chaos theory, and advanced multivariate statistics) allow scientists to explore complex ecological relationships and processes in a rigorous way (Ruitenbeek and Cartier 2001). Shifts in ecological thinking that highlight the importance of disequilibria where ecosystems are continually influenced by disturbances and the flow of materials and individuals across system borders are raising new challenges of direct consequence to landscape planning in agricultural regions (Pulliam and Johnson 2002).

- *New understanding of the potential for local adaptive management of natural resources.* Models for agricultural technological innovation developed in the mid-twentieth century depended on a central source of innovation, primarily in agricultural research centers. While such centers continue to be important, the importance of other sources of innovation—including local knowledge, innovation, and adaptation—is now widely recognized. New approaches to innovation that emphasize the role of community-based farm or forest research are increasingly used for complex agricultural and natural resource management systems (Buck et al. 2002; Franzel and Scherr 2002; Salafsky, Margoluis, and Redford 2001). These approaches are lowering the cost of developing and adapting new management systems.

- *New tools for data collection and analysis.* New techniques of remote sensing, such as diffuse reflectance spectrometry, and new measurement methods, such as small, computerized hydrology monitoring equipment, enable managers to monitor environment and agricultural systems in ways unimagined only a decade ago. Computerized geographic information systems make it possible to store and use the data from these new tools in highly dynamic ways that can directly inform ecosystem planning. Advances that have been made in genetic analysis and biochemistry will allow us to understand much better the complexities of predator-pest relations and thus develop better methods of pest control. For example, a computer mapping tool called FloraMap, originally designed to help plant breeders and managers of germ-plasm banks at CIAT and CIP predict new collection sites for wild relatives of domestic crops, is now being used to identify suitable locations for culti-vating promising wild species, and to identify natural habitats that could serve as living gene banks for wild plants (Russell and Jones 2000).

• *New agroecological models and planning tools.* Computer hardware and software advances in the past decade have introduced a whole new potential for build-ing complex, data-based agroecological models for use in exploring agricul-tural and biodiversity management options. Economic and institutional vari-ables can be built into these models, increasing their value as support tools for land-use planning among multiple stakeholders. The development of user-friendly, computer-based, interactive landscape planning methods that can rep-resent spatially explicit land uses and material flows and application of scarce resources, such as labor and agricultural technology, are revolutionizing our capacity to involve diverse land users in landscape planning and management. Such methods have been adapted successfully for use in low-income countries.

• *New institutional models for local technology development and diffusion.* New, low-cost tools for participatory landscape assessment by local farmer groups and new institutional models for effective community-based natural resource management have been developed. Combined with low-cost telecommuni-

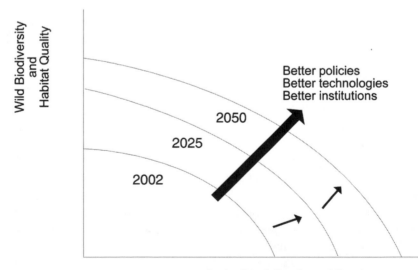

Figure 5.4. Expanding the Agriculture Plus Biodiversity Production Possibility Frontier

cations and new low-cost methods for on-farm research, these make it possible to disseminate ecoagricultural approaches widely to poor rural farmers in developing countries.

A rich cross-disciplinary exchange of such ideas has begun, making it possible for the first time to apply a sophisticated science-based approach to design and manage landscapes for both production of food and the conservation of wildlife and other ecosystem services. By applying these tools and new approaches, it should be possible to move beyond the present sustainable production possibility frontier to a far less constraining set of trade-offs between biodiversity and productivity (see Figure 5.4; Scherr 1999b).

Ecoagriculture

Ecoagriculture places food security and rural livelihoods at the center of strategies for biodiversity conservation and ecosystem management. It is intended to include and enhance—not to replace—protected area systems, which play an essential role in ecosystem health. But even if the extent of public protected areas were to double—a prospect few consider to be feasible in the short to medium term in most low-income or densely populated countries—land use would still be dominated by crop, livestock, and forest production systems. By strengthening the conservation values and capacity of those most directly responsible for rural land husbandry, ecoagriculture could help to strengthen public support for biodiversity conservation, including support for allocating more protected areas. As described here, ecoagriculture has particular relevance:

- to enhance habitat quality of buffer zones and other land uses around nationally and globally important protected areas;
- to conserve and increase wild populations in agricultural regions with important remaining biodiversity; and
- to restore biodiversity and ecosystem functions to agricultural regions with seriously depleted ecosystems.

Ecoagriculture has several key features that distinguish it from similar approaches. The first is its intentionality: ecoagriculture conceives biodiversity conservation (along with other ecosystem services) and food and fiber production as *joint outputs* of crop, livestock, forest, and fisheries production systems. Second, ecoagriculture needs to be developed at a relatively large scale involving different forms of land use that together explicitly consider the ecological and spatial requirements for effective habitat. Third, ecoagriculture embraces a wide range of technologies and management practices that enable agricultural producers to raise output while also improving biodiversity values.

It draws and builds upon the knowledge being generated in all of the sustainable agriculture and ecosystem management approaches, including integrating agroecological management, high-tech components and management tools, and wildlife biology. These approaches have diverse philosophical roots and conceptual frameworks that sometimes conflict, but it is too early to judge which elements will prove to be most effective and sustainable over the long term. Thus it makes sense to take an eclectic approach to ecoagriculture design at this time.

Ecoagriculture systems may include diverse crop, livestock, forest, and fisheries systems, both intensive and extensive, so long as these are produced by natural ecosystems. Greenhouse farming and hydroponics are not considered ecoagriculture systems. Ecoagriculture is concerned primarily with conserving wild biodiversity. But agricultural species and system diversity are often essential to enhance the quality and sustainability of wildlife habitat or the level and sustainability of agricultural production.

Large-scale, high-tech agricultural producers have considerable scope for ecoagriculture. Given the dominant ownership of farmland in many countries by such producers, pursuing these opportunities could have important benefits for wildlife and wildlife habitats. However, the most widespread social and economic benefits will come from promoting ecoagriculture among the millions of low-income farming households in ecologically degraded regions. Increasing wild biodiversity in ecoagriculture systems can potentially help the poor to increase yields, increase yield stability, increase dietary diversity, reduce the risk of crop failures, and make human habitats more healthy. Scarce resources for agriculture, conservation, and poverty reduction may often be invested more efficiently and effectively through ecoagriculture.

Strategies for Ecoagriculture

In the many agricultural landscapes that are still rich in wild species, ecoagriculture can help to maintain or enhance wild biodiversity. Where wild biodiversity has been greatly diminished, systems can be promoted in ways that enhance the effectiveness of protected areas and recover locally valued species and habitats. Ecoagriculture system design draws from six basic strategies. The first three strategies make space for wildlife within agricultural landscapes. In areas of the landscape that are not used for production farming, high-quality natural habitats or semi-natural habitats can be restored and managed more effectively for biodiversity conservation. The last three strategies enhance the habitat value of productive areas. Agricultural fields (or grazing lands, production forests, or fisheries) can themselves serve as habitat or corridors, and they can be managed to reduce the impact of agricultural production on adjacent areas of wild habitat. These are the strategies:

1. *Create biodiversity reserves that also benefit local farming communities.* New biodiversity reserves can be established in agricultural regions, where the environmental or other services they provide clearly benefit farmers and their communities in the surrounding lands.
2. *Develop habitat networks in nonfarmed areas.* The many nonfarmed areas in agricultural landscapes—on both private farms and public spaces—can be integrated into networks of high-quality habitat for wild species that are compatible with farming. Where possible, these can be targeted to link existing protected areas.
3. *Reduce (or reverse) conversion of wild lands to agriculture by increasing farm productivity.* Increasing agricultural productivity and sustainability on lands farmers already use can slow or reverse expansion into wild habitats, increasing the likelihood that Strategies 1 and 2 can be implemented.
4. *Minimize agricultural pollution.* Agricultural practices can be modified to minimize pollution of wildlife habitat through more resource-efficient methods of managing nutrients, pests, and waste, and through installation of farm and waterway filters.
5. *Modify management of soil, water, and vegetation resources.* Farmers can modify their management of their critical resources (soil, water, and vegetation) to enhance habitat quality in and around farms.
6. *Modify farming systems to mimic natural ecosystems.* Economically useful trees, shrubs, and perennial grasses can be integrated into farms in ways that mimic the natural vegetative structure and ecological functions, to create suitable habitat niches for wildlife.

Different farming conditions will call for different combinations of these key strategies. Each agricultural setting has its own particular features, making it difficult to generalize about what will work where. Tables 5.2 and 5.3 indicate the relative importance of the different strategies for biodiversity conservation. The tables are fairly speculative, given the relatively modest hard data on agriculture-biodiversity interactions.

Table 5.2 suggests potential benefits for wild biodiversity from using the six strategies in different types of agricultural systems. Promoting wild biodiversity in nonfarmed areas seems widely applicable across production systems. In intensive cropping systems, the greatest benefits to biodiversity are likely to come from minimizing pollution and modifying management practices. For expanding protected areas for conserving biodiversity, the lowest opportunity costs (i.e., where the alternative land uses are least economically attractive) would typically come from farming systems in lower-quality lands, forest fallow systems, extensive grazing and forestry systems, and fisheries. However, if ecosystem services and products provided by such reserves are significant (e.g., if conserved wetlands provide important supplies of fish), reserves may be

Table 5.2. Potential Benefits For Wild Biodiversity from Interventions in Agricultural Regions

Strategy	1.	2.	3.	4.	5.	6.
Land-Use System	Create wild biodiversity reserves to benefit local people	Develop habitat networks in nonfarmed areas	Increase agricultural productivity to reduce land conversion	Minimize agricultural pollution	Modify resource management	Modify farming systems to mimic natural systems
Intensive Irrigated Cropping	*	**	*	***	***	*
Intensive Annual Rainfed Crops on High-Quality Land	*	**	*	***	***	**
Intensive Annual Crop Systems on Marginal Lands	***	**	**	**	**	**
Perennial Tree Crop Systems in Rainfed Lands	**	**	**	***	***	***
Extensive Fallow-based Cropping Systems	***	***	***	*	**	***
Pastoral and Ranching Systems	***	***	**	*	***	***
Intensive Livestock Systems	X	*	*	***	***	**
Agroforests and Forest Fallows	***	***	***	*	**	***
Forest Plantations	**	***	*	***	**	***
Natural Forest Management	***	***	***	**	***	***
Aquaculture	*	***	*	**	***	X
Natural Fisheries Management	***	***	*	*	***	X

Notes:

***   High benefit for wild biodiversity.

**   Medium benefit for wild biodiversity.

*   Modest benefit for wild biodiversity.

X   Negligible benefit for wild biodiversity, not feasible in this system, or not relevant to this system.

Table 5.3. Types of Biodiversity Most Likely to Benefit from Interventions in Agricultural Regions

Strategy	1.	2.	3.	4.	5.	6.
Type of Biodiversity	Create wild biodiversity reserves to benefit local people	Develop habitat networks in nonfarmed areas	Increase agricultural productivity to reduce land conversion	Minimize agricultural pollution	Modify resource management	Modify farming systems to mimic natural systems
Soil Organisms	***	**	**	**	**	**
Beneficial Insects Including Pollinators	***	**	**	***	**	***
Herbs, Shrubs, Grasses	***	***	**	*	**	**
Trees	***	**	**	*	*	***
Fish	***	**	**	***	**	*
Reptiles	***	**	**	*	**	**
Amphibians	***	**	**	**	**	*
Birds	***	**	**	**	**	***
Small Mammals	***	***	**	**	*	***
Large Mammals	***	**	**	**	*	**

Notes:

*** High benefit for wild biodiversity.

** Medium benefit for wild biodiversity.

* Low benefit for wild biodiversity.

valued in more intensive systems as well. Perennial tree, long-fallow, forestry, and pasture systems offer the greatest potential (at least in the short term) to mimic natural habitats.

Table 5.3 suggests the types of wild species most likely to benefit from interventions in agricultural regions. While large mammals and some tree species will benefit most by creating reserves and expanding and utilizing nonfarmed areas (Strategies 1, 2, and 3), abundant and diverse types of smaller wildlife can flourish in suitably modified farming systems. Both biodiversity and human society are likely to benefit most when agriculture is sustainably managed.

The process of moving from low-biodiversity agricultural landscapes toward a goal of full ecosystem management takes time. In keeping with the "adaptive management" strategy of learning by doing, and to respect the livelihood needs of local people, it may be useful to aim first for a transitional system. Habitat sites that most support both agricultural production and biodiversity may be established first. Other areas can then be established to cover additional habitats and to form wildlife corridors for allowing processes such as migration to continue. Where new agricultural lands are being developed, networks of protected areas should be an integral part of the land-use plan from the very beginning.

Intrinsic to the success of ecoagriculture is regular monitoring and evaluation of key indicators of habitat conditions and wild biodiversity in and around agricultural areas considered part of the broader ecosystem. An example of criteria developed for the western Australia wheat belt is shown in Table 5.4. The presence of healthy populations of particular wild species can serve as

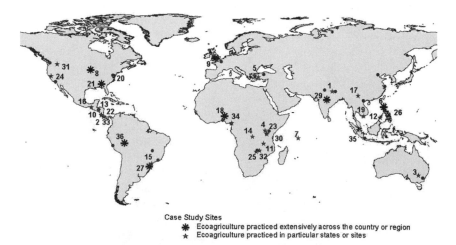

Case Study Sites
✳ Ecoagriculture practiced extensively across the country or region
★ Ecoagriculture practiced in particular states or sites

Map 5.1. Map of Case Study Locations

Table 5.4. Ecological Parameters for Assessment of Biodiversity—
An Example from the Western Australia Wheat Belt

Parameter	Indicator	Unit of Measure	Present	Transitional	Desirable
				Level	
1. Diversity of plant species and forms (native vegetation)	Species count by plant form: Trees Shrubs Herbaceous perennials Perennial grasses Annual grasses Annual herbs	Proportion of species originally present			
2. Representation of plant functional groups	Proportion of area occupied by Trees Shrubs Herbaceous perennials Annual pasture Crops	Percent	3 5 2 45 45	5 10 5 35 45	10 10 10 30 40
3. Diversity of native fauna	Key species count: Birds Mammals Other	Proportion of species originally present			
4. Soil biological activity	Population density of key species	Number of organisms per cubic cm			
	Level of metabolic activity in the soil	Respiration rate			
	Rate of organic matter and surface litter breakdown	Kg per hectare per year			
	Acidity/alkalinity	Change in pH (acidity) per year			
5. Conservation status	Congruence between protected remnants and minimum subset	Percent congruence	20	50	100
6. Connectivity	Percentage of remnants connected by corridors		0	50	100

Source: Adapted from Lefroy, Salerian, and Hobbs (1992), Table 8.4.

indicators that other environmental services, such as soil biological activity, are in good supply.

Case Studies of Ecoagriculture

While preparing this study, many different practices and land-use systems that could qualify as ecoagriculture were identified. In far fewer cases, however, was the impact of introducing the new system or practice well documented on all three key criteria: agricultural yield, farmer income, and wild biodiversity. The following chapters present thirty-six cases where these impacts have all been studied (Map 5.1). They include only examples where wild biodiversity clearly increased (or major threats were significantly reduced), and where agricultural yields and farmer income either increased or the impact was neutral. Thus all of the cases reflect situations where it was possible to "expand the agricultural production possibility frontier" in ways that enhance biodiversity.

Social impacts were not assessed for these cases, largely due to a lack of documentation. While all of the cases have been analyzed for their effects on farmer income (using diverse methods), most did not have data available specifically on demand for labor or the investment resources required to bring about these system changes. Future analyses should include these factors explicitly.

Chapter 6

Making Space for Wildlife in Agricultural Landscapes

This chapter describes three strategies for making more space available for wildlife in agricultural landscapes. In and around existing farmlands it is often possible to identify spaces that can be maintained as protected areas, either as larger reserves or as habitat networks within production areas. Under some conditions, increasing agricultural productivity on existing farmlands will reduce the expansion of farming onto new lands, or even encourage the contraction of production areas to leave more wild habitat. Each of the strategies is illustrated by several case studies that demonstrate how the strategy is being applied in practice.

Strategy 1: Create Biodiversity Reserves That Benefit Local Farming Communities

Large protected areas are a central feature of ecosystem management for biodiversity. Large mammals, birds, and trees require large territories for effective reproduction, as do interior species, that is, those that do not flourish on the edges or in fragments of habitat but are dependent on extensive areas of closed-canopy forests. Some conservationists have established the goal of maintaining at least 10 percent of each major habitat type in fully protected reserves. This figure has already been exceeded in some habitat types. A recent study of ninety-three parks in twenty-two countries found that 83 percent had been largely successful in controlling encroachment, especially land clearing.

In more than 60 percent of the parks studied, hunting and grazing pressures were better controlled than in surrounding areas (Bruner et al. 2001). A larger sample might not give such a positive result, but it is clear that even less-affluent countries with low per-capita gross domestic product (GDP) have embraced the value of protected areas.

Finding ways for local people to benefit from protected areas has become standard practice, as a strategy to compensate them for lost access to resources or to ensure their continued willingness to comply with access restrictions (Borrini-Feyerabend 1997). While the model remains widely appreciated, there has been considerable debate about how to make these "integrated conservation and development projects" (ICDP) effective in delivering both livelihood and conservation benefits (Wells, Afiff, and Purnomo 1992; Barber, Brandon, and Hannah 1995). The source of many problems is that local people still have little say in establishing the boundaries and rules of protected areas, and many reserves are established in areas over which local communities have prior claims. This ecoagriculture strategy builds on lessons learned in successful ICDPs, as described below.

Identifying Locally Beneficial Areas for Protection

Ecoagriculture places local food security and livelihood concerns at the center of its conservation strategies. Thus, in establishing new protected areas, it emphasizes choosing sites jointly with local people, and in places where there are clear benefits to local agricultural populations. While some individual landowners may be disadvantaged, the benefits to others should be sufficiently substantial to justify their support for protection, and even to provide some compensation. This is likely to be the case where:

- the planned site clearly helps to make farming more productive or sustainable (for example, by protecting valued pollinators);
- the reserve helps to protect locally valued environmental services (for example, good water quality);
- the site offers attractive alternative livelihood options (for example, by enhancing fishing income or attracting tourists);
- farmers are adequately compensated for the loss of land or are helped to make the transition to an equally attractive livelihood option (e.g., through payments for habitat protection); or
- local communities themselves value the aesthetic, cultural, or recreational aspects of the habitat or of particular species (leading them, for example, to protect sacred groves from development by outsiders).

Most protected areas have been established in and around lower-intensity rainfed agricultural systems, where land values and productive potential were relatively low. Even in these areas, however, the value of the agricultural land for local people may be significant, and without their "buy-in" the boundaries between agriculture and conservation may not be respected. Biodiversity conservation initiatives are increasingly being targeted at lands with much higher value for agriculture. In such cases, a much clearer analysis of trade-offs is needed, and compelling evidence of potential benefits of conservation for the surrounding farmers must be rigorously produced. Otherwise, mechanisms must be found to fairly compensate local people for giving up valuable resources.

For example, even some areas that are highly valued for agriculture because of ready access to irrigation are more suitable as protected areas than might be expected. While the value of production per hectare in irrigated systems is typically high, costs of production and of infrastructure construction and maintenance per hectare are also high. These costs are often not reflected in the price of the water provided, just as the economic values of biodiversity and habitat preservation are not reflected in decision-making about land use. In particular, the value of natural fisheries, an important source of food in many rural areas, is ignored when allocating water to irrigation. Also, while some irrigated agricultural areas have been sustainably farmed for millennia, others are at high risk due to water scarcity (from groundwater depletion or competition from nonagricultural demands), or due to salinization (usually from poor water management or naturally saline soils). Strong economic, as well as ecological, arguments may thus be mobilized for retiring some of these lands from irrigated farming (Ahmed and Hirsch 2000).

Protecting Agricultural Habitats

In some cases, wild species actually require the open habitat found in settled farming or the rotational land patterns of shifting cultivation. Yet abandonment of agriculture for economic reasons may convert open space to forest or woodlands (OECD 1998), while other farmland habitats are threatened with urban or residential development. Conservationists in many countries are working to provide such farmlands with some type of protected status. For example, Plieninger and Wilbrand (2001) have proposed that the dehesa system of Spain (see Box 4.2) be protected as part of a biosphere reserve project. The Nature Conservancy, an international conservation organization based in the United States, has been working with farmers and ranchers to protect prime agricultural lands in the state of Virginia and ranching ecosystems in the states of Arizona and New Mexico from urban development (OECD 1997a).

Other agricultural areas are being protected for in situ conservation of genetic diversity. Such conservation will help conventional breeding and biotechnology improve crop characteristics. Traits such as pest and disease resistance, for example, are to be found in wild relatives of domesticated plants.

Although large *ex situ* gene banks play an important role in conserving such germplasm, in situ conservation allows species to continue to evolve in relation to their natural environment and their pests.

Recognition of the need for in situ efforts to conserve wild relatives has led to the establishment of protected areas that include working farms (Amaral, Persley, and Platais 2001). The basic principle behind genetic reserves of this sort is to conserve enough diversity to enable the species to co-evolve with associated pest and other species. Gadgil, Hemam, and Reddy (1996) developed landscape-based conservation plans for several important taxa of plants. Reserves currently exist for corn in Mexico, wheat in Israel, and a countrywide program funded by the Global Environment Facility (GEF) in Turkey (Hodgkin and Arora 2001). India has established a "gene sanctuary" in the Garo Hills for wild relatives of citrus; additional sanctuaries are planned for banana, sugarcane, rice, and mango (Hoyt 1992). The Chatkal Mountain Biosphere Reserve in Kirgizstan conserves important wild relatives of walnuts, apples, pears, and prunes. These programs seek to preserve farming areas and nearby wildlands, usually with some restrictions on management and harvest to protect wild biodiversity.

Enhancing Benefits from Protected Areas for Local Farmers

Local willingness to support protected areas will depend in large part on the perceived benefits. Thus, planning for protected areas should include local people in decision-making about boundaries. The participation of local people will ensure that they maintain access to those resources they consider most important economically and culturally, and that they negotiate reasonable compensation. Such compensation may include opening up access to unprotected sites for hunting and gathering or farm settlement, or providing alternative employment. In most cases these agreements will not compromise the main conservation goals.

Planning requires an understanding of the threshold values of the intensity of disturbance, beyond which genetic diversity is rapidly lost. For example, research from Malaysia, Thailand, and India found that imposing total prohibition on access and use of tropical forests was not necessarily the best way to benefit species or conserve resources, at least partly because it was so difficult to enforce. Significant logging did reduce populations of climax species (those species characterizing fully rehabilitated natural vegetation), but it had no effect or a positive effect on the pioneer species (those that move in quickly after disturbance). Some of the negative impacts of logging on the diversity of the logged and nonlogged species were only transitory. While heavy harvesting of nontimber forest products (NTFPs) can threaten their species' populations, others are little disturbed. Thus sustainable exploitation regimes can be developed for the high-value NTFPs, and genetic diversity can be maintained even with resource use by local people (Boyle 1997).

Communities themselves sometimes take the lead in establishing wildlife reserves. Tonda Wildlife Management Area in Papua New Guinea contains a unique savanna interspersed with tall forest along the rivers that provides a globally significant wintering ground for migratory species of waterfowl. It covers 590,000 hectares under customary tenure (recognized by local people, though not enshrined in national legislation), with about 1,200 inhabitants distributed among sixteen villages. It was established in 1975 at the request of the customary landowners for the conservation and controlled utilization of wildlife and other natural resources. The population is mobile, and shifting cultivation with a fallow period of fifteen to thirty years remains an important part of the local economy. Customary landowners are allowed to hunt freely, but they have agreed that the portion of the reserve between the Bensbach and Morehead Rivers will be closed to hunting, and that no vehicles or boats can be used for hunting anywhere within the site (Alcorn 1993).

In Indonesia, community conservation agreements have been developed to give rights of use to local communities for buffer zone areas while allowing them to prevent outsiders from opening new lands within Sumatra's Kerinci National Park. Villagers are encouraged to grow indigenous tree species after harvesting existing tree crops within the national park; they are also granted twenty-year rights of use to the land in order to slow further encroachment (Barber, Afiff, and Purnomo 1995). In Cameroon, community wildlife management systems provide commercial hunting rights to local people, under their own management.

Six Case Studies

Below are six examples in which local farmers have actively cooperated in the management of state protected areas, established community-managed protected areas, or designated areas of their own farms as protected for wildlife. In all of the cases, both wild biodiversity and the community have benefited.

EXAMPLE 1. BUFFER ZONES TO PROTECT RHINOS AND TIGERS IN A NATIONAL PARK IN NEPAL

Royal Chitwan National Park is located on Nepal's border with India, along the floodplains of the Rapti, Reu, and Narayani Rivers. It covers 93,200 hectares, and its ecological value is greatly extended by the adjacent Parsa Wildlife Reserve (49,900 hectares). The alluvial terraces laid down by the floodplains provide a very productive habitat for grazing mammals, most notably the Indian rhinoceros (population around 450) and several species of deer that provide prey for Chitwan's tiger population (now estimated at 107).

The riverine/tall grass habitat there is more extensive than that reported anywhere else in Asia. The 1991 census recorded 275,000 people in thirty-six Village Development Committees settled around the park's vicinity, and conflicts
between the park and local people were a major management problem. Three
to five local people were killed each year by rhinos and tigers in the park. In
addition, domestic cattle constituted up to 30 percent of tiger kills in settled
areas on the periphery of the park, rhinos sometimes caused significant damage to rice and other crops, and local villagers cast covetous eyes on the park's
rich resources (Mishra 1982). People living in the buffer zones around Royal
Chitwan National Park derived up to 80 percent of their needs for firewood
and fodder from the forest, still leaving some of their needs unmet. The resulting competition between humans and rhinos, tigers, and other large mammals
caused significant conservation problems.

Pioneering legislation in 1993 empowered the Government of Nepal to
declare areas surrounding Chitwan as a buffer zone; it also permitted local forest user groups to use 30 to 50 percent of park revenues for managing community forests (Sharma and Wells 1997). In the locally managed forests within
the park, one community group constructed nature trails for elephant-back
safaris through the riverine grasslands plus a wildlife viewing tower where
tourists can stay overnight to view rhinos and other wildlife. Within the first
six months of operation, nearly 8,000 tourists visited the Baghmara wildlife
viewing area, generating nearly U.S.$200,000 in revenues and providing
enough money to enable the Baghmara Group to refurbish three schools and
a health clinic. An area that had been largely deforested and supported little
wildlife prior to this conservation investment has now become one of the
most popular tourist destinations in Nepal (83,000 visitors per year). Moreover, the village-managed forests to the north of Chitwan help to protect villagers from Rapti River floods and provide a barrier against rhinos raiding
their crops. Some 1,240 hectares of park land have been given by the Forest
Department to local forest user groups for production of fodder, fuel, and
building materials. Villagers receive income of 10,000 rupees per elephant per
year (U.S.$200) plus $3 per tourist trip. The income has enabled them to build
biogas plants and smokeless stoves, provide training to local women's groups,
and carry out numerous other activities that reduce human pressure on the
park. The local villagers are now convinced that rhinos are a critical tourism
attraction, and they claim to do whatever they can to support conservation
(McNeely 1999).

EXAMPLE 2. COSTA RICAN ORANGE PLANTATION COOPERATES WITH CONSERVATION AREA

Del Oro is a Costa Rican company focused on the growing, production, and
processing of oranges, and the marketing of orange juice concentrate, mostly

sold on the international market. The company owns 7,000 hectares of land, of which 3,339 hectares are planted with oranges and other citrus fruits. The remainder, bordering on the orchards, includes a variety of habitats from grassland to primary forest. Each year the company produces an estimated 3 million boxes of orange juice concentrate weighing 90 kilograms each. The Del Oro property shares approximately 12 kilometers of common boundary with the 120,000-hectare Guanacaste Conservation Area (GCA). A large portion of the Del Oro plantation is interspersed with patches of rare dry tropical forest lying within the farm property or adjacent to the GCA.

The presence of forest has contributed to keeping the population of insect pests at levels low enough that, unlike most orange plantations, little or no pesticide applications are required. Pesticide use is limited to controlling leafcutter ants (*Atta cephalotes*) in specific areas and controlling outbursts of citrus leaf miner (*Phylloenistis citrella*) in nurseries. Del Oro benefits from reduced costs of pesticide application, and from the ECO-OK certification of their concentrate by the Rainforest Alliance. By differentiating their product as environmentally friendly, Del Oro obtains added value in the market.

An innovative contract was agreed upon between Del Oro and GCA whereby the farm purchased a number of environmental services provided naturally by the state conservation area over a twenty-year period. Specified services included biological control, water supply and watershed protection, natural decay of organic material from orange residues discarded in forest areas (avoiding expenses for waste removal or burning), and research support. The company committed $480,000 to purchasing these services. The environmental services were paid for by the transfer of 1,200 hectares of remnants of rare dry tropical forest patches from Del Oro to the GCA, thus expanding the area and biodiversity quality of the reserve (Gamez 1999). Due to a series of legal challenges, ironically brought on by conservation interests as well as commercial competitors, this partnership has not been able to function as envisaged and, therefore, was unable to live up to its full potential.

EXAMPLE 3. AUSTRALIAN LANDCARE GROUPS PLAN FOR
BIODIVERSITY GOALS

The Landcare movement in Australia is premised on farm planning that keeps both production and conservation goals in mind (Campbell 1991). Groups of farmers support one another in seeking land improvements and coordinate their actions when necessary at a landscape scale involving numerous farms and nonfarmed areas (Campbell 1994). To date, around 4,500 active community groups are working in partnership with government, nongovernmental organizations, and corporations to address soil, water, and biodiversity degradation. In the late 1990s, several provinces established Community Nature Conservation Extension Networks to assist landholders and community groups with wildlife conservation planning and management, under programs such as "Bushcare," "Land for Wildlife," and "NatureSearch" (Millar 2001).

The Genaren Hill Landcare Group, for example, includes fourteen farming families in the heart of the wheat/sheep belt of New South Wales. The group's activities focus on nature conservation, but landscape modification goals also address farm profitability, soil structure decline, gully erosion, feral animals, and introduced weeds. With community and government support, the group erected an 8.4-kilometer-long fox- and cat-proof fence around an area of good-quality remnant native vegetation. All livestock and introduced predators were removed, and two marsupial species were reintroduced to the area—the threatened brush-tailed bettong (*Bettongia pencillata*) and the endangered Bridle nail-tailed wallaby (*Onychogalea fraenata*). Another 85 kilometers of fencing are being laid and 35,000 trees planted across a 50,000-hectare farmscape that will strategically link existing remnants of wildlife habitat. Covenants have been negotiated with government agencies to secure commitment to long-term conservation use (Sutherland and Scarsbrick 2001).

EXAMPLE 4. LIVESTOCK AND WILDLIFE COEXISTING IN A
WORLD HERITAGE SITE: THE NGORONGORO
CONSERVATION AREA, TANZANIA

The grazing ecosystems that are favored by livestock are also often those especially productive for large mammals (and their predators). The Ngorongoro Conservation Area in Tanzania, one of the few protected areas in eastern Africa established explicitly to promote multiple land uses including grazing of domestic stock, demonstrates that coexistence is not only possible, but productive. Inscribed on the UNESCO World Heritage List in 1979, this vast area of 828,800 hectares is contiguous to the much larger Serengeti National Park and to Maasai Mara National Park in Kenya. One of the great wildlife spectacles in the world is the annual migration of about 1.8 million wildebeest, 300,000 zebra, and 450,000 gazelles (though the actual numbers vary considerably from year to year), along with their predators (lions, hyenas, and cheetahs). The Ngorongoro Crater attracted over 200,000 visitors in 2000, including 125,000 foreigners. Some farms adjacent to the site host tourists for bird-watching, while the rest of the site is shared by wildlife and livestock (cattle, sheep, and goats) tended by some 40,000 Maasai herders. Pastoralism is an ancient practice in the Ngorongoro region, stretching back at least 2,000 years, though the Maasai have lived in the area for only two centuries (Homewood and Rodgers 1991).

Maasai settlements dot the region outside the crater. The site is specifically designed under the Ngorongoro Conservation Area Ordinance of 1959, revised in 1975, to ensure that appropriate benefits are provided to the Maasai, along with the benefits of conserving wildlife populations and promoting tourism. While permanent agriculture is discouraged, grazing is actively supported. The herdsmen are allowed to actively protect their livestock from predators. The forested parts of the site outside the crater are also legally protected to ensure permanent water flow to downstream farmers who grow high-value crops such as coffee.

To ensure that the Maasai have a voice in the management of the protected area, a reasonable understanding of its management objectives, and a platform for presenting their interests, the conservation area agency set up an extension unit and a community development department. As a result of negotiations, food security has been improved by subsidizing grain sales, veterinary services have been provided, water resources have been further developed or rehabilitated, employment has been provided by tourism agencies (including as guides for walking tours), and income from tourism has been returned to the Maasai. A livestock marketing system, dairy industry, and tsetse-fly eradication programs have been established, all of which are highly popular with the Maasai. This has greatly improved the relationship between the protected area management and the local people, who now themselves help to control poaching, though land ownership by the Conservator of Ngorongoro remains an issue for the Maasai. Lessons learned from the Ngorongoro Conservation Area on joint management of wildlife, tourists, indigenous people, and domestic livestock are already being applied elsewhere in Tanzania, including around the Selous Game Reserve and the Ruaha National Park (Leader-Williams et al. 1996).

Example 5. Agricultural Gene Sanctuaries Protect Wild Biodiversity in Turkey

Agriculture was born over 10,000 years ago in the Fertile Crescent, which encompasses modern-day southeastern Anatolia, the Asian part of Turkey. Today, more than 8,700 species of vascular plants are found in Turkey, and about 30 percent of these are endemic to the area. In the early 1990s a project funded under the auspices of the GEF was established to conserve plant genetic resources in their natural habitats, that is, in situ. In situ conservation maintains interactions between plants and their natural pests, predators, and environmental conditions, supporting continued co-evolution, and is thus crucial to efforts to provide resistance to new pest and pathogen mutations as they arise. The GEF project in Turkey was the first of its kind in the world to protect multiple wild crop relatives—both woody and nonwoody—using an integrated multispecies, multisite approach.

A key feature of the project was the establishment of Gene Management Zones (GMZs) based on ecogeographic surveys and inventories of state-owned land. Protected areas with specific management requirements adapted to individual plant species and environmental conditions, GMZs serve as reserves for one or more endangered or economically important plant species;

they are large enough to encompass considerable genetic variation within populations. The GMZ concept was first used in California in the 1960s, but it is a new concept to most of the rest of the world. Based on findings on genetic diversity, project planners designated twenty-two GMZs. Kazdagi National Park was home to ten GMZs covering five target species, including wild plum, chestnut, Turkey red pine, Anatolian black pine, and Kazdagi fir. Seven GMZs were designated at Ceylanpinar State Farm, containing five species of wild wheat relatives. The Bolkar Mountains contained five GMZs covering Anatolian black pine, Turkey red pine, two types of Taurus fir, and Taurus cedar.

A vital element of GMZ management is local community participation, which preserves local people's access to the GMZ and enables them to practice traditional activities important to local livelihoods. Grazing in many cases can continue with some modifications. During some parts of the year, grazing animals actually enhance a GMZ's desired vegetation pattern by shattering the seed and trampling it into the soil for germination the following year (one form of "natural seeding"). Similarly, the local practice of harvesting chestnuts was incorporated into the management plan for the GMZs for this target species. Lessons learned in this project are informing the development of a large GEF biodiversity project in Turkey and other projects elsewhere (*Diversity* 2000).

EXAMPLE 6. MARINE RESERVES HELP BOTH FISH AND FISHERMEN IN THE PHILIPPINES

Fisheries scientists have learned that banning fishing completely in certain areas can lead to an increase in the overall fish catch in adjacent areas. A survey of 100 "no-take" reserves around the world with complete bans on fishing found average increases of 91 percent in the number of fish, 31 percent in the size of fish, and 23 percent in the number of fish species present within the reserves. Those increases occurred within two years of starting the protection scheme. The beneficial effects spilled over into areas where fishing was still permitted. In St. Lucia, for example, a third of the country's fishing grounds were designated no-take areas in 1995. Within three years, commercially important fish stocks had doubled in the seas adjacent to those reserves. A benefit of the reserve approach is that rules are simpler to enforce than with traditional regulations, such as harvesting limits, as inexpensive global positioning systems can be used to monitor compliance (*Economist* 2001).

This concept has been applied in community-level marine source management in several island villages in the Philippines, where overexploitation of coral reef fisheries has become a major problem. In 1985, Marine Management Committees (MMCs) were formed to design new coral reef protection and management schemes that reflected the interest of local people. As marine reserves began to function and illegal fishermen were repelled, community support for the MMCs increased. Apart from village-based patrolling of the coral reefs, activities included growing giant clams in the fish sanctuary areas for the community to manage and harvest. MMC members were trained to manage tourism and to establish alternative income schemes such as mat weaving and sea cucumber mariculture. Marine reserve guidelines were refined into a legal document adopted by the municipal town councils. Each of the three protected areas now has a fish-breeding sanctuary and a surrounding buffer area for ecologically sound fishing. Destructive fishing methods, such as using dynamite, cyanide, or small-mesh gill nets, have been effectively banned. Species diversity and abundance have significantly increased for certain families of fish, especially the favorite targets of fishermen. Mean percentage increases in species diversity ranged from 20 to 40 percent, while increases in the numbers of all food fishes ranged from 42 to 293 percent over the three sites. Total fish yield for the fishermen also increased, providing them important economic benefits (Savina and White 1986; McNeely 1988). Recognizing the success of such community initiatives, the government of the Philippines has decentralized fisheries management, giving municipal and city governments responsibility for their marine waters out to 15 kilometers off shore. The MMC model has spread throughout the Philippines and into Indonesia (White and Vogt 2000).

Potential Application

The examples above are drawn from a wide range of ecological zones and agricultural systems. In the Tanzania example, extensive livestock activities continued within the protected areas, as did extensive grazing and food-gathering in Turkey, and low-impact forestry in Nepal. Economically important local ecosystem services motivated local people in the more intensive production systems in Costa Rica, Australia, and (through fish reproduction) the Philippines. Providing nonagricultural income opportunities were important supplements in Nepal and Tanzania. The success of these systems depended upon proactive and participatory management by the institutions governing the protected areas. In the Landcare and fisheries examples local institutions took the lead, while elsewhere national conservation organizations did so. All these project areas have been subject to political pressures from within local communities, from their economic competitors outside the project, and/or from within conservation organizations. Land ownership issues are sensitive and may need to be re-negotiated. The costs to establish and maintain protected areas themselves are sometimes fairly high. Evolving technical knowledge has enabled a high degree of compatibility between these relatively extensive food-producing activities and conservation. These examples suggest that the model of designing protected area management systems to benefit local people is a viable one that could be extended widely.

Strategy 2: Develop Habitat Networks in Nonfarmed Areas

In most agricultural landscapes, even those with intensive farming systems, considerable land area is devoted to nonagricultural uses. Some obvious forms these uses take include farm wetlands, woodlots, and windbreaks, as well as often-ignored ones such as schoolyards, temple grounds, and graveyards (Table 6.1). More wild biodiversity, and scope for protecting it, usually exists in these areas than most people realize. Thus a second major strategy to promote biodiversity in agricultural regions is to modify the use of these "in-between" spaces so that better ecological conditions enable wild biodiversity to thrive. These spaces can then be linked into networks that enable wildlife to move between important food and water sources, find mates, and have access to protected areas. This strategy draws heavily from the work on biological corridors and large-scale landscape planning.

The Biodiversity Value of "In Between" Spaces

No matter how carefully they are protected, small reserves will progressively lose their most distinctive species if they are surrounded by a hostile landscape. But if the surrounding matrix is managed with biodiversity in mind, agricultural areas can make a positive contribution to biodiversity (see Box 6.1). The best method for meeting biodiversity conservation goals is to establish habitat according to a pattern within and across farms that reflects ecosystem planning on a fairly large scale. Different types of niches in agricultural landscapes may support different types of biodiversity, depending upon their size, shape, and location. Nonfarmed areas can be utilized to provide patches of certain types of habitat, or to form corridors that link protected areas and enable species to maintain genetic contact between populations that otherwise would be isolated. This strategy may involve protecting remnant native vegetation or reestablishing wild species, often the essential or "keystone" plant species that provide microhabitats for associated species. Remnants may include both biological communities that depend on a continuation of traditional land-use practices, and surviving areas of preagricultural vegetation. Through various kinds of linkages with the surrounding landscape, protected areas can avoid fragmentation and degradation and instead be used more effectively in conserving biodiversity. For example, research in Costa Rica found that isolated, small (20- to 30-hectare) fragments of remnant forest had far fewer butterflies than a large 227-hectare patch less than a kilometer away. Establishment of corridors between these fragments and the larger forest—even under highly modified management—would support larger butterfly populations (Daily and Ehrlich 1995).

While much remains to be learned about ecological relationships between wild species and agricultural habitats, some general principles are developing. For example, since many vertebrate and insect species use and require two or three habitats during the course of a day, in different seasons, or over their life

Table 6.1. Noncultivated Areas in Agricultural Lands—Potential Habitats for Wild Biodiversity

AROUND WATER RESOURCES:

- Riparian forests and ecosystems
- Natural waterways
- Irrigation canals
- Watershed areas to promote water harvesting and management
- Ditches and other drainage along roads or on farms
- Drainage water areas used for fish habitat or production
- Stream filter strips (using strips of perennial vegetation to catch sediment and chemical runoff before it enters waterways)

IN AND AROUND FARM FIELDS:

- Conservation reserve areas taken out of farming
- Uncultivated strips within crop fields as habitat for wild relatives of crop plants (especially in areas known to be centers of origin or diversity for crop plants)
- Windbreaks
- Border plantings or live fences between plots or paddocks, or between farms
- Irrigation dikes in crop fields
- Vegetative barriers to soil and water movement within crop fields
- Areas taken out of production to control soil or water salinity, or abandoned as a result of salinity
- Little-used or low-productivity croplands
- Little-used or low-productivity grasslands

IN AND AROUND FOREST AREAS:

- Farm or community woodlots
- Farm, community, government, or private natural woodlands or forest
- Private industrial tree plantations

OTHER SITES:

- Homesteads
- Along roadsides
- "Sacred groves" in communal lands, churchyards, or graveyards
- Schoolyards
- Industrial or hospital sites
- Agro-ecotourism sites
- Public or private recreational parks
- Special sites conserved for local, national, or international cultural value

Box 6.1. Biodiversity in Agricultural Landscapes in Costa Rica

To preserve some species, large patches of habitat are needed. For example, forest butterflies in southern Costa Rica require unexpectedly large patches, although heavily managed systems of largely exotic plants (a botanical garden, in this example) could effectively serve as corridors for many of these species (Daily and Ehrlich 1995). A study of a 227-hectare forest fragment and four surrounding agricultural habitats (monoculture coffee, shade-grown coffee, pasture, and mixed farms) found no significant difference in the richness and abundance of moth species among the agricultural habitats. However, agricultural sites less than 1 kilometer from the forest fragment had significantly higher richness and abundance than sites more than 3.5 kilometers away, and species composition differed according to distance as well (Ricketts et al. 2000).

A substantial proportion of the native bird fauna occurs in the agricultural habitats densely populated by humans, almost half a century after extensive clearance. From 1 to 9 percent of the possible number of original bird species has been extirpated since deforestation began. Of the 272 bird species found today in this small-scale coffee-cattle-crop-forest mosaic, 55 percent occur in forest habitats only, 23 percent in open agricultural habitats only, and 22 percent in both. Thus while the agricultural landscape in Costa Rica has the capacity to maintain considerable bird biodiversity, this capacity depends upon proximity to forest cover (Daily, Ehrlich, and Sánchez-Azofeifa 2000).

---

cycle, the proximity and access to such habitats is critical (Forman 1995). Networks of natural vegetation are particularly effective for maintaining populations of species that require edge habitats, and for connecting breeding stocks in dispersed protected areas. Such networks could potentially meet a significant portion of the habitat needs for many types of species, even without large protected areas nearby. In western Australia researchers found that even modest increases in native vegetation of 7 to 10 percent, strategically located, significantly improved habitat value (Carl Binning, CSIRO, personal communication, 2001).

Even small fragments of native habitat can help migratory animals at sites that provide food and shelter for specific periods of the year. Many migratory species of birds, for example, will find these relatively small areas of habitat along their migratory routes sufficient to meet their transitory needs. Recent studies of insect-eating birds in isolated fragments in Brazil indicate that the rapid establishment of tall secondary forests around isolated small fragments, linking them to more extensive primary forest areas, greatly accelerates the recovery of the avian insectivore community to something close to the pre-isolation situation. Thus small fragments can provide a safety net for a significant number of species and their genetic diversity, and a breathing space for conservationists to plan strategies for preventing the loss of the species concerned. Intervention management can then be focused on species that are particularly sensitive to fragmentation, such as large carnivores, large trees, and epiphytic orchids. Cowlishaw (1999), for example, concludes that 30 percent of forest primate fauna will be lost even if deforestation is controlled, unless corridors to connect protected areas are established.

Finding Landscape Niches for Biodiversity

Simple, low-cost landscape interventions can help biodiversity. Farmers can locate and orient crop production lands, corridors, and barriers to minimize species loss, promote dispersion of certain species, and prevent dispersion of harmful populations. For example, farmers managing fallow-based systems can leave strips of natural vegetation when clearing land for temporary farm fields. Where these strips are left along the contour of steeper slopes, they also markedly reduce the erosion caused by clearing and cultivation. Especially in extensive agricultural systems, natural ecological disturbance patterns (such as fire and floods) can be used, wherever possible, to play their normal role in biodiversity conservation.

Many opportunities are also available to improve habitat in and around farms in more densely populated and intensively cultivated farming areas. Farmers already recognize the potential role of perennial vegetation in "in between" places to stabilize agricultural production. In heavily populated parts of Burundi, Kenya, and Uganda, where farm size is very small, the number of trees on farms is increasing because farmers recognize their value for fodder, fuelwood, boundary delineation, and food production (Sanchez, Buresh, and Leakey 1996). But on these small farm holdings, farmers must carefully screen perennials, trees, shrubs, and grasses planted around farm fields to ensure that they are compatible with agricultural crops, and that any nonnative species do not threaten the local ecosystem or farm production.

Many farmers are interested in wildlife conservation, if it does not entail financial loss or livelihood risk. For example, farmers have worked to recover native or endemic species now rare in the landscape by converting low-value unfarmed areas to native vegetation, or by preserving biodiversity-rich wetlands. In ranching systems, landowners and community groups have allocated marginal grazing lands to help conserve wild species. In Ontario, Canada, a farm survey found that in 1999 77 percent of farmers felt wildlife was "very or somewhat important as a necessary part of the balance of nature," and that in the same year farmers had invested a total of almost $8 million in enhancing wildlife habitat (Ontario Soil and Crop Improvement Association 2001). A 200-hectare row- and field-crop farm in central California, for example, planted more than fifty locally adapted species of native perennial grasses, wildflowers, sedges, rushes, shrubs, and trees into various parts of the farm— on poor-quality lands, roadsides, irrigation canals, natural marshes, ponds, and hedgerows. The farm has 3 hectares of 5- to 10-meter wide multispecies hedgerows that serve as year-round habitat corridors for deer, fox, bear, coy- otes, and many other animals whose populations have dramatically increased since the hedgerows were established. They act as a web connecting the native habitat patches on the farm, as well as supporting beneficial insects that con- trol pests in adjacent row crops. While the farmer faces additional costs for seed and plant materials, special equipment, and increased transportation, cost savings are achieved from reduced pesticide use, labor, and tillage on the parts of the farm devoted to standard cultivation. Field studies demonstrated no meaningful difference in crop yields between farmed areas with and without hedgerows on this farm, and implementing the conservation practices in unfarmed areas has caused little or no reduction in the land available for crops (Anderson et al. 1996).

In selecting species to use in habitat networks, it is often desirable to include species that produce locally valuable products for cash sale or house- hold consumption. By supplementing the natural vegetation growing between farm fields with nutritious food species, the nutritional status of local people can be improved. Native vegetation established in nonfarming areas, such as roadsides or schoolyards, can include food or fuel plants to be harvested by the poor. Even if not all vegetation in these "in between" sites is native, the increased below- and above-ground biodiversity will often be ecologically valuable (assuming that none of the introduced species are invasive). Inclusion of noninvasive exotic species that provide products of value to local residents can encourage community participation in biodiversity conservation. Adding such species should be considered wherever their establishment represents a net improvement in overall habitat quality and the risk of the species becom- ing invasive appears very low.

Whole Farm and Landscape Planning

Programs in various parts of the world have begun to support individual farm- ers and local community groups in their efforts to undertake whole-farm and landscape planning, often with explicit biodiversity goals. These activities can identify the most strategic sites for biodiversity conservation. Maps of local resources developed together with the local community, supported by aerial photograph analysis, can be used to design vegetation patterns that cross multiple farms to achieve biodiversity goals. Plans may identify needed modifi- cations to infrastructure and other physical features that restrict animal movement or expose critical habitats to pollution. A key element is strategic placement of perennial native vegetation in and around agricultural production areas, as shown in Figure 6.1. Networks and corridors of perennial vegetation can be designed to link remnants of native habitat with protected areas. Vegetation networks can be designed, in collaboration with farmers and other resource users, to improve local hydrological regimes, and in dry areas to increase water availability in recharge areas. For example, tree belts might be arranged in association with earthworks that are designed to intercept runoff water on sloping lands. The earthworks could direct the intercepted water to the tree belts and to annual crops or pastures growing in broad-contoured areas between the wide-spaced tree belts. In high-rainfall areas, establishment of native vegetation could reduce water erosion and waterlogging of farm fields, if strategically situated to improve rain infiltration, slow water movement across slopes, and channel drainage water away from farm fields (Lefroy, Salerian, and Hobbs 1992).

Habitat networks in nonfarmed areas could potentially be developed on a large scale, covering thousands of hectares even in intensively managed agricultural regions (Center for Applied Biodiversity Science 2000). These networks would require that an institutional framework for planning and coordination be in place, with a mix of instruments to facilitate long-term commitments and local respect for those commitments. Changes in zoning rules may be required for agricultural production, landscape corridors, and management of resource use and waste to optimize the quality of habitats for wildlife in the larger system. Networks may also require intensive on-the-ground facilitation by individuals trusted by local farmers. Conservation groups, such as The Nature Conservancy, have garnered considerable experience in making such programs work, even in areas where farmers were initially quite hostile. Large-scale efforts now under way include the Yellowstone-to-Yukon initiative in North America and the Mesoamerican Biological Corridor in Central America (Miller 1996). Agroecosystems can even be managed as "agrolandscapes" that reconnect agricultural and urban landscapes to achieve biodiversity objectives, as in the Atlantic forests around Rio de Janiero, Brazil (Barrett, Barrett, and Peles 1999).

Four Case Studies

Four cases illustrate how wildlife habitat can be created in between agricultural production areas to the mutual benefit of farmers and wild species. In the first three cases, private farmers have agreed to designate areas of their land for biodiversity conservation, with minimal productive use. NGOs or government agencies have taken responsibility for planning their integration into biological networks or corridors. In the last case, farmers worked together to establish

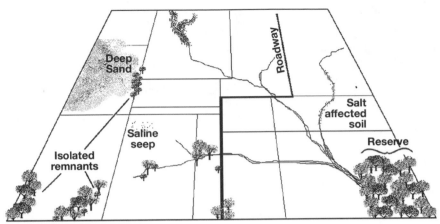

A. The Base Plan
showing drainage, structures, vegetation, major problem areas, and aspect

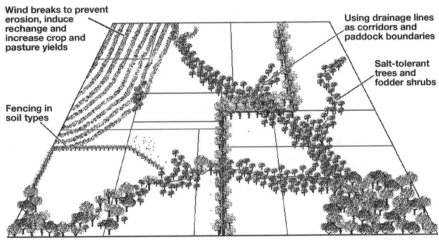

B. A Long-term Revegetation Plan

Figure 6.1. Enhancing Biodiversity through Whole Farm Planning.
Source: Adapted from Lefroy et al. 1992.
Notes: A stylized representation of a central wheat-belt farm illustrating the transition from the present landscape (A) to a mosaic of agricultural, natural, and seminatural systems (B). Key elements of the transition are: 1. Reinstated lines used as conservation corridors and field boundaries; 2. Coincidence between soil types and management unit; 3. Protection of existing remnants in vegetation corridors; 4. Productive revegetation of land affected by secondary salinity; and 5. Alley-farming: a network of widespread shelterbelts (10 to 25 times their height apart) to prevent wind erosion, increase recharge of ground water, increase crop and pasture yield, and act as conservation corridors.

a community-scale windbreak to protect their farms; this windbreak also func-
tioned as a wildlife corridor. Major wildlife conflicts have not yet been
reported in any of the cases.

Example 7. Private Landowners Help Protect Habitat of Amazonian Saki

Two subspecies of bearded sakis (*Chiropotes satans satanas* and *C. s. utahicki*) are
among the most threatened monkeys in Brazil's southeastern Amazonia,
because of a combination of deforestation, hunting, and intolerance of habitat
disturbance. Intense habitat fragmentation has been exacerbated by hunting
pressure, and local populations have been exterminated. But the saki can be
relatively abundant in isolated forest fragments. Well-protected forest fragments
of greater than 5,000 hectares can offer good potential for the protection of
these populations.

Conservation-oriented private landowners in the region have recognized
the importance of retaining forested areas, both to provide habitat to the sakis
and to protect watersheds and provide a range of forest-based products (such
as Brazil nuts, latex, medicinal plants, and many others). They are strongly
motivated and have the means to protect their properties from agricultural
encroachment, and therefore can be important partners for conservation.
Conservation on these large landholdings depends very much on the good
will of the landowners and on their ability to control illegal settlements on
their lands (Ferrari et al. 1999). Many of the private landholders derive their
main income from the land, but mostly from cattle that graze in pastures
cleared from forest. They do not depend as much on forest products as small-
scale farmers do.

Example 8. Protecting Wetlands on U.S. Farms

The Convention on Wetlands of International Importance, also known as the
Ramsar Convention (see Box 2.1), has some 115 state parties that have identified nearly 1,000 wetlands for the Ramsar List of Wetlands of International
Importance. This set of protected areas covers more than 70 million hectares.
Globally, some 25 percent of the Ramsar sites contain agricultural lands within
them. In Latin America, half the Ramsar sites contain agricultural land, while
in Africa and Eastern Europe the figure is about 40 percent. Within Ramsar
sites, agricultural lands are to be managed in ways that are sustainable and
"wise"—do not deplete critical resources.

By enlisting farmers whose land surrounds important wetland areas in the
effort to conserve biodiversity, entire wetland ecosystems can be protected. For
example, in various parts of the United States, programs have been devised to
solicit the voluntary participation of local farmers in wetland protection

efforts. Farmers participate because wetland protection enhances the economic value of farmland used for recreation, hunting, and fishing. A variety of conservation agreements between farmers and state agencies and/or conservation organizations underlies this wetlands protection program. With the 1985 Farm Act, wetlands were no longer to be drained without jeopardizing a farmer's subsidies, but former rigidities in environmental regulation were relaxed, encouraging greater farmer participation.

These programs permit sustainable local use of wetlands and continuous profitable use of areas around the wetlands. When fields are converted back to wetlands, public assistance helps farmers invest to improve agricultural productivity in fields outside wetlands (Frazier 1999; Considine, Roe, and Willard 1999). The U.S. Wetlands Reserve Program, formally established in 1990, has protected over 375,000 hectares, and another 100,000 hectares have been proposed for enrollment. In 2000, requests for participation were running 5 hectares for every 1 enrolled (Audubon 2000). Between the mid-1970s and 1992, the average rate of wetland conversion for agricultural production plummeted from about 165,000 hectares per year to about 13,000 hectares per year, and it is expected that the United States will soon experience a net increase in total wetland areas in agricultural landscapes (OECD 1997a).

EXAMPLE 9. BIODIVERSITY SET-ASIDES ON U.K. FARMS

The European Union's Common Agricultural Policy has sought to reduce overproduction and support farmer incomes. Financial payments have been made to farmers under the condition that some agricultural land be set aside, essentially taken out of agriculture. In the United Kingdom the 600,000 hectares of set-aside land became the third-largest land-use type in the lowlands after grass and cereal production (Sotherton 1998). Farmers producing certain crops or with certain types of land are required to set aside at least 15 to 18 percent of their land. The set-aside policy has been most successful in benefiting wildlife habitat and plant diversity where the land has a short history of intensive cropping and the seed supply in the soil can still generate a diverse local flora. In other areas, seeding a selected plant cover for birds or game may be more beneficial. Freed of crops or planted in appropriate species, set-aside land provides valuable winter-feeding and nesting habitat for farmland birds. Particularly important has been the introduction of a nonrotational set-aside scheme that has enabled farmers to plant specifically designed seed mixtures to create wild bird habitat, and to plant such mixtures in relatively small strips and plots distributed strategically around the farm, thus avoiding large blocks of set-aside land. These small set-aside strips and patches have allowed the creation of designed game and wildlife habitats. It appears that small blocks of wild habitat scattered across the farm considerably increase the capacity of set-aside land to benefit wildlife (OECD 1997b).

EXAMPLE 10. FARM WINDBREAKS AS CORRIDORS IN COSTA RICA

In 1989, the Conservation League of Monteverde initiated tree-planting activities with farmers in a wet, mountainous region of northeast Costa Rica of high natural biodiversity value. The project worked in nineteen communities and helped farmers establish more than 150 hectares of windbreaks. The windbreaks, a mix of indigenous and exotic tree species, were designed to protect coffee trees and dairy cows from the negative impacts of high winds.

The economic returns from the windbreaks are very high, even without considering the timber products, because wind protection results in higher coffee and milk yields, reduced calf mortality and morbidity, and higher herd-carrying capacity of pastures. Nearby farmers have also planted high-value horticultural crops in the protected fields. Damage to coffee from wild parakeets has been reduced, because the parakeets prefer the fruit of a native tree known as colpachi, one of the species used in the windbreaks. Furthermore, farmers who received benefits from the windbreaks have been more receptive to efforts to protect the remaining natural forest on their farms (Current 1995).

Research has shown that the planted windbreaks serve as effective biological corridors connecting remnant forest patches in the Monteverde area. These corridors are especially useful for the migratory species of songbirds that are an essential part of agroecosystems in North America during their summer breeding season. The windbreaks also dramatically increase the deposition of tree and shrub seeds within the agricultural landscape. A careful study of annual "seed rain" patterns in the windbreaks and adjacent areas found that seeds deposited in the windbreaks represented 174 species and at least 53 plant families. Trees accounted for a third of all species. Epiphytes and trees were primarily bird-dispersed, whereas herbs were dispersed by wind, gravity, or explosive mechanisms, and shrubs by a combination of mechanisms. These windbreaks were only 3 to 7 meters in width, yet they increased seed deposition by birds more than 95-fold relative to the pastures. They were effective despite consisting primarily of exotic, nonfruit-bearing species that offered no food to birds. If native, fruit-producing trees were incorporated into the windbreaks, it is likely they would enhance the seed rain and species richness still further (Harvey 2000).

Potential Application

The four examples of habitat networks described here are from diverse types of tropical and temperate forest. The examples in Costa Rica, the United Kingdom and the United States are from intensive farming systems. All of these involve some element of external funding: in Costa Rica a small subsidy for windbreak establishment from a conservation organization and in the

United Kingdom and United States significant government payments. The Brazilian example demonstrates the potential to involve large landowners with extensive farming systems involving pastures and agroforestry in conservation planning, without the need for financial compensation. In Costa Rica there also were significant positive effects on agricultural productivity documented to result from habitat networks, although economic benefits downstream were to be expected from the U.S. wetlands reserves. All were designed with some strategic planning for biodiversity impact at a regional/ecosystem scale, although implementation was often spotty, depending on landowner interest. The U.K. and U.S. models are likely to be too expensive for most developing countries, although their benefits quickly became apparent over a relatively large scale. The inexpensive Costa Rica and Brazilian approaches are more realistic models for poor countries, but they require strong, ongoing commitment by locally based conservation organizations.

Strategy 3: Reduce Land Conversion by Increasing Farm Productivity

As shown in Chapters 2 and 4, agricultural land conversion is a leading threat to wild biodiversity. Pressure for agricultural expansion sometimes results from incentives to expand profitable production systems. But in many cases, expansion is due to stagnant agricultural productivity in the face of rising market and population pressures, lack of alternative employment that induces the landless to seek unexploited lands, and degradation from unsustainable intensification and subsequent abandonment of lower-quality lands. In these latter situations, increased agricultural productivity and sustainability may help to slow or reverse land conversion. Indeed, productivity increases stimulated by Green Revolution technology almost certainly helped to slow land conversion in the developing world in the late twentieth century (see Box 6.2).

Role of Technological Change in Stabilizing Agricultural Area

The Green Revolution and other science-based strategies to increase agricultural productivity begun in the 1960s were targeted primarily at areas already under permanent (often irrigated) agriculture in long-settled communities, especially in South and Southeast Asia and Latin America. More recent international research initiatives have sought to increase productivity in lower-quality agricultural lands and make more intensive production systems more sustainable. All of the sustainable agriculture approaches discussed in Chapter 5 have among their main objectives—at least in developing countries—to increase agricultural yields on existing farmland.

Box 6.2. Global Impacts of Agricultural Productivity Growth on Land Conversion

Several recent studies have assessed the global effects of productivity increases on agricultural land conversion. They estimate the land that would be required to produce today's level of food supply with the average yields obtained in 1961. One study calculated that the yield increases resulting from crop productivity improvements since 1961 may have forestalled the conversion of an additional 3,550 million hectares of habitat globally to agricultural uses, including 970 million hectares for cropland (Goklany 1999). Another study estimated that land savings in the tropics from productivity-increasing research in seven staple food commodities and three livestock products during the same period were roughly 170 million hectares (Nelson and Maredia 1999). This estimate is more conservative than the first but still equivalent to the current area under production in India.

The difference in the two estimates stems from three sources: (a) Goklany includes all cropland plus permanent pasture, whereas Nelson and Maredia include only the seven major crops; (b) Goklany's estimate of cropland saved includes 1.4 billion hectares of cropland in all crops, as compared to Nelson and Maredia's use of 714 million hectares in 1990 under seven major crops; and (c) Goklany assumes that after 1961, cropland would have expanded proportionally to population, though this was not likely, due to increasing cost and declining availability of land, especially in Asia (M. Maredia, personal communication, 2000). Both studies ignore the possible dampening effect of restricted food supply on human population and the likely reduced use of grain to feed livestock.

While unplanned negative impacts on wild biodiversity resulted from many of these technologies, such as high agro-chemical use and encouragement of monocultures, they also had unplanned positive impacts, such as reducing the need for land conversion. By lowering commodity prices, increased crop yields also reduced incentives for production in frontier areas, which typically have higher production and marketing costs than long-settled areas (Southgate 1998). More productive or lower-cost systems in higher-quality lands may encourage migration or abandonment of farming in environmentally sensitive (and less modified) lands. In more marginal lands, technology that enables farmers to replace shifting cultivation with permanent, higher-yielding fields may allow former fallow land to revert to natural forest, woodland, and grassland, or prevent further clearing.

Historically, productivity increases in more favored production regions of developed countries were indeed commonly associated with contraction of agricultural area elsewhere. But recent research concludes that the easy availability of off-farm employment was a critical factor in reducing agricultural conversion (Mather 2001; Rudel 2000). Cases from Honduras, the Philippines, and Zambia are described below. Elsewhere, the incorporation of relatively labor-intensive coffee into colonization settlements in northeast Ecuador allowed colonists to maintain large portions of their area in forest some twenty to thirty years after settlement (in a context of limited in-migration). Incorporation of kudzu into fallows in Peru reduced primary forest clearing, although it increased secondary forest clearing and introduced a potentially devastating invasive alien species into the ecosystem. Incorporation of rubber into farming systems in Borneo led farmers to shift from annual crops to agroforestry land uses (Angelsen and Kaimowitz 2000).

Increases in the productivity and harvesting practices of timber plantations and in forest-product processing have also slowed the harvest and conversion of natural forests for timber production in some parts of the world, aided by the decline in natural forests that are economically accessible for logging. While one ton of paper and paperboard in 1970 was made up of more than 80 percent wood pulp, by 1997 this figure had dropped to 56 percent, in part due to increased use of recovered papers (Bazett 2000).

However, the conditions under which productivity increases do result in a slowdown or reversal of agricultural land conversion appear to be limited. Efforts by conservation groups to reduce forest conversion by helping farmers to adopt more sustainable and intensive agricultural systems were fairly successful in Mexico, but not in Guatemala, although in both places conservation benefited from reduced agricultural burning (Margoluis et al. 2000). In synthesizing the results of a large number of well-documented case studies involving technological improvements in developing countries, Angelsen and Kaimowitz (2000) conclude that technological progress in the intensive agricultural sector is good for forest conservation, unless it substitutes capital for labor and expels the displaced labor to the agricultural frontier. Technological change in intensive agricultural systems can shift resources away from the frontier by increasing wages or lowering agricultural prices. Labor-intensive technological progress will tend to reduce forest clearing because it absorbs labor, while labor-saving technologies will have the opposite effect. Yield-increasing, labor-saving innovations in extensive agriculture may reduce deforestation if producers are largely subsistence oriented, labor markets function poorly, or prices are set locally and labor is unresponsive to wage changes. Forest conservation is likely where technologies are labor intensive, opportunities for in-migration are limited, and demand for agricultural products is unresponsive to local production. From a biodiversity perspective, the worst type of technological progress at the frontier has involved mechanization and other labor-saving, capital-intensive improvements in export crops that have fixed world market prices, and where large-scale in-migration occurs.

Some case studies document increased land conversion resulting from technological change, often through rather complex processes. In Côte d'Ivoire and Indonesia, low-cost production of plantation crops allowed farmers to ignore any international price reductions their activities induced, while a constant inflow of migrants from other regions dampened pressure on wages. Massive deforestation resulted. In Indonesia, replacing traditional shifting cultivation with more profitable, intensive small-rubber plantations has stimulated the clearing of additional forest (Tomich et al. 2001). The dissemination of new soybean technologies encouraged farmers in Paraná, Brazil, to convert from labor-intensive coffee cultivation to mechanized soybean production. As a result, land tenure became more concentrated, employment fell, and 2.5 million rural people left the state in the 1970s, many of them moving to forested areas in the Amazon where they were major actors in deforestation (Diegues 1992). Field studies in the tropical agricultural frontiers of Nicaragua, Brazil, and Indonesia found that under land-abundant conditions, productivity increases for locally produced outputs accelerated land-clearing by increasing economic incentives for new production (Vosti et al. 2001; Faris 2000). The development of crop varieties adapted to growing conditions in lower-quality lands has often accelerated agricultural intensification in those areas and threatened remaining wild biodiversity. Of course, where such technologies replace demonstrably resource-degrading patterns of intensification, they may have positive effects on biodiversity.

Impacts on land conversion also depend in part on the geographic relation between areas undergoing intensification and extensification. In heterogeneous landscapes, where individual farmers or neighboring villages have lands of varying quality, labor-using intensification in better-quality lands often halts or reverses land conversion. But when communities with lower-quality lands have weak economic links because of poor transportation systems and fragmented labor markets, intensification in distant regions will have little impact on local land-clearing. Local or economy-wide changes in the relative costs of labor and inputs that substitute for labor will influence land-clearing practices. Assessment of the impact of productivity-increasing intensification is further complicated when longer-term impacts are considered. In the Zambia example described below, technological change greatly reduced deforestation in the short term but promoted a far higher population than that which the new agricultural systems could support, and indirectly led to additional deforestation later on.

We draw two conclusions from this evidence. First, agricultural innovations to increase productivity and sustainability have been, and will continue to be, *necessary* for controlling land conversion and protecting wild biodiversity. However, they are clearly not *sufficient* in themselves to protect biodiversity. Other proactive measures are needed as well.

Potentials to Increase Agricultural Productivity

The discussion above already suggests that to be most effective in protecting biodiversity, productivity- and sustainability-increasing research should focus on labor-intensive innovations and yield-increasing technologies for lower-quality lands that already support high populations.

Genetic improvements have played a major role in increasing agricultural production and will continue to do so in the future. More focus is needed on conventional crop breeding to produce satisfactory yields in marginal environments where farmers are already practicing intensive agriculture. Genetic improvement can improve land- and water-use efficiency of crops, enhance productivity under intercropping systems, increase biomass production, and reduce the need for agrochemicals. Development of shade-tolerant food crop cultivars is essential for expansion of polyculture systems. Genetic improvement can develop breeds of livestock that are well adapted to local conditions and resistant to local diseases, which will help to ensure that domestic livestock do not serve as a reservoir of diseases that might spread to wild species. Selective breeding also has great potential for improving forestry production. Only a modest portion (less than 140) of the more than 50,000 estimated tree species worldwide is being utilized in forestry, and present breeding efforts cover far fewer.

Conserving genetic diversity is an essential part of any program of genetic improvement; wild species are often an important source of new genetic traits. Methods and strategies from crop genetic conservation are being adapted to study and conserve genetic diversity of forest resources, with particular attention to in situ conservation (Koskela and do Amaral 2001). However, relatively few tree species are currently receiving such attention. Biotechnology may raise the potential yield of crops (Mann 1999). Specific biotechnology tools, such as marker-assisted selection and new markers for assessing genetic diversity, can accelerate conventional breeding efforts (Amaral, Persley, and Platais 2001).

Ecological and agronomic research on crop, livestock, forest, and fisheries management will be equally important to achieve sustainability in higher-yield production systems. New strategies for lower-cost and more effective pest control, soil enrichment, and water utilization are urgently needed in diverse agroecosystems. Research will need to anticipate changes in various environmental, social, and economic parameters, and build the capacity to adapt to the unpredictable. For example, research on deep-water rice, including floating rice, may assume greater importance as the sea level rises with global warming. If genetic engineering were able to breed salt tolerance into rice, then many brackish water areas could remain in production, or irrigated lands that have been salinized might be brought back into production. However, care must be taken that new technology does not lead to converting areas to agriculture that today provide crucial habitat for wild biodiversity.

Genetic and ecological research are closely interlinked. In Vietnam, for example, IPGRI researchers have been studying the genetic diversity of three threatened timber tree species: *Pterocarpus macrocarpus, Xylia xylocarpa* and *Albergia oliveri*. Their work shows that conservation strategies require assessment of tree species composition and natural regeneration potential; relations between the target species and other dominant species in the remaining forests; historical land use changes; information on biological characteristics and ecological processes regulating genetic variation, such as seed production and dispersal and regeneration patterns; and the impact of human activities on genetic processes (Boffa, Petri, and do Amaral 2000).

New aquaculture technologies can also be developed to protect biodiversity. For example, farming precious black pearls from cultured oysters *(Pinctada margaritifera)* for a lucrative export market was identified by ICLARM as an alternative economic activity for Solomon Island fishermen, whose resources are being depleted by overfishing. Researchers had to find concentrations of baby oysters ("spat") to cultivate and then had to develop production techniques that would protect the oysters and keep them safe from predators. Now when the oysters are sufficiently large, a small hole is drilled through their hinges and they are suspended in vertical lines above the ocean floor. Eighteen months later, the oyster is "seeded" by inserting a small bead into its tissue, which stimulates the oyster to coat the bead with layers of mother-of-pearl. The first crop of cultured pearls from the Solomon Islands was auctioned in September 2000, and local people are now being trained to seed the oysters (Child 2000).

Six Case Studies

Among the six well-documented examples described below, in three cases agricultural productivity increases led to a contraction in agricultural lands and reversion to wild vegetation. In a fourth case, large-scale habitat clearing was avoided by an environmentally friendly pest control technology. All of these examples took place in farming systems on marginal lands that relied on short fallows. Intensification of production on the better (irrigated or more fertile) lands permitted farmers to withdraw from (or slow expansion into) the more extensively managed fallow areas. The last two cases describe how assistance to improve farm productivity enabled farmers to take some of their crop or pasture lands out of production and commit them to forest reserves.

EXAMPLE 11. INTRODUCTION OF CASSAVA IN NORTHERN ZAMBIA
REDUCES DEFORESTATION

Most of northern Zambia is the sparse, open deciduous woodland known as *miombo*. The nutrients in these soils are mostly leached and acidic, not suited for continuous grain cultivation. Using low-input systems, farmers can only cultivate these lands one out of every four years. With medium and high levels of fertilizer or other inputs the frequency rises to two and three out of every four years, respectively. The rest of the time farmers must leave the land fallow to restore its nutrients, organic matter, and soil structure, and to control weeds, pests, and diseases. Early in the last century, the *chitemene* shifting cultivation system dominated most of northern Zambia. In this system, farmers chop down a large area of trees, pile the trunks onto a smaller area, and burn them. Then they grow crops in the ash for a few years. The fire releases the nutrients in the woody biomass and makes them available for crops and provides a seedbed free of weeds. The heat also affects the soil structure, leaving a fine seedbed for finger millet (*Eleusine coracana*), the first crop. In more populated areas, fallow periods were too short for woodlands to regenerate, so the forests were gradually replaced by grasslands, and farmers used a modified grass–mound system. In this system, grass turf collected from surrounding areas was piled into mounds. Finger millet and beans were planted on the spread-out mounds after the organic matter decomposed. Uncontrolled fires were common in the grass–mound system as well; these helped suppress regrowth of woody vegetation and affected the composition of nonwoody species.

In the first half of the twentieth century, the British introduced cassava (*Manihot* spp.), a starchy root crop from Brazil, to northern Zambia, with the objectives of controlling deforestation and permitting the land to support more than the 2.2 to 2.5 persons per square kilometer permitted by the chitemene system. Cassava cultivation was labor intensive and greatly improved food security. Farmers typically planted cassava on ridges or mounds as the main crop in "cassava gardens," often alongside other crops during the first year. Although farmers initially resisted the new crop, they discovered its advantages and it eventually became most people's main staple, producing good yields even in poor soils where most other crops failed. Household models suggest that cassava boosted the number of people the land could support by two to six times relative to the chitemene system, and that households could meet their food requirements with 40 percent less labor input, giving them much larger surpluses for sale. The introduction of cassava reduced deforestation by increasing land productivity in a context where farmers mainly sought to meet their subsistence requirements (Holden 2001).

EXAMPLE 12. INCREASING YIELDS OF LOWLAND RICE REDUCES
EXTENSIVE HILLSIDE FARMING IN THE PHILIPPINES

Population growth in the Philippine frontier province of Palawan has been particularly high (4.6 percent per year), and as a result agriculture there has expanded into marginal and environmentally sensitive areas, promoting acute upland deforestation. The area's main staple is rice and its main cash crop is corn. Farm size averages 2.6 to 5.1 hectares. To intensify and raise agricultural production, the Philippine National Irrigation Administration constructed or upgraded a number of small-scale communal irrigation systems in Palawan. These are in the lowlands, but most are adjacent to inhabited upland forest areas.

Household surveys show that irrigation allowed farmers to increase cropping intensity to 1.9 crops per year, whereas rainfed farms held them to 1.2. Farmers used less family labor and less total labor during each cropping season, but more hired labor. By creating better-paying jobs, intensification of lowland irrigated cropping induced upland farmers to participate less in lower-paying forest-clearing and forest product extraction (hunting, charcoal making, resin collection). After the irrigation systems were installed in the lowlands, annual forest clearing by upland households declined by 48 percent. Average wage income rose nearly threefold among upland households that traveled to lowland farms to work. These positive impacts resulted in part from the fact that the upland area is physically adjacent to the lowland irrigated area, reducing the cost of traveling to work. Also, lowland farmers made relatively little investment in labor-saving technology, which allowed them to hire more labor (Shively and Martinez 2001).

EXAMPLE 13. REGENERATING NATIVE PINE FOREST HABITAT IN HONDURAS THROUGH IMPROVED CROP TECHNOLOGY

The central region of Honduras covers about 890,000 hectares, of which more than 90 percent is rugged hillside. All of it was originally forested, but today only about half of the area is covered by native pine forest, with scattered deciduous forest stands. Significant deforestation occurred prior to the mid-1970s, due to overlogging and frontier agricultural settlement. Since then, commercial logging has been sharply controlled. However, conversion of forest to farmland has continued as a result of a 2.3 percent annual rural population growth rate, agricultural demand from the even faster-growing capital city nearby, and widespread erosion and nutrient depletion in steep fields used for low-value staple food crops. Forest habitat and wild populations of deer, agouti, raccoon, various squirrels (which have traditionally provided an important source of animal protein for local diets), and other native fauna and flora have declined sharply.

But a different pattern of land use has emerged in some of the region's communities as a result of research and extension by the National Coffee Program of Honduras and the local Pan-American Agricultural School of Zamorano. In the 1980s the Zamorano School identified many fruit and vegetable varieties suitable for local steep slopes. The school also developed integrated nutrient and pest management strategies and sprinkler irrigation and conservation practices. The Coffee Program encouraged coffee-growing communities to intensify production of basic grains, in order to free up farmland to expand the area for shade-grown coffee, and to replace traditional coffee with higher-yielding varieties. In the late 1980s and early 1990s, many communities (occupying a third of the area of the central region) adopted and adapted these new technologies. Higher incomes from vegetables and coffee enabled farmers to purchase fertilizers to replenish soil nutrients both in their commercial fields and in subsistence staple food crops, thus nearly doubling corn yields on permanent fields. The increases in income also allowed farmers to abandon marginal fallowed fields, which reverted to forest. Aerial photograph analysis shows that the net area under forest cover remained stable during this period in the coffee-growing communities and declined only slightly in the horticultural communities. This contrasts with a decline in the forest cover of at least 13 percent, and in some cases as high as 20 percent, in the communities growing corn or other grains without new technologies. Unlike these extensive grain farming communities, the former did not report a decline in wild game over the period (Pender, Scherr, and Durón 1999; Scherr 2000c).

Example 14. Biocontrol of Cassava Mealybug in Africa

Though originating from Brazil, cassava (*Manihot* spp.) is grown over an area of Africa one and a half times the size of the United States. It is the staple food of more than 200 million Africans. Most cassava in Africa is grown by small-scale, semi-subsistence farmers, who have little access to agricultural chemicals. The cassava mealybug (*Phenacoccus manihoti*) was first discovered in Zaire in 1973, but it rapidly spread through the cassava belt. This pest can cause yield losses of up to 80 percent and thus poses a severe threat to African food security. A multinational collaborative research project involving IITA, CIAT, and CIBC was established in 1981 to combat the pest through a classical biological control approach, that is, to introduce natural enemies of the mealybug from its place of origin in Latin America. The parasitoid wasp *Apoanagyrus* (*Epidinocarsis*) *lopezi* (*A. lopezi* for short) was found on a mealybug from Argentina and introduced experimentally into Africa. It dramatically reduced the mealybug threat, maintaining numbers of mealybugs below levels that cause economic damage (Glass 1988). From 1981 on, *A. lopezi* was released in about 150 sites in twenty countries. The impact on cassava was slow, and stable biological control was achieved only after several years. But by the end of the decade, the wasp had spread to all major mealybug infestations in twenty-seven countries and had brought the pest under control in 95 percent of all fields—at a relatively low cost to the public sector, at no cost to farmers, and without any use of chemical pesticides. A study of the economic benefits of cassava mealybug biocontrol over a forty-year period estimated a benefit–cost ratio of about 200 at world market prices, and 370 to 740 when inter-African prices were considered (Zeddies et al. 2001).

This pest control method had positive overall impacts on wild biodiversity. Ecological studies undertaken after the introduction of the parasitoid wasp indicate only transient effects on indigenous competing predators and parasites of the wasp. A food web study of 135 species found that the parasite was specific to cassava and did not affect other plants. The biocontrol measures are considered to have had a large, though as yet unmeasured, impact on habitat protection, by precluding the need for farmers to clear large areas of additional land to compensate for mealybug destruction of cassava fields (Neuenschwander and Markham 2001).

EXAMPLE 15. SAVING BRAZIL'S ATLANTIC FOREST THROUGH IMPROVED DAIRY FARMING

Brazil's Atlantic Forest, a unique type of humid, subtropical forest, is one of the most threatened in the world. The area's noteworthy biodiversity includes endemic lion tamarin marmosets, hundreds of endemic birds, and a rich flora, particularly of rare orchids and bromeliads. Only 7 percent of the original forest cover remains—the result of five centuries of population growth, land-clearing for coffee, livestock production, and uncontrolled fire used in pasture management. Dairy farming is one of the most important economic activities of the rural areas today, but it has very low productivity, requiring ever-expanding conversion of forest to low-quality pasture.

To reverse the destruction, an NGO called Pro-Natura began in the early 1990s to work with dairy farmers living around one of the largest remnants of protected Atlantic Forest in the state of Rio de Janeiro: the Desengano State Park. Pro-Natura made a deal with the farmers: they would provide technical assistance to improve dairy-farm productivity and incomes; in exchange, the farmers would commit to reforest and regenerate part of their land and maintain it as a conservation easement. Pro-Natura helped farmers to invest in genetic improvement of their dairy herds, use mineral supplements, improve fodder, and produce silage. As a result, milk yields of participating farmers tripled, and incomes rose by more than 100 percent. The improved pastures were sufficient to meet farmers' forage needs, so the area in pasture could be reduced. More than 60 hectares of pasture on sixteen farms in northern Rio de Janeiro have already been converted back to forest, serving as a buffer between agricultural and protected forest areas. Many additional pastures are now candidates for reforestation, including parts of the Capelinha settlement of the Landless People's Movement (MST), a former sugarcane plantation. Farmers and local communities supported the reforestation in part for its other ecosystem services, especially water flow regulation and erosion control. Seedlings raised in two tree nurseries managed jointly by Pro-Natura and municipal governments have been planted on farms and in rural communities (see newsletter at www.pronatura.org).

EXAMPLE 16. INCREASED BIODIVERSITY FROM
"CARBON FARMING" IN MEXICO

The Scolel-Té ("growing trees") Pilot Project for Community Forestry and Carbon Sequestration was established in the highland and lowland ecoregions of Chiapas, Mexico. It is being developed as a prototype scheme for sequestering carbon from the atmosphere to reduce the threat of climate change. The project's strategy is to assist farmers in shifting from unsustainable, low-income land-use patterns—mainly extensive fallow systems that involve regular forest clearing—to sustainable forestry, agroforestry, and agricultural systems. The initial phase of the land-use transition is financed by revenues from an international greenhouse-gas-mitigation agreement with the International Federation of Automobiles, which is committed to offsetting the carbon emissions resulting from sponsored car races (Wilson, Moura Costa, and Stuart 1999).

The project involves ten indigenous Mayan communities with 5,000 inhabitants; it was developed and implemented jointly by a local farmers' organization, with support from U.K. university researchers. New systems, including live fences, enriched fallows, coffee with shade trees, and reforestation, are being established on about 2,200 hectares (Totten 1999). So far, 300 small-scale farmers have sold about 170,000 tons of carbon sequestered in natural and planted forest areas at U.S.$10 to $12 per ton. As much as 75 percent of carbon revenues have gone directly to farmers to cover establishment costs of new, more profitable agricultural systems (Smith and Scherr 2002). It is estimated that without the changes in farming systems made possible by the carbon project, continued practice of extensive fallow farming would have resulted in a 2 to 3 percent annual loss of forest. Wild biodiversity conservation results from both protecting defined areas of biodiversity-rich natural forest from agricultural land clearing, and enhancing the habitat value of permanent cultivated area through agroforestry.

Potential Application

These six cases present diverse approaches to agricultural intensification. The Honduras, Mexico, and Zambia cases represent a shift from extensive to intensive cropping systems that enabled farmers to retire lower-quality land from production. The Brazil example represents a similar shift from an extensive to an intensive dairy system. The Philippine example shows how intensification can reduce pressure for land-clearing in a neighboring region by offering more lucrative employment. The mealybug example shows how technology to control pests (which improved yields) can avert unnecessary agricultural expansion. In cassava farming in Zambia, carbon farming in Mexico, and dairy farming in Brazil, the land freed up by intensification was explicitly designated for conservation use; in the latter cases this was part of the "deal" for farmers to gain access to improved agricultural technology. In the other three cases, biodiversity conservation was not a primary objective of agricultural research, investment, or extension, but a fortunate consequence. All of these cases relied on improved agricultural technologies developed by research institutions. These cases and other evidence above suggest that yield-increasing research (even on technologies that are not recognized as ecoagricultural practices) can enable biodiversity conservation initiatives to succeed.

Chapter 7

Enhancing the Habitat Value of Productive Farmlands

Interventions directly in productive croplands, grazing lands, forests, and managed fisheries can make a big difference to the habitat value of agricultural landscapes. On the one hand, agricultural pollutants and sediments can be reduced through more efficient use and by using physical and vegetative conservation structures as filters to protect nonagricultural resources. These practices generally leave the production system itself largely the same. Public agricultural programs and conservation projects have encouraged the development and adoption of such systems throughout the world, though more often to protect water supply and quality for human use or to reduce downstream sedimentation rather than to protect biodiversity.

But, on the other hand, advances in agricultural and natural systems ecology are also allowing researchers to explore ways in which agricultural systems could be designed to mimic the structural and ecological functions of natural ecosystems more closely and thus avoid some of the ecological problems of conventional agriculture. New patterns of crop and associated natural vegetation—at field, farm, and landscape scales—can restore water flows, close the nutrient cycle as much as possible to minimize nutrient loss, improve energy use efficiency, and increase biodiversity. Such production systems begin to blur the distinction between agricultural land and nature reserves by having, at the same time, nature-conservation values, functional ecosystem values of benefit to agriculture, and direct improvements in agricultural production. These approaches had their origin in the 1970s and 1980s in organic farming, agroecology, permaculture, agroforestry, permanent agriculture, and natural systems agriculture.

This natural-ecosystem approach to agriculture is of particular interest in the tropics and subtropics, where low-quality soils and harsh climates make land and water resources highly vulnerable to degradation under continuous annual crop cultivation. Unlike the Middle East and Europe, where temperate grain agriculture evolved, many ecosystems in the tropics and subtropics have narrow ecological breadth, so they cannot stray far from the original vegetative structure without paying a penalty (Lefroy and Stirzaker 1999). In agricultural systems based on ecosystem models, monocultures would be used sparingly, and where used would appear in smaller patches (large fields, for example, would be divided by hedges). Scientists hypothesize that by mimicking the structure of native habitat and incorporating key native species, tropical and subtropical systems could greatly increase their biodiversity potential, benefit hydrological regimes, and improve agricultural pest control. Different agricultural products could be produced in suitable niches and environments within the landscape.

This chapter discusses specific strategies that have already been developed to enhance habitat quality of productive farmlands by minimizing agricultural pollution, modifying resource management to make production more compatible with wildlife, and modifying farming systems to mimic natural ecosystems. With continued advances in technology and marketing to develop these alternatives, it may be possible for such systems to become commercially viable on a large scale.

Strategy 4: Minimize Agricultural Pollution

The earliest recognized threats to wild biodiversity from modern intensive agriculture were pollution from toxic pesticides applied to crops, high nutrient loads from overfertilization, and organic and inorganic farm wastes, as described in Chapter 4. Such intensive, high-input crop, livestock, and aquaculture systems are widespread in developed countries, but they are generally limited to high-value cash crop and livestock operations in developing countries. Fertilizer pollution is a minimal concern in many low-income countries and marginal-land cropping systems, where boosting yields may require increasing both inorganic fertilizer use and organic matter levels in poor soils.

Reducing agrochemicals in high-input systems can greatly benefit wildlife. For example, these chemicals typically kill the insects and weeds that constitute the food base for insect- and grain-eating species, depressing bird populations. The use of conservation headlands kept free of pesticide applications can dramatically enhance bird reproduction and population viability (O'Connor 2001).

This ecoagriculture strategy builds on major advances in organic agriculture, integrated pest management, and soil conservation. Proven pest- and nutrient-management strategies have been developed for both small and large-scale farms. The development of improved strategies has had various motivations. Government regulatory controls and local community concerns have driven major efforts by the research and farming communities to reduce pollution in agroecosystems and improve management of the pollutants and wastes that remain. Concerns about water and air quality for humans have often been important in motivating action. Low-income and risk-averse farmers have actively sought technologies that would reduce their dependence on expensive chemicals. While some technologies are more labor-intensive than chemical control, others—such as precision farming techniques that use computer programs to apply fertilizers tailored to site-specific soil nutrient needs—are capital-intensive. This strategy is composed both of techniques to reduce the use of pollutants and of improved management techniques to reduce their impact on biodiversity.

Reduced Use of Pollutants

Major advances have been made in methods to reduce and improve the efficiency of agrochemical use. In the case of chemical fertilizers, excessive use has been reduced through farmer education and research on optimal application levels for specific farm conditions. Improved nutrient efficiency has reduced overall usage, mainly through better timing and methods of application; the benefits are immediately apparent to farmers in lower costs. For example, spraying fertilizer on leaves results in rapid absorption and translocation through the plant. It has cut applications on some vegetable crops by 25 percent, relative to applications in the soil. Some ammonia losses can be reduced simply by incorporating fertilizer more thoroughly into the soil during land preparation (Conway 1997). Agrecological studies have shown that complex tropical irrigated rice systems control pests so that insecticides are not needed (Settle 2001).

With greater understanding conferred by modern cellular and molecular biology, chemical companies have begun to search for tailor-made pesticides. Some have used compounds that mimic the effects of juvenile hormones in insects to disrupt the transition from one life-cycle stage of an insect to another (preventing caterpillars from becoming moths, for example). Natural plant compounds that have been traditionally used by farmers for pest control are also getting more attention. These compounds include custard apple, turmeric, croton oil tree, neem, and chili pepper (Conway 1997).

Pesticide pollution has declined significantly in areas where new forms of pest control have been introduced. Integrated pest management (IPM) allows farmers to "scout" for pests and apply pesticides strategically only when necessary to prevent major losses, rather than according to a preset schedule. Again, farmers derive immediate benefits from reduced pesticide costs. Biological controls have been developed for several major crop pests, involving the introduction of parasitic or predator species that control pest populations. Another method of pest control is through cultural control, the use of management practices to create habitat conditions adverse to pests. For example, farmers can plant more diverse mixtures of crops to encourage natural enemies of pests and slow the spread of pests. Methods for controlling pests and agricultural diseases have even been applied to maintaining threatened wild species, as described in Box 7.1.

High-yielding and profitable organic farming methods have been developed for a wide range of crops (Oelhaf 1979; Welsh 1999). Certification of organic farming usually requires that for some time period, such as three years, no chemical inputs at all be applied. Organic sources of soil nutrients are utilized, such as animal manures, green manures (crops, especially legumes, that are grown for incorporating into the soil), and composts. Farmers also rely on certain bacteria and other microorganisms present in or added to soils to convert nitrogen from the atmosphere into ammonia in the soils that can nourish crops.

Box 7.1. Biological Control Saves Endangered Tree Species

The isolated island of St. Helena in the south Atlantic ocean lies 1,850 kilometers from the western coast of Africa. Only 122 square kilometers in size, it contains a highly distinctive flora. An endemic species of gumwood (*Commidendrum robustum*) dominated much of the extensive woodland that once covered the higher regions of the island. The island's national tree, it is now so threatened that it has become restricted to two stands of around 2,000 trees. Extinction seemed a real possibility when an invasive species of scale, *Orthezia insignis,* native to South and Central America, was introduced into St. Helena in the 1970s or 1980s. The scale started feeding on gumwood in 1991, killing 400 trees by 1993. To fight this threat, a species of predatory beetle, *Hyperaspis pantherina,* was obtained from Kenya, where it had been introduced to control *Orthezia* on jacaranda, a popular flowering ornamental tree. Assessment of the St. Helena fauna showed that no related indigenous species were present, so scientists concluded that introducing this predator would be safe in terms of effects on nontarget organisms, and that it would be likely to control the *Orthezia* scale. In 1993, the beetle was imported, cultured, and released. It rapidly established itself and did indeed control *Orthezia* on gumwoods, saving the national tree from extinction. This is probably the first case of biological control methods developed for domesticated plants being used against an insect to save a wild species of plant from extinction (Wittenberg and Cock 2000).

The higher rate of microbial activity in the soil and an increase in earthworm populations that result from organic techniques both help maintain soil quality as well as soil biodiversity (Duelli, Obriast, and Scmatz 1999; Pfiffner and Niggli 1996). Organic farms also support a much greater abundance of birds than conventional farming (Christensen, Jacobsen, and Nohr 1996; Fuller et al. 1998). As a result of these benefits—as well as perceived benefits to human health—organic agriculture is growing in popularity. In 1997 the organic markets in the United States, Europe, and Japan totaled about U.S.$11 billion, and demand has been increasing by 20 to 30 percent per year (Welsh 1999).

Breeding pest and disease resistance into plants using conventional means has worked with some, but not all, types of pests (Conway 1997). Genetically modified organisms (GMOs) offer new opportunities to build in resistance by drawing on a wider array of genetic resources. Fungicide use could potentially be reduced by "switching on" genes for fungal resistance currently found only in the seed, but not in the adult plant. Insecticide use could be reduced by changing the physical characteristics of crops to better resist pests, for example by increasing hairiness or thickening the outer protected layer of cells on the plant. It may be possible to reduce fallow periods by changing the growing season or time of harvest of crops. Crops, including trees, could be developed that tolerate high levels of competition from wild plants while still producing well (Johnson 2000). New ecological and biochemical research techniques are revealing an unexpected sophistication of host–pest relations (described in Box 7.2) that could revolutionize agricultural pest control in the future.

Improved Management of Pollutants

The problem of agricultural pollutants and wastes from intensive systems has led to important advances in management and disposal. Whole-farm planning addresses three features of the farm to minimize agrochemical pollution of aquatic systems. The first is the pollution source, such as barnyard areas, silage systems, and sheds containing chemicals. If a source is managed in a way that eliminates or reduces the release of contaminants, risks to water quality are diminished. The second is the farm field over which contaminants may be transported to a watercourse following rain and runoff. Transportation may be reduced by applying or releasing contaminants only when there is no rain expected, to minimize runoff; by reducing the amount released; or by managing soil and crops to absorb more of the applied materials and therefore minimize the effects of runoff. The third opportunity is at the watercourse itself, where buffer zones may be established on stream margins to prevent contaminants from reaching the water (Coombe 1996). Many technologies are being developed for on-farm processing of livestock wastes and their conversion to valuable crop manures.

Box 7.2. Allelochemicals: A New Frontier for
Crop Pest Management

The interaction between plants and the natural enemies of the herbivores that
attack them is more sophisticated than previously realized. Certain chemicals—
allelochemicals—carry information within or between plant and animal species.
Recent research shows that when some plant species are damaged by insect pests,
they emit allelochemicals that attract natural predators of those insects, essentially
calling in help from another species. Tobacco, cotton, and corn plants each produce distinct volatile chemical signals in response to damage by plant-eating
insects, attracting a wasp that is parasitic on the insects (De Moraes et al. 1998).
Using spider mites (*Tetranychus urticae*) and predatory mites (*Phytoseiulus persim-
ils*), researchers have shown that not only the attacked plants but also their neighbors produce this chemical response. So they too become more attractive to the
predatory mites and less susceptible to spider mites. The mechanisms involved in
such interactions remain elusive. But one study showed that uninfested lima bean
leaves activate five separate defense genes when exposed to volatile signals from
leaves infested with *T. urticae,* but not when exposed to signals from artificially
wounded leaves (Arimura et al. 2000). Research has shown that predators and
parasitoids (insects that lay their eggs inside larvae of other insects, which are
then consumed from within) learn to associate the volatile chemicals with the
availability of prey, and that plants usually produce the signals as a response to
chemicals released in the saliva of the pest species. The larvae of many moths feed
on leaves of crop plants, and at least some species of plants release volatile compounds at night that are highly repellent to female moths (De Moraes et al.
1998). If the natural defenses of plants could be more effectively developed into
nontoxic forms of pest control, safe and effective crop protection strategies could
be designed that would significantly minimize the negative side effects of the
current generation of chemical fertilizers.

Species of plants vary in the amount of attractants they produce, suggesting
that conventional plant breeding could lead to varieties that could be very effective in resisting attacks from certain herbivorous insects. New technologies, such
as DNA micro-assays, can determine patterns of gene expression in plants
responding to herbivore attack, enabling improved varieties to be produced relatively quickly. Clearly, this approach depends on a nearby reservoir of predators
and parasitoids to come to the defense of the plant that signals it has been
attacked by a herbivore. Farmers trying to use natural enemies will need to
ensure that the necessary reservoir of natural enemies is available nearby, argu-
ing for maintaining biodiversity in the system. Thus the success of approaches to
pest control that use plant-predator communication will depend upon the existence of habitat that harbors populations of predators and parasitoids; such habitat includes hedgerows, set-aside areas, and other unplanted areas. Mixtures of

(continues)

Box 7.2. *Continued*

crops that can be grown together to encourage predators and parasitoids might also be effective, and such strategies are being developed in Kenya, Uganda, China, and India, promoted by the International Centre of Insect Physiology and Ecology (ICIPE).

A variety of integrated soil nutrient and integrated pest management techniques, crop rotations, and vegetative filter strips along waterways have been developed to reduce runoff of agricultural pollutants into water sources (Franzluebbeers, Hossner, and Juo 1998). For example, in Flevoland, the Netherlands, farmers collaborated on a program designed to achieve a more ecologically conscious mode of streambank and field-margin management, in order to reduce agrichemical pollution and create new habitats. The program successfully reduced deposition of fertilizers and agrochemicals during the growing season, and the number of plant species increased by 33 percent per farm. The techniques developed through this initiative are easily integrated into overall farm management, though weed control requires additional time and energy (Witteveen 2001).

Financial incentives for livestock production currently favor large herd sizes and concentration near urban markets, but this may change as the real economic costs of pollution are reflected in agricultural policy. The location and size of polluting livestock or on-farm agricultural processing operations may need to be regulated, and requirements may need to be set to protect water quality (Reijtntjes, Haverkort, and Waters-Bayer 1992). One response to the environmental problems of intensive livestock operations is to promote free-range operations with lower livestock densities. A variety of methods have also been proposed to reduce methane emissions from ruminants (animals that chew their cud, like cattle), including improved diet quality and nutrient balance, increased food digestibility, and food additives.

Seven Case Studies

Numerous cases exist where farmers have reduced their use of agrochemicals or improved their management of farm wastes. Often the initial goal is reducing cost rather than conserving wild species, but the benefits to wild biodiversity are an important result. The seven cases described here represent a wide range of technological approaches, all in intensively managed cropping systems.

EXAMPLE 17. PEST CONTROL THROUGH INTERCROPPING IN
YUNNAN, CHINA

While monocultures are convenient—easier to plant, harvest, market, and
identify varieties of a crop—agroecological evidence comparing diverse
and simple agricultural systems shows that pest populations are dramatically
lower in mixed-species than in single-species plantings (Pimm 2000). In
Yunnan, scientists sought to apply this principle to manage fungal rice blast
(*Magnaporthe grisea*). This disease has been a serious problem for cultivation of
glutinous or "sticky" rice (32 percent losses) and upland rice (20 to 50 percent
losses) in the province (Zhu Youyong 2001). The characteristics of this blast
made breeding of genetic resistance into rice ineffective; new pathogenic races
emerged within a few years.

Chinese researchers, working with IRRI, tried a new strategy of mixing
standard blast-resistant rice with a more valuable, but lower-yielding sticky
rice. Their approach was based on an observed farmer practice of dispersing a
single row of glutinous rice between groups of four rows of hybrid rice, to
meet local demand for glutinous rice. The unusual trial was done at a large
scale, with thousands of farmers, in order to evaluate the epidemiology of the
disease. In 1999, the trial included all rice fields in five townships, on 812
hectares, and in 2000 all rice fields in ten townships, on 3,342 hectares. By
mixing the two varieties, total rice yields increased by 89 percent and incidence of fungal rice blast declined by 94 percent. Yields of the more valuable
sticky rice rose because they had sunnier, warmer, and drier conditions in the
mixed stand than in the pure stand. These conditions discouraged growth of
the disease. Fungicidal sprays were no longer needed, as the greater distance
between plants limited spread of the blast's airborne spores. With farmers' standard practice of hand harvest, it was easy to separate the different varieties of
rice. By 2000, the practice was being used by farmers on 42,500 hectares
throughout Yunnan province. Ten other provinces in China are beginning to
test the technique (Mew 2000; Wolfe 2000; Yoon 2000; Zhu et al. 2000).

EXAMPLE 18. SAVING STORKS AND SONGBIRDS WITH A
NATURAL BIOCIDE IN WEST AFRICA

The plague of the desert locust has been feared since biblical times. From West
Africa to India they attack once every ten to twenty years, when weather conditions in their breeding grounds around the Red Sea and on stopovers on
their invasion routes are suitable for them to multiply. Their less well-known
cousin, the ordinary grasshopper, is present on a more regular basis and thus
causes even more economic harm. These pests cause millions of dollars in crop
damage, with consequent widespread hunger and disease. In recent decades
farmers have used a variety of pesticides to combat both pests. In 1986 alone,
U.S.$200 million worth of broad-spectrum insecticides were used to combat

a major locust invasion. But these insecticides also kill beneficial insects, birds, and small mammals and are often dangerous to people. The yearly spraying of insecticides over areas through which migratory birds pass is now blamed for large decreases in the numbers of migratory flocks in Europe. Especially hard hit were storks and songbirds.

Between 1989 and 1999, an international group of scientists from CABI and IITA pursued a nonchemical approach to controlling these pests (Lomer, Prior, and Kooyman 1997). After studying more than 160 strains of fungi and other pathogens, they identified a strain of an environmentally benign fungus (*Metarhizium anisopliae*) that grows naturally under African conditions and is deadly to both pests (Langewald et al. 1997; Cherry et al. 1999). An economical way was found to produce large quantities of the fungus in an oil-based carrier. One scientist, impressed with the power of the new biocide, dubbed it "Green Muscle," and that name is now used for the commercial version. Green Muscle has several advantages over traditional insecticides. Its cost is similar, but it requires only one application as compared to at least three for others. The fungus, a living organism, can be stored for up to a year without refrigeration—a distinct advantage in tropical countries. Finally, the new biocide is environmentally benign: it does not damage other insects, plants, animals, or people (Smits, Johnson, and Lomer 1999).

EXAMPLE 19. MAKING WATER SAFE FOR AQUATIC LIFE IN VIETNAM

In the intensive irrigated rice systems of Southeast Asia, rice productivity has increased greatly but landscapes have become highly simplified. This has increased vulnerability to pests and diseases, leading to some of the world's highest levels of pesticide use. Pesticide pollution has directly reduced biodiversity in and around paddy fields, and indirectly reduced it through contamination of fresh water, thus affecting the entire food chain. Pesticide-resistant insects and pathogens have developed as a result. In many parts of Asia, scavenging birds such as vultures and some hawks that play an important ecosystem role in removing dead animals and controlling rodent populations have virtually disappeared from agricultural areas, largely due to pesticide accumulation.

Research has found that cost-effective levels of pesticide use are much lower than farmers (influenced heavily by information provided by chemical companies) thought were needed. In Vietnam, research on insect infestations and farm-level financial returns from pesticide spraying led to new recommendations disseminated by radio dramas and leaflets in 1994 in Long An Province. The program spread to eleven other provincial governments by 1997, reaching about 92 percent of the Mekong Delta's 2.3 million farm households. By 1999, farm surveys showed that insecticide applications had fallen from 3.4 per farmer per season to just one, a decrease of 72 percent. At the same time, the gross paddy output of the Mekong Delta increased from 11 to 14 million tons per year. Scientists believe that through IPM systems now under development, insecticide use can be reduced by another 50 percent without reducing rice production (IRRI 2000). Reducing pesticide use will benefit the many species of frogs and fish that also inhabit the rice fields, not to mention the people who depend on these species as a source of protein. It will likely allow the return of predator bird species and their rodent-controlling benefits.

EXAMPLE 20. RESTORING FISHERIES IN THE CHESAPEAKE BAY

The Chesapeake Bay on the eastern coast of the United States is one of the richest natural fisheries in the world. Since the late 1800s, however, wildlife of the bay has been faced with serious threats. First, large-scale dredging for oysters led to habitat degradation in the ocean bed and dramatic declines in various species. During the period before World War II, the total area of the bay was reduced by large-scale sedimentation resulting from land clearing and construction. Following World War II, urban growth led to high levels of pollution from sewage, while growth in intensive vegetable and livestock production systems surrounding the bay delivered high levels of agricultural nutrients to the bay through runoff. These pollutants caused human health problems, local wildlife extinctions, and dramatic declines in aquatic resource harvests and wildlife. Initiatives to address sewage pollution have brought this problem more under control, so that agricultural sources now pose a larger share of the remaining problem, accounting for about a third of total nitrogen and two-fifths of total phosphorus runoff into the bay. The Chesapeake Bay Agreements of 1983, 1987, and 1992 commit the state of Maryland to restore the bay to its former health and productivity by reducing the flow of major pollutants into the bay. Agricultural sources of nitrogen and phosphorus are required to decline by 24 and 21 percent, respectively, relative to estimated 1992 levels.

Key strategies to reduce agricultural pollution focus on improving the efficiency of nutrient use, reducing runoff through conservation tillage and conservation practices, improving livestock waste management, and establishing riparian buffer strips. Buffers are areas of perennial vegetation that filter chemicals from runoff water moving across the ground or entering streams. They may include filter strips, establishing permanent vegetative cover in critical areas, and special plant mixtures selected for wildlife habitat. Various programs encouraged adoption of these practices through cost-shares and rental payment for conservation easements, including two programs specifically meant to restore wetlands and improve wildlife habitat on private land. By 1995, almost half of Maryland farmers used some form of conservation tillage and 40 percent had improved nutrient management practices. More than a quarter of those with livestock now have animal waste storage facilities and use vegetative buffer strips. As a result of these and other nonagricultural inter-ventions, point-source emissions of phosphorus were cut by 56 percent from 1985 levels, while point-source emissions of nitrogen were cut by 35 percent (Lichtenberg 1996). Populations of several aquatic species began to recover as Bay water quality improved.

EXAMPLE 21. INTENSIVE DAIRY GRAZING IN THE UNITED STATES
CONSERVES WATER QUALITY FOR WILDLIFE AND PEOPLE

During the 1970s, favorable tax policies and outside advice encouraged U.S. dairy farmers to intensify production in response to high crop and milk prices. Many farmers incurred high debt by investing in machinery and facilities to confine cows for milk production, and by expanding cultivation of feed crops into previously uncultivated erodible land. When commodity prices tumbled in the 1980s and machinery costs proved to be much higher than expected, many dairy farmers went out of business; others sought an alternative system. At the same time, increased environmental regulations to protect urban water-sheds and coastal resources from nutrient loads began to pose significant new costs for conventional livestock producers. Since dairies near or adjacent to cities or urban counties currently produce 76 percent of all milk in the United States, pollution reductions and a positive interface with urban and suburban neighbors are essential.

One alternative, management-intensive grazing, is a method of controlling the frequency and intensity of grazing to improve livestock carrying capacity on pasture. The approach was developed originally by farmers and researchers in France and New Zealand. This low-capital strategy relies on pasture graz-ing to feed dairy cows, allowing conversion of lands used for feed crops back to pasture and avoiding the financial and environmental costs of concentrated livestock waste disposal. Dairy farmers in Vermont (Murphy et al. 1996), North Carolina (Washburn et al. 1996), and Maryland (Barao 1996) who have con-verted to this system increased net income per cow and total farm profits sub-stantially (30 to 80 percent in cases cited in the various studies). Their higher incomes resulted from decreased costs for grain, concentrated supplements, equipment and labor for feeding and manure handling, machinery use and repairs, and fuel. More surprising savings came from lower veterinary costs; longer productive life of the cows; and, in some cases, a premium price for higher-quality milk. Studies suggest that the dairy grazing system significantly reduces nitrogen and phosphorus runoff, leaching, volatilization, and soil ero-sion. This not only reduces costs of public regulation of water quality and farmer compliance, but neighbors appreciate the improved landscape and reduced odor (Petrucci 1996). These pasture-based systems also greatly enhance local wildlife habitat. One study in Maryland found the following increases in observed wildlife following conversion to the dairy grazing system:

deer, 65 percent; fox, 27 percent; waterfowl, 30 percent; and turkeys, 70 percent (Barao 1996). The system has spread to most of the southeastern and upper midwestern states, including Ohio, Kansas, Minnesota, Nebraska, and Wisconsin, and northeast into Vermont.

EXAMPLE 22. ORGANIC COCOA FARMING STABILIZES COSTA RICAN BUFFER ZONE

Worldwide consumption of chocolate is worth U.S.$75 billion annually. Currently, organic cocoa is only a small niche market—1 percent of U.S. sales in 2000. However, interest by the chocolate industry in organic cocoa has grown significantly. Despite rising demand, cocoa productivity levels are declining in many parts of the world, a result of persistent problems with diseases in the humid lowland tropical environment where this crop is grown. Even when prices are attractive, farmers are beginning to feel that the risks are too high. Major buyers of cocoa, such as M&M Mars, believe it likely that the most promising option for long-term, sustainable cocoa production is with small-scale, organic systems. Unlike conventional cocoa production, which uses high levels of pesticides and fungicides, new systems for disease control are management-intensive rather than chemical-intensive, requiring more complex manipulation of cocoa plants, shade trees, soil, and pest species. For example, researchers of the industry-funded American Cocoa Research Institute have identified new pruning techniques and modified harvest schedules that greatly reduce the incidence of key pests.

The Nature Conservancy is drawing upon these scientific advances to help protect one of their top conservation priorities in Central America—the Talamanca/Bocas de Toro reserve, an area that spreads from mountains more than 4,000 meters high to the sea's edge. The reserve encompasses more than 30,000 hectares of private lands, indigenous reserves, and coastal zones in Costa Rica, and 14,000 hectares of coral reefs, mangroves, and lagoons in Panama. More than 358 species of birds, of which over 100 are North American migrants, depend on this intact forest for their survival. The Talamanca Corridor in Costa Rica is a major buffer zone for the reserve. Along one side of the Corridor are 400 small-scale producers of shaded cocoa who, for lack of capital, were unable to apply many purchased chemical inputs to their crop before The Nature Conservancy began its program. Their incomes had dropped, and some farmers had begun to cut down their cocoa and shade trees to plant annual crops. To halt this land-use change in the buffer zone, The Nature Conservancy program promoted organic cocoa production. Financing for the program came from the EcoEnterprise Fund, a venture capital fund co-financed by The Nature Conservancy, the Inter-American Development Bank, and private investors for environmentally compatible businesses. This investment is enabling 1,500 farmers to produce for the more lucrative organic cocoa markets using new, higher-productivity organic management techniques. Farmers, mostly of the BriBri and Kekoldi indigenous peoples, are paid 85 percent more than other cocoa growers. As a result, farmers have not only stopped cutting down their trees, they are now expanding the area under cocoa. A study of bird populations in the organic cocoa buffer zone found a higher number of species than in the forested areas because the cocoa plantations create the forest edge habitat preferred by many bird species (Alfaro 2000; Niiler 2001; The Nature Conservancy 2000b).

EXAMPLE 23. SAVING NATIVE FISH IN LAKE VICTORIA THROUGH REMOTE SENSING TECHNIQUES

Lake Victoria, the world's second largest freshwater lake and the chief reservoir of the Nile River, serves as a major source of employment for some 30 million people, as well as a source of tourism and wildlife in the region. The lake is considered a biodiversity hotspot because of its species richness (especially fish) and the level of threat to it. Blooms of blue-green algae caused by the runoff of agricultural chemicals from surrounding lands are starving fish and plankton of oxygen and sunlight and reducing the diversity of important aquatic plants. The lake is losing dozens, perhaps hundreds, of species of fish because the increased turbidity constrains color vision and so interferes with mate choice (Hecking 2000). This process adds to the already serious pressure from an introduced species, the Nile perch, which has exterminated numerous species of native fish. Eutrophication of the lake has also enabled the population of another invasive species, the water hyacinth (*Eichhornia crassipes*), to explode, causing lake water to stagnate and making the choked shoreline a breeding ground for mosquitoes that spread malaria and snails that host bilharzia, a human parasite. The hyacinths also block waterway traffic and have greatly damaged the lakeside economies (Nkuba 1997). Action to address the problems has been hindered by poor understanding of the source of the nutrients that feed the algal bloom. Hypotheses included atmospheric deposits from forest burning and wind erosion, agricultural runoff carried into the lake by streams, and wastes from livestock watering at the lake shore.

Scientists discovered in 1999 that satellite remote sensing technology could be used to identify sources of the nitrogen- and phosphorus-rich sediments that feed the algae and water hyacinths and cause turbidity. In this method, each soil type has a unique spectral "fingerprint" that can be detected by satellite, using standard analytical spectrometry techniques. The analysis indicates that the sediments are made mostly of two soil types. These soils come from gullies created by soil and water erosion on agricultural land, on the human and livestock paths through those lands, and in riparian zones where vegetation has been removed. Testing soil particles in the areas where the sediment plume appeared, scientists were able to determine which parts of the valley system were producing the most nutrients. Part of the solution is simple, inexpensive, and targeted. It mainly involves reintroducing vegetation on critical sites most responsible for sediment deposit detected by satellite imagery. Having identified the exact sources of the problem, researchers, government extension agents, and local NGOs are now working with communities to target rehabilitation of key microwatersheds, including noncultivated areas under communal property rights.

Potential Application

These seven examples illustrate various strategies for reducing agricultural pollution, for both large commercial producers and small semi-subsistence farms. New crop management systems that greatly reduced or eliminated the need for agrochemicals were developed for rice production in Yunnan and Vietnam, and cocoa in Costa Rica. Nontoxic pest management was developed for West Africa. Entirely new systems were developed for dairy production in the United States. Advanced methods for tracing nonpoint source pollution in Lake Victoria enabled highly targeted conservation investment. Ecological filters were used to reduce the impact of agrochemicals in farming around the Chesapeake Bay. For most cases, finding strategies that control pollutants without reducing yields or farm incomes was research-intensive and crop-specific. In all of these cases, interventions were developed largely by research organizations, but often with the urging and support of conservation agencies.

Approaches like natural pest controls and new planting patterns may be quite simple and low-cost for farmers to adopt. While the transition to less polluting systems such as intensive dairy grazing, organic cocoa, or integrated pest management may require several years, they ultimately reduce farmer costs significantly. These systems often require site-adapted application, for which farmers need special training. Where intensive production systems must instead rely on investments to control or process their wastes, costs are much higher and in the United States and Europe have required public subsidies. This suggests that research to reduce the use of agrochemicals and the production of wastes has the greatest long-term potential to control agricultural pollution.

Strategy 5: Modify Management of Soil, Water, and Vegetation Resources

The habitat quality of farmlands for wild biodiversity may be improved by changing the management of farm resources to improve agricultural production. Key aspects may include water management, soil management, tillage practices, management of natural vegetation, fire management, cover crops, agrobiodiversity, and co-management of livestock and wildlife. This ecoagriculture strategy builds on the contributions of agroecology, regenerative agriculture, agroforestry, and sustainable rangeland and forest management, as well as wildlife biology and ecology. Many of these investments in land, water, and perennial vegetation resources serve to increase farmers' natural capital, and thereby long-term flows of farm outputs.

Water Management

There is great scope for increasing use efficiency of both rainwater and irrigation water in agriculture, thus making more water available for wetlands, wildlife, and humans (Penning de Vries et al. 2002). Better management of drainage water in irrigation systems can prevent salinization of soils and water and the consequent harm to habitat quality; this is appealing to farmers because salinization also significantly reduces crop yields. Water conservation measures can help to slow the velocity of water moving across the surface, encouraging better percolation through the soils and availability of water for noncrop plants. Reduced velocity also helps the farmer by reducing soil erosion, reducing nutrient removal from fields, and reducing water requirements of crops. A variety of water-harvesting systems can help sustain agriculture in drylands, while also supporting wildlife in the dry season (Agarwal and Sunain 1999). Improved crop varieties have been developed by ICARDA and ICRISAT, for Asian and African drylands, that are more efficient in water utilization. WARDA is seeking to develop methods for irrigated rice production in the inland valleys of West Africa that will meet human health requirements as well as ecological needs.

Soil Management

Soil microorganisms include macrofauna (earthworms, termites, and ants); decomposers, pests, and pathogens (fungi, mesofauna, bacteria, and nematodes); and microsymbionts (nitrogen-fixers and mycorrhizas). These are pivotal to sustainable agriculture, because they make soil nutrients available to crops by breaking down organic matter and fixing nitrogen, suppress soilborne diseases and pests, detoxify waste materials, and maintain good soil structure, improving absorption of rainfall or irrigation water. They also constitute a high proportion of all wild biodiversity. By modifying the frequency and intensity of tillage practices, preventing soil erosion, and maintaining soil organic matter, habitat for these microorganisms can be enhanced, with beneficial effects on crops (Swift 2001). Some soil organisms are essential for healthy plant growth. The soil bacteria that fix nitrogen in association with legume plants and mycorrhizae are essential for the survival of many plant species, so the establishment of natural forest species in agricultural fallows may be improved by inoculating the soil with mycorrhizae species (Styger, Fernandes, and Raktondramasy 2001).

Tillage Practices

In agricultural regions where cropped land accounts for a high proportion of total land use, frequent tillage not only damages or destroys soil structure, but also significantly reduces soil biodiversity. This in turn reduces the long-term sustainability of soil, as soil organic matter and microorganisms contribute to maintaining good soil physical structure and water- and nutrient-holding capacity. Minimum or zero tillage practices were originally stimulated by the oil crisis of the early 1970s, as farmers sought to lower farm energy costs. Agricultural engineers developed new machinery permitting seed to be drilled into untilled soil. Herbicides were developed that could control weeds in the seedbed without tillage. Minimum-tillage systems are now used on more than 100 million hectares of grains and legumes worldwide, particularly in mechanized temperate farms of Argentina, Australia, Brazil, Canada, and the United States. While some biodiversity concerns have been raised where herbicides are misused, minimum tillage has had a major positive impact on soil biodiversity. In many places, minimum tillage is combined with other soil-building practices. Other benefits to farmers include reduced soil erosion, significantly higher levels of organic matter in soils (with all its associated benefits), and improved water-holding capacity.

There may also be biodiversity benefits from leaving crop stubble rather than turning it under at the end of the season. A study of the northeast region of the United States has shown that the number of wild turkeys, Canadian geese, deer, raccoons, skunks, and possums has increased where farmers have left more crop residues in autumn and winter (Mac et al. 1998).

Management of Natural Vegetation

Natural vegetation in farmlands can be better managed for both habitat quality and economic production. In much of the tropics, agricultural production cannot be sustained with only inorganic fertilizers and pesticides, even where both are available and used in ecologically sensitive ways. As organic material decomposes rapidly in tropical climates, and many soils are naturally low in organic matter, to maintain adequate levels of organic matter in the soil requires additional organic biomass application. Natural vegetation growing on or near the farm is one source of such biomass.

Moreover, to control pests in regions with no winter or pronounced dry season, lands must be left out of production—that is, left fallow—for periods from one season to several years. Thus natural grass, shrub, and tree fallows remain important for production, while serving as habitat for diverse wild plant and animal species.

Cover Crops, Green Manures, and Improved Fallows

Biodiversity-friendly measures such as cover crops and green manures are widely used, even in large commercial systems. Cover crops are high-biomass crops, such as alfalfa, that are grown after the main crop is harvested so as to maintain effective ground cover, thus reducing soil erosion due to wind and rain. When used as green manures, they are tilled into the soil prior to the subsequent cropping season in order to enrich soil organic matter and nutrient content. For example, large-scale commercial wine-grape plantations in both developed and developing countries are now using a mixture of cover crops between rows of vines. The cover crops are mowed and incorporated into the soil several times a year (Thrupp 1998), thereby benefiting both biodiversity and the crop. Shrubby cover crops such as velvet bean (*Mucuna* spp.) have been widely adopted by small-scale farmers in Central America, to provide organic matter and maintain soil cover, and CIMMYT research has confirmed their beneficial effects on yields, soil quality, and profitability (Buckles, Triomphe, and Saín 1998).

As agriculture has intensified around the world, traditional medium- and long-term fallows are disappearing. It used to be common wisdom that fallows would have no role in the permanent agriculture of the future. Over the past decade, however, researchers working together with farmers have developed short-duration (six-month to three-year), improved woody fallows for many tropical agroecosystems. Because they reduce farmers' cash costs for purchase of fertilizers and produce a range of valuable products for household use or sale, the practice has spread rapidly, even on small farms. Short fallows using trees, shrubs, or herbaceous plants can enhance wild biodiversity by reducing agro-chemical pollution and providing suitable habitats. Fallow systems provide mosaics of spatially interacting fallow and cropped plots (van Noordwijk 1999). These can be an important part of broader land use mosaics to enhance wild biodiversity. To provide the benefits of fallows, functional and structural biodiversity seem more important than species richness, because of their effect on nutrient recycling and retention, reduction of risks from climatic variability or pests, and economic outputs from the fallows themselves (Buresh and Cooper 1999).

Fire Management

Fire is an important management practice in shifting cultivation and grazing systems to control weeds and pests and to stimulate growth of useful plants. But fire can also threaten many wild species. Careful fire management can ensure that timing and scale of fires are appropriate to the ecosystem, and can improve crop and livestock systems (de Haan, Steinfeld, and Blackburn 1999). Potentially invasive species of plants can be prevented from establishing themselves, and controlled through other means where they have become a problem. Weeds can sometimes be better managed through shade, soil cover, and other means that avoid use of the herbicides that can have negative impacts on biodiversity.

Crop and Livestock Diversity

Increasing the diversity and spatial mix of crop, tree, and livestock components on the farm can greatly enhance habitat value. Deliberately diversifying farming systems by planting a wider range of crop species and cultivars (for example as a mosaic of small plots, as crop rotations, or in fully integrated agroforestry systems) can result in much greater wild biodiversity as various forms of wildlife move in to occupy the expanded ecological niches. Underutilized niches on the farm can be filled with economically valuable indigenous species. It is estimated that 20 percent of the world's farmers already rely on mixed polyculture systems. Small-scale farmers in various parts of Africa have ingeniously used crop diversity to control land degradation, maintain yields with minimum external inputs, and exploit their variable environment (Tengberg and Stocking 2001). Even where economic forces continue to strongly favor farm specialization, it is often possible to intercrop minor species strategically on the farm, in particular native species that do not compete with economic crops. Field, farm, and landscape diversity can also enable farmers—especially those in economically marginalized areas—to cope with abrupt and transformative changes in their biophysical and socio-economic environments (Piñedo-Vasquez, Padoch, and Coffey 2001).

Mixing species may benefit farmers. In the Philippines simple intercrops of corn (maize) and peanuts help to control the corn-stem borer. Springtails found under the peanut plants are eaten by young spiders, which prey on stem borers as adults. The mechanisms of mixed-crop benefits may be more subtle. Aromatic odors from the intercropping of cabbages and tomatoes repel the diamondback moth; the shading effect of mung beans or sweet potatoes grown with corn reduces weed growth (Conway 1997). It may also be possible to increase the diversity of animals on farms by domesticating selected game species, using native vegetation for feed. These animals may provide a more reliable protein source for local people than traditional livestock, especially where common domestic species are poorly adapted to local ecologies (National Research Council 1983, 1991).

Co-managing Livestock and Wildlife

In many dry tropical lands, livestock play a central role in local livelihoods, both for cultural reasons and because crop yields are so precarious in those environments. The establishment of large wildlife reserves in traditional grazing areas, with sharp restrictions on local rights to graze or to defend cattle against wild predators, has caused conflict and exacerbated poverty. In response, new paradigms have been developed for co-managing domestic livestock and wildlife (Bourn and Blench 1999; IFAD 2001; Kiss 1990). Research has shown that livestock and wildlife exploit different (but overlapping) ecological niches in time and space, and have evolved different physiological and behavioral strategies to reduce competition. Thus, on rangelands, annual and perennial grasses and legumes can be mixed in ways that both increase the sustainability of graze and browse for domestic livestock, and improve feed sources and protection for wildlife.

Predation and infrastructure damage from large wild mammals can be limited through improved physical protection. Some experts now advocate mixed livestock raising and harvesting of wild herbivores as the most economic use of low-rainfall rangelands, thus maintaining the full natural biodiversity (Western and Pearl 1989). The new strategies not only help pastoralists maintain income and use values from livestock, but also benefit them economically by integrating wildlife into their livelihood strategies, which include earning income from ecotourism, safari hunting, park revenue sharing, cash compensation for the risks of wildlife damage, and sale of rangeland products to tourists. For example, the CAMPFIRE community-based wildlife management program in Zimbabwe has increased incomes in communal areas by an estimated 15 to 25 percent, though household level income increases may be less. Research in Ghana, Kenya, Zimbabwe, and Namibia showed significantly higher economic rates of return on wildlife ranching than from cattle, though the income from tourism, trophy hunting, and wild meat is subject to market saturation (Bojö 1996).

Similar principles may apply to aquaculture. By introducing noncompetitive, nonpredatory species with complementary food needs, polyculture systems in aquaculture can raise productivity, reduce effluents, diversify output and maintain a wider range of wild aquatic species. In China, for example, four types of carp are produced in the same pond to take advantage of food resources in this unique ecosystem: silver carp (which feed by filtering phytoplankton), grass carp (which feed on plant-eating microorganisms, common carp (an omnivorous bottom feeder), and bighead carp (which feed by filtering zooplankton) (Naylor et al. 2000).

Seven Case Studies

Seven cases of improved resource management resulting in biodiversity benefits are highlighted here. While they are drawn from very different agricultural systems, a common element is that the innovations beneficial to wild biodiversity were introduced primarily to increase the profitability or yield of crops, livestock, or forests. Biodiversity benefits were collateral and might reasonably be expected to increase with greater investment and attention.

Example 24. Managing Flooded Rice Fields for Wildlife Habitat

Flooded crop fields may provide foraging habitat equivalent to seminatural wetlands and, because of reduced predation threat, may provide a safer habitat for water birds. Thus, if managed appropriately, even one of the world's dominant forms of intensive agriculture can provide valuable habitat. Flooded rice fields in California are used by numerous aquatic birds during winter. Since this habitat functions like a natural wetland, increased flooding may help replace the extensive wetlands that existed in the region prior to agricultural development (Elphick 2000). Researchers compared the habitat value of flooded rice fields and seminatural wetlands for several species of aquatic birds. The availability of invertebrate species used by birds for food did not differ among habitats. Seminatural wetlands had less rice grain but more seeds from other plants than the two rice habitats. Predators passed over a feeding area less often in flooded fields than in unflooded fields or seminatural wetlands, and birds fed more often in flooded fields.

Such results are relevant in many parts of the world. In the Sacramento and San Joaquin Valleys of California, farmers working together in the Valley Care program have instituted minor management changes in flooded rice production that greatly increased their value for tropical migrant shorebirds and waterfowl. These methods were pioneered by Ducks Unlimited, a conservation and hunter's organization.

After rice is harvested, rice stubble and straw are rolled and crushed, and then flooded over the winter as an alternative to burning them. The system accomplishes the grower's objective of decomposing waste straw and controlling weeds and diseases, while providing winter habitat and food for water birds. Rolling rice straw is economical in comparison with other agronomic methods that do not offer the same wildlife benefits. It also eliminates air pollution due to burning, which is now tightly regulated. Some restored natural wetlands are being managed jointly with agricultural lands to provide year-round wildlife habitat. Species that benefit include waterfowl (such as ducks), wading birds, shorebirds, and cranes. Shorebirds include dunlins (*Calidris alpina*), dowitchers (*Limnodromus scolopaceus*), killdeers (*Charadrius vociferus*), and other sandpipers. Ducks include northern pintails (*Anas acuta*), American widgeons (*A. americana*), and even mallards (*A. platyrhynchos*) and northern shovelers (*A. clypeata*). Snow geese and Ross's geese also commonly benefited (Payne, Bias, and Kempka 1996).

The rice cropping system in the upper coast of Texas creates a heterogeneous mosaic of flooded-rice wetlands, grazed fallow lands, and plowed fields that has dramatically increased use by migratory birds including the lesser snow geese, the greater white-fronted geese, and Canada geese. More than 20 million waterfowl and geese winter on the upper Texas coast, with the bulk of these using freshwater wetlands associated with rice agriculture (Lacher et al. 1999).

EXAMPLE 25. DAMBO IRRIGATION COMPATIBLE WITH
WETLAND PRESERVATION

Despite widespread low rainfall and drought, less than 15 percent of Africa's irrigation potential has been exploited. But concerns about environmental sustainability, high costs for infrastructure development for conventional irrigation, and lack of managerial capacity for large systems have prevented significant irrigation expansion. Instead of relying on "conventional" (and especially large-scale) irrigation, small-scale irrigation systems under farmer management—which already comprise an estimated 47 percent of irrigated area in sub-Saharan Africa—are generally more suitable. Dambo irrigation offers particular promise. *Dambos* are shallow, seasonally waterlogged depressions at or near the head of a drainage network. Wetland forms like dambos are found in Zimbabwe (*bani*), Malawi and South Africa (*vlei*), Rwanda (*marai*), Sierra Leone (*boli*), and Nigeria (*fadama*). Dambo farmers fence a plot and dig a series of water channels between beds, often with a shallow well in one part of the plot. Water is applied through these channels as subirrigation in the root zone, with buckets, hoses, and watering cans for supplemental application. Investment costs are low, making dambos more affordable than conventional irrigation. Returns on investment are high. Zimbabwe has 1.28 million hectares of dambo landform (3.6 percent of the total land area). Approximately 15,000 to 20,000 hectares of dambo gardens are under productive cultivation, and up to 80,000 hectares have potential for sustainable production; these are mainly in communal areas. However, concerns about erosion and downstream water flows led to 1970s legislation banning dambo cultivation. Though often circumvented, these laws have restricted research, extension, and credit for these systems.

Researchers studying dambos in Zimbabwe found that yields per unit of land and water were approximately twice as high as in formal irrigation systems. Hydrologic measurements show that cultivation on the dambos with indigenous methods (that is, no deep drains or mechanical pumps) is environmentally sustainable; it does not dry up the dambo, mine the groundwater, or reduce downstream flows. This is because gardens do not consume much more water than native vegetation on the dambos, and the water that is not used by the crops flows through to other fields or to a stream below the field. Indeed, dambo cultivation is less damaging, in terms of erosion and water releases (because of dense dambo fencing), than dryland cultivation of the watershed above the dambo (Rukuni et al. 1994). Impacts on habitat and wildlife appear relatively benign. On dambo fields farmers often retain some native vegetation, have high crop diversity in small plots, plant diverse native tree species as live fencing, and leave part of the dambo uncultivated. Dambo cultivation alleviates pressure from livestock grazing and the resulting compaction.

Dryland cultivation is the main alternative production system of small-scale producers on communal lands, but it is only a tenth as productive as the dambos. Thus dambo production can relieve pressure on upland resources, including wildlife habitats (Meinzen-Dick and Makombe 1999). Responding to evidence of the economic contributions and benign environmental effects of dambos, policy-makers in southern Africa are rethinking the potential for wise use of dambos.

EXAMPLE 26. SOIL EROSION BARRIERS WITH NATIVE PLANTS IN THE PHILIPPINES

Contour hedgerows are rows of perennial plants established along the contours that have been promoted on steep lands to reduce erosion and produce organic matter for soil improvement. Most contour hedgerows have used exotic grass or shrub species, requiring special nursery development to provide planting materials and considerable labor for establishment. In the early 1990s, researchers from the International Centre for Research in Agroforestry (ICRAF) working in the Philippines became frustrated at farmers' low adoption of hedgerow technology. They began a series of studies to identify the most cost-effective approach to contour planting of perennials. They discovered that natural vegetative strips—contour rows left uncultivated during plowing, so that natural vegetation could grow there—were not only the least expensive method for erosion control (zero cost for planting materials and establishment), but also nearly as effective as planted shrub or exotic grass hedgerow technologies. Studies found that rows as far apart as 2 to 4 meters in elevation distance served nearly as well for erosion control as more closely spaced rows, while removing much less area from production (Mercado, Stark, and Garrity 1997). Further research developed a low-cost method for laying out initial contour lines and for enriching the natural vegetative strips with high-value fruit trees and timber species from which farmers could earn cash income.

First introduced to natural vegetative strips in 1996, thousands of Philippine farmers have now adopted this low-cost technology in the densely populated steep farmlands of northern Mindanao (Figure 7.1). The strips are not only valuable for maintaining soil fertility on farms and protecting local watersheds but also provide important habitat for wild biodiversity. A study of floral composition and community characteristics of fields with natural vegetative strips confirmed the high diversity of native plant species, while the presence of untilled areas provided good habitat for native fauna (Ramiara-

Figure 7.1. Natural Vegetative Strips Protect Watershed and Biodiversity. *Source:* ICRAF, Philippines.

manana 1993). Economically profitable timber and fruit-tree species in the strips further expand their habitat value for wildlife.

EXAMPLE 27. PRESERVING SOIL ORGANISMS THROUGH NO-TILLAGE SYSTEMS IN BRAZIL

In many intensive, mechanized annual cropping systems, the land is tilled several times each season to improve the seedbed for crop germination and early growth. While providing short-term benefits, this process tends to break down organic matter in the soil and physically disturbs the soil habitat, leading to soil erosion, sharp declines in soil numbers and species diversity of soil organisms, and a shift in types of microorganisms. Heavy machinery causes compaction in many types of soils, not only destroying the worm and root channels that create the porosity essential for rainfall absorption and healthy plant growth, but also reducing pore spaces in which bacteria and their protozoan predators live (Bamforth 1999).

No-tillage systems were first developed in the 1970s in the United States, in response to high energy prices, and began to be widely adopted in Brazil in the 1980s. While conventional minimum-tillage systems rely on heavy use of herbicides, smaller-scale farms in Brazil adopted more sustainable and lower-cost systems using crop rotations and cover crops. Even as energy prices declined, farmers found that no-till greatly reduced their costs of production. In Paraná state in Brazil, for example, no-tillage farm plots yielded 13 to 34 percent more wheat and soybean than conventionally ploughed plots, and reduced soil erosion by up to 90 percent. An added advantage is their suitability for small farms, using modified animal-drawn implements (Calegari 2000). Thus farmer adoption has been rapid and widespread. By 2001, no-till covered over 13 million hectares in Brazil, mostly in the southern region and more recently also in the Cerrado. The spread of no-till was driven by farmer-to-farmer contacts. In 1992 a group of farmer organizations formed the National Federation for Direct Planting in Crop Residues (FEBRAPDP), which provided country-wide facilitation of technological change. The farmers' groups were supported by integrated efforts of the private sector, NGOs, and government agencies (Landers 2002).

The use of no-tillage systems can also enhance soil macrofauna populations and their contribution to soil function. Conventional plowing incorporates crop residues into the soil profile to produce homogeneous soils that favor bacteria, protozoa, and bacteria-eating nematodes. By contrast, new no-till systems leave organic residues on the surface and a rich organic layer near the surface, thereby enhancing the fungi and earthworm portions of the underground food web. Recent studies of Brazilian no-till systems found that millipedes, earthworms, arachnids, insect larvae, and beetles tend to be far more abundant, while ants and sometimes termites tend to be more abundant in conventional-till systems. The number of large scarab beetle-grub holes can be up to ten times greater in no-till systems. These organisms play an important role in soil porosity and infiltration, especially when there are heavy rain showers (Brown et al. 2001).

EXAMPLE 28. TREES IN COSTA RICAN PASTURES FEED FOREST BIRDS

Central America has more than 9 million hectares of land in silvipastoral systems (pastures with scattered trees). Depending on farm size and where a particular plot of land is in the fallow cycle, dispersed trees on farms may be established from an improved fallow, through strategic clearing of a mature fallow, or (where no fallow is used) through direct planting or protection of naturally regenerated trees (Kowal 1999). Farmers keep the trees intentionally in their pastures because of significant benefits to associated livestock, subsistence needs, and income benefits. It is estimated that in Central America as a whole, trees in pastures could produce 18.4 million cubic meters of timber per year, given adequate support for timber marketing and assistance to farmers in tree management (Beer, Ibrahim, and Schlonvoigt 2000). Research has begun to show that such isolated trees also play a critical role in maintaining wild biodiversity, by serving as nesting, feeding, and roosting sites for a variety of bird and bat species, many of which are forest species, and by providing transient habitats for many neotropical migratory birds. Remnant trees in pastures often retain rich communities of epiphytic plants that would not otherwise be present in the agricultural landscape. They assist in forest regeneration, both because they produce seed locally and because the birds and bats that visit their canopies regurgitate or defecate seeds of forest plants while perched in the trees. In addition, the microclimatic conditions (for example, low light levels and high humidity) beneath tree crowns help facilitate the germination, survival, and growth of forest plants.

A survey of twenty-four small-scale dairy farms near Monteverde, Costa Rica, found a mean density of 25 trees per hectare; primary forest trees accounted for 57 percent of all species and a third of all individuals. Many of the tree species are important locally for humans as sources of timber (37 percent), firewood (36 percent), or fence posts (20 percent). Timber, shade for cattle, fruits for birds, and fence posts were the most commonly cited of nineteen reasons for keeping the trees. More than half the farmers interviewed thought that the presence of trees in their pastures improved their farm productivity, whereas 32 percent felt the trees had a beneficial effect in the dry season (conserving moisture) but could be negative in the wet season if they produced too much shade. The remaining 13 percent thought the trees had no effect on productivity. The biodiversity impacts were significant. More than 90 percent of the species are known to provide food for forest birds, bats, and other animals. Their presence is especially important for forest birds (such as three-wattled bellbirds, resplendent quetzals, and keel-billed toucans) that seasonally move from the Monteverde Reserve Complex down to the Pacific lowlands. Since most of the landscape surrounding the Monteverde Reserve Complex is highly deforested, isolated trees in pastures provide essential food sources at various elevations along the birds' migratory path (Harvey and Haber 1999).

EXAMPLE 29. REGENERATING NATIVE FORESTS IN INDIA THROUGH
JOINT FOREST MANAGEMENT

Following independence in 1947, Indian forest policy emphasized nationalization and commercial utilization of the country's national forests, which account for nearly 23 percent of the nation's land area. The national government, however, was unable to manage all these resources effectively, instead promoting exotic tree species plantations to meet fast-growing demand. By the 1980s, less than half of this "forest" land had good tree cover, and subsistence forest products had become scarce. In response to this degradation, thousands of communities, primarily in eastern India's tribal forest tracts, took action to protect their degrading forests. They organized hamlet-based forest protection groups that halted cutting and grazing, which often resulted in the regeneration of natural forests. Researchers and NGOs, recognizing the potential of natural forest regeneration, began to support these village initiatives by developing methods to accelerate natural regeneration (by encouraging seed germination, for example) and to manage sustainable product extraction. After a new National Forest Policy was passed in 1988 that identified the need to motivate forest communities to develop and protect their forests, several states allocated partial public forest management authority to forest communities. The National Joint Forest Management Resolution in 1990 supported the rights and responsibilities of forest communities (Poffenberger and McGean 1996).

Today, an estimated 30,000 village-level committees are protecting and regenerating 10 million hectares of forests (Poffenberger 2001). Village forests are managed to provide a flow of subsistence products such as fuelwood, medicines, fodder, and condiments, and some income from commercial timber sales. Their biodiversity value is enhanced many-fold relative to their previous degraded state, while biodiversity is also enhanced in adjacent areas because the environmental services of forests are recovering. The extent of habitat improvement for many types of wildlife in these multiuse forests may be greater than that provided by a number of India's official protected areas. Although there are still major issues to be resolved in fully empowering local communities in joint forest management, local benefits from the local share of timber and nontimber forest product income and direct use values are substantial in some communities. In Madhya Pradesh, for example, local people in over 2,000 forest villages now have rights to all intermediate products and nontimber forest products, estimated to be worth U.S.$125 million per year, or $280 per household (World Bank 2000).

EXAMPLE 30. VETERINARY VACCINE COMBATS LETHAL DISEASE OF
KUDU AND WILDEBEEST

Veterinary research suggests that diseases need not be a major constraint to wildlife and livestock coexistence in the semiarid rangelands. The incidence and transmission of disease can be mitigated through habitat management to discourage breeding habits of disease vectors. In some cases, domestic livestock veterinary practices can have positive spillovers for associated wildlife. Indeed, such is the case for rinderpest, a highly contagious and lethal viral disease of cattle and cloven-hoofed wild animals. It was first introduced into Africa about 100 years ago, along with Asian cattle brought in to feed the Italian army, which at the time was invading Ethiopia. Almost all ruminants, wild and domestic, were highly susceptible to the virus and a disastrous pandemic ensued, sweeping eastern and southern Africa. Hundreds of thousands of cattle, buffalo, eland, kudu, wildebeest, giraffe, and warthog died, as did many pastoralists who depended for their livelihood on domestic cattle.

In the mid-1960s a rinderpest vaccine was developed. Millions of cattle throughout East Africa were vaccinated annually, almost eliminating rinderpest from Africa. By removing the source of rinderpest virus, the wildebeest population in the Serengeti Plain rose more than sixfold. Unfortunately, the vaccination effort was not completed and the virus reemerged. A rinderpest outbreak in Kenya in 1994–95 killed 50 percent of the buffalo and 80 percent of the lesser kudu in Tsavo National Park. In 1997, a massive cattle vaccination campaign was organized, and susceptible wild animals are now regularly sampled throughout sub-Saharan Africa. Wild animals are being used as sentinels to track the movement of the virus. An improved, heat-stable vaccine for rinderpest is now under development. It allows naturally infected cattle to be distinguished from those that received the vaccine, thus removing the need to continuously monitor the virus in susceptible wildlife species. The Food and Agriculture Organization of the United Nations (FAO) predicts that rinderpest can be eradicated—for both domestic livestock and wildlife—for as little as U.S.$3 million (Woodford 2000b), with wildlife being secondary beneficiaries.

Potential Application

These seven cases represent a small fraction of the diverse types of possible resource management innovations to enhance wildlife habitats in farm fields and production forests. All of the crop-related examples are from intensive commercial or semicommercial cropping systems; all contributed to quite substantial increases in productivity (Philippine vegetative strips; Zimbabwe dambos) or cost reductions (minimum tillage in Brazil; flooded rice in the United States). The Zimbabwe system was compatible with a fragile wetlands habitat; the other systems enhanced wildlife populations. In the low-input grazing system of Costa Rica, forest system in India, and grazing/wildlife system in East Africa, modest changes in vegetation management by local people enhanced their incomes, while enabling wildlife to flourish. The veterinary vaccine developed for domestic grazing herds in East Africa served to protect wild species as well. Some of these practices may reduce economies of scale or increase farmers' monitoring costs; such practices are more feasible for smaller-scale farmers and for producers in ecologically heterogeneous areas, where variable farm management across diverse landscape niches is necessary anyway. However, other management practices like minimum tillage are more financially attractive for large-scale farmers, while scattered trees in pasture benefit farms of all sizes.

The technique of minimum tillage and veterinary vaccines represented considerable investment in formal research, and their positive impacts on wild biodiversity were largely unplanned collateral benefits. By contrast, the dambo, flooded rice, trees in pasture, and forest-management systems were largely local innovations, and the vegetative strips were adapted by researchers from local practice; all of these were quite low-cost. But for these practices to be most effective requires that farmers design and manage them with an understanding of wildlife habitat requirements. Thus, for this strategy to have a major impact will require farmers to have a commitment to biodiversity conservation and for researchers (wildlife biologists, ecologists, and agricultural scientists) to identify the most strategic management practices from both production and conservation perspectives.

Strategy 6: Modify Farming Systems to Mimic Natural Ecosystems

The premise of "natural systems agriculture" is that agroecosystems modeled after natural ecosystems should exhibit many of the functional attributes and processes that stabilize natural systems, including vegetation adapted to the local climate, closed nutrient cycling so that nutrients are not lost to the system, soil preservation, and methods of crop production (Jones, Sieving, and Jacobson 2001). For example, by planting mosaics with patches of different tree species, or multistrata mixtures of crop species, the structure of natural forest habitats can be imitated. Soil disturbance would be minimal compared to conventional crop cultivation or forest harvest, and the canopies, stems, and root systems of perennial plants would provide habitat niches for many wild species. In temperate areas, effective integration of farm and ecosystem management has been demonstrated for diverse prairie and wetland systems (Jackson and Jackson 2002).

Growing interest in agroforestry and other integrated systems since the 1970s reflects, in part, a desire to capture the environmental services that perennial trees, shrubs, and grasses provide. These services include soil and water conservation, control of water flow, soil nutrient cycling, wind protection, and production of organic mulch in the form of leaf litter, which reduces soil erosion. The actual contribution that perennials make to these services depends very much on their temporal and spatial configurations, location relative to key landscape features (such as waterways or remnant forests), species choice and mixtures, density, and management—all elements of landscape design.

The science for designing agricultural systems that provide the ecological functions of natural systems is still in an early phase of development. In particular, research is needed to establish which ways of modifying the spatial arrangements of natural and domestic plant species and other landscape elements can really increase crop and pasture productivity and sustainability (Lefroy, Salerian, and Hobbs 1992; Lefroy et al. 1999). This section discusses how farming systems that mimic natural forest and grassland ecosystems can contribute to sustainability and to wild biodiversity, and the associated challenge of making more perennial species in such systems economically viable.

Forest Ecosystems

In humid and subhumid forest ecosystems, perennial plants can be incorporated into farming systems in numerous ways to enhance ecosystem function.One general approach is to strategically integrate woody plants into conventional annual-crop based farms, through various agroforestry systems.Trees and shrubs can be used to improve soil and water management and provide farm products or income. On sloping land, strips of economically productive perennials can be planted along the contour between strips of erodible annual crops.Within a few years, natural processes can build up the level of soil behind such contour strips, forming terraces. Patches of natural vegetation can be allowed to regenerate to protect waterways and alleviate severe gullying.There is often unrecognized scope to integrate trees even into intensive irrigated cropping systems. In the Northwest Frontier Province of Pakistan, 67 million trees have been planted in irrigated fields at an average rate of 72 trees per hectare. Species including poplar and *Dalbergia sissoo* thrive in rice paddy conditions and compete minimally with the crop (Vergara 1997).

Improved fallows can be designed to mimic natural habitats, by using species mixtures of relevant size and structure. In Southeast Asia, improved fallows based on indigenous fallow management strategies are shown in Figure 7.2.These range from single-species fallows that quickly restore soil fertility for annual crops (shrub-based accelerated fallows, viny legumes as seasonal fallows), to simpler, mixed-species fallows (improved woody fallows; retention or promotion of preferred volunteer species), to complex mixtures of economic plants (such as in agroforests or in perennial-annual crop rotations) (Cairns and Garrity 1999). Fallow enrichment can both improve the speed and quality of soil rehabilitation, and provide an economic incentive to maintain forest fallows for longer periods.

Multistrata systems, such as agroforests or complex home gardens, can be developed in several ways. The first is by enrichment planting within logged or degraded forests.The second is by planting perennial trees in cleared forest land, as an alternative to shifting cultivation, and by leaving other wild plants to grow among them.The third way is by intercropping among existing plants of either upper- or middle-story species (Leakey 1999a). Annual food crops may be planted along with trees to provide quick benefits to the soil, and to produce food until the trees mature enough to yield fruits, resins, medicines, nuts, and high-grade timber (De Foresta and Michón 1994). Agroforest systems can be highly productive and profitable for farmers, providing a secure and low-risk livelihood (Anderson 1990).

Even monoculture tree plantations, if sited appropriately and managed to minimize agrochemical use and maintain ground cover, provide far superior biodiversity values than most annual cropping systems.They are ideally planted in smaller plot sizes, in mosaics containing other land uses.

Agroforests	Perennial-Annual Crop Rotation (cyclical tungya system)	Interstitial Tree-Based Improved Fallow	Promotion of Preferred Volunteer Spp.	Shrub-Based Accelerated Fallow	Viny Legumes as Seasonal Fallow

LATEX BASED:
■ *Havea braziliensis*
RESIN BASED:
■ *Shorea javanica*
  - Krui, Sumatra, Ind.
■ *Paraserianthes falcataria*
  - Mindanao, Phil.
■ *Toxicodendrom vernicifera /*
■ *Pinus yunnanensis*
  - west Yunnan, China
■ *Styrax tonkinensis /*
■ *S. benzoides*
  - N.E. India
■ *Styrax benzoin /*
■ *S. paralleloneurus*
  - North Sumatra
FRUIT/NUT BASED
■ *Durio zibethinus*
  - Kalimantan, Ind.
■ coconut palm
  - Menado, Sulawesi Ind.
  - widespread
■ commercial fruit orchard
  - widespread
OTHERS:
■ *Amomum compactum*
  - northern Laos
■ *Amomum subulatum*
  - Himalayan foothills
■ *Piper nigrum* (pepper)
  - widespread in Ind.
■ *Camellia sinensis*
  - S. China/ N. Thailand
■ *Coffee* spp.
  - widespread
■ *Zanthoxylum limonella*
  - northern Thailand
■ *Borassus sundaicus*
  - Roti & Savu Is., Ind
MIXED SYSTEMS:
■ Kenyah / Iban fallow enrichment
  - Kalimantan, Ind.
■ Ifugao woodlots
  - Ifugao, Phil.
	TIMBER-BASED:
	▲ *Cunninghamia lanceolata*
	  - southern China
	▲ *Paraserianthes falcataria*
	  - Mindanao, Phil.
	▲ *Melia sp.*
	  - N.W. Vietnam
	▲ *Tectona grandis*
	  - Laos
	▲ *Gmelina arborea*
	  - widespread
	▲ *Eucalyptus spp.*
	  - widespread
	▲ *Pinus wallichiana*
	  - Bhutan
	▲ *Santalum sp.*
	  - central Laos
	▲ cedar
	  - Kwangtung, China
	▲ *Alnus nepalensis*
	  - S/S.W. Yunnan, China
	NON-TIMBER
	▲ *Cinnamomum burmanil*
	  - Sumatra, Ind.
	▲ *Aquilaria sp.*
	  - central Laos
	▲ *Broussonetia papyrifera*
	  - northern Laos
	▲ *Giganthochlos levis*
	  - Mindoro, Phil.
	▲ other bamboo species
	  - Timor, Ind.
	  - southern Vietnam
	  - northern Vietnam
	▲ *Calamus caesius*
	  - Kalimanatn, Ind.
	▲ *Calamus sp. /*
	▲ *Plectocomia himalayane*
	  - Southern China
	▲ 'Talun kebun' mixed
	  -West Java, Ind.
		SIMULTANEOUS:
		● *Alnus nepalensis*
		  - Nagaland, India
		  - S/S.W. China
		● *Leucaena glauca/ Gliricidia sepium*
		  - Naalad, Cebu, Phil.
		● *Leucaene leucocephala*
		  - Amarasi, Timor, Ind.
		  - Sikka, NTT, Ind.
		  - Mindoro, Phil.
		● *Sesbania grandiflora / Leucaena leucocephala*
		  - Sumba/Flores/Timor
		● *Erythrina sp./Desmodium sp. / Hybiscus sp.*
		  - Flores, Ind.
		● *Albizzia chinensis*
		  - Sumba, Ind.
		● *Acacia villosa*
		  - Timor, Ind.
		● *Ficus spp.* (fodder)
		  - eastern Bhutan
		● *Pinus keslys*
		  - northern Italy
		SUCCESSIVE:
		● *Tephrosia purpures*
		  - north Vietnam
		● *Sesbania spp.*
		  - Isabela/Cagayan, Phil.
		  - Yap, S. Pacific
		● *Alnus nepalensis*
		  - southwest China
		● *Casuarina oligodan*
		  - New Guinea
		● *Alnus japonica*
		  - N. Luzon, Phil
		● *Parasponia rugosa / Schieinitzia novo-guineansis*
		  - Papau New Guinea
		● *Hibiscus tiffaceous*
		  - Yap, S. Pacific
			ECONOMIC UTILITY:
			Food
			▶ bamboo shoots
			▶ native vegetables & other wild food plants
			Fiber
			▶ construction materials, e.g., planting *Corypha uish* Lan. & other palm spp. Before abandoning swidden to provide roofing materials for field hut construction in next cropping phase
			▶ harvest of poles useful for house or hut construction
			  - thatch-grass for roof construct.
			Fodder
			▶ *Imperata cylindrica* & other native forages
			Fuel
			▶ Medicinal Herbs
			▶ Stimulants
			▶ *Nicotiana tabac.* (tobacco)
			▶ *Piper beetle* (beetle leaf) + spp. providing shade, pleasant smells, nectar for honey product, attracting wildlife for hunting, etc.
			  -all widespread insubsistence
			▶ **ECOLOGICAL UTILITY:**
			  -selective felling to retain 'mother trees'& speed up recovery of secondary forest
			  -protect existing coppices: limit cropping period, fire management & avoid tillage
				NON-N FIXING:
				● *Compositae spp.* (N-accumulating?)
				+ *Austrosupatorium inulifolium*
				  - West Sumatra, Ind.
				+ *Tithonia diversifolia*
				  - Mindanao, Phil.
				+ *Chromolaena odorata*
				  - Luang Prabang, Laos
				  - Nusa Tenggara, Ind.
				  - Kalimantan, Ind.
				  - Yunnan, China
				  - northern Thailand
				  - widespread below 1000 meters
				Other
				+ *Mallotus barbatus*
				  - northern Thailand
				+ *Ricinus communis*
				  - Timor, Ind.
				+ *Tecoma stans*
				  - Timor, Ind.
				N-FIXING:
				+ *Mimosa invisa*
				  - Leyte, Phil. (spiny)
				  - northern Thailand (spineless)
				+ *Cajanus cajan*
				  - Mindoro, Phil.
					LEGUME ROTATIONS:
					> *Phaseolus calcaratus*
					  - northern Vietnam
					  - northern Thailand
					> *Amphicarpaea linearis*
					  - Hainan Island, China
					> *Flemingia vestita*
					  - N.E. India
					> *Dolichos lablab*
					> *Vigna sinensis*
					  - northern Thailand
					> *Calopogonium mucunoides*
					  - Leyte, Phil.
					> *Pachyrhizos tuberosis*
					  -northern Vietnam
					
					- Increasing integration of legume components into cropping sequence + ruminant livestock

Figure 7.2. Continuum of Fallow Types in Southeast Asia. *Source:* Cairns and Garrity 1999

Savanna Ecosystems

Farming in dryland areas, where the native vegetation is savanna or woodland-savanna mixture, might be more sustainable if designed to mimic the structure of native perennial-grassland mosaics. For example, productive ecoagriculture in the Sahel region of Africa ideally would have 10 to 15 percent tree cover and include a mixture of perennial grasses and shrubs to make the best use of limited water and minimize wind erosion. The dehesa system of southern Spain and Portugal described in Box 4.2 is a human-engineered ecosystem that resembles a natural savanna. The improved soil structure, improved rainfall absorption, and reduced evaporation underneath the trees benefit adjacent pastures and fields of grain. These agroecosystems have adjusted over the long term to achieve sustainable production of grains and livestock in an area of limited rainfall (Joffre, Rambal, and Ratte 1999).

Analysis of dryland ecosystems in Australia suggests that efficient use of natural rainfall will be achieved only if the total leafy area of plants approaches that of their natural state. This would require revegetation of most or all parts of the watershed with either native trees or plants with similar hydrological characteristics, although few such systems are yet profitable (Hatton and Nulsen 1999). The greatest short-term opportunity for ecoagricultural systems in the semiarid tropics is to plant trees and other perennials into niches in the landscape where resources are currently underutilized by crops. For example, they can be planted along footpaths and near houses. This mix would integrate crops with boundary plantings, scattered trees with vegetable gardens, and crops with fodder banks protected by hedges. In this way, the farming system would mimic the large-scale patch dynamics and successional progression of the natural ecosystem (Ong and Leakey 2002).

Benefits of Farm Perennials for Wild Biodiversity

Less intensive management, less soil disturbance, permanent biomass, deep roots, shade, canopy cover, and provision of food sources through much of the year all contribute to the habitat value of areas with perennial vegetation. While agroforestry systems cannot substitute for protected areas as habitat for large animals and plants or for species that are agricultural pests (van Schaik and van Noordwijk 2002), they can provide critical natural elements and refuges for diverse wildlife in agriculture-dominated bioregions. Research has demonstrated how retaining even low tree densities in farmland can enhance wild biodiversity. Isolated trees in pastures, for example, greatly increase the seed rains produced by fruit-eating bats and birds for pioneer and primary species of trees, shrubs, herbs, and epiphytes (Galindo-Gonzalez, Guevara, and Sosa 2000). Perennials thus play a crucial role in maintaining the capacity of agricultural lands to regain their historical levels of plant diversity, should they ever go out of production or into fallow. Longer, woody fallows mimic the natural forest habitat, and they can maintain many wild species that help forests recover after a cropping cycle or land abandonment. Bird populations in the United States respond dramatically to the spatial distributions of cropped and noncropped landscape elements. At more local scales, the distribution, relative vegetative cover, and juxtaposition of agricultural landscape elements (cropped fields, pastures, uncropped areas) can determine avian community structure (Jones, Sieving, and Jacobson 2001).

In mature complex rubber agroforests in Sumatra, some 300 plant species per hectare have been reported, where adjacent natural forests have 420 plant species per hectare, although species diversity decreases as intensity of planting increases. While bird species in the agroforests may number only half as many as those found in mature forests, virtually all mammal species are present (Nobel and Dirzo 1997; Sibuea and Herdimansyah 1993). In Gabon agroforests, species such as the gray parrot and certain hornbills were more abundant where cocoa agroforests accounted for a greater portion of the overall landscape than where native closed-canopy forests dominated. This was due to the high densities of indigenous fruit tree species, such as the African plum (*Dacryodes edulis*) and oil palm (*Elaeis guineensis*), which bear the favorite foods of several of the bird species found in the agroforests (Gockowski 2000).

Increasing the Economic Value of Perennial Crops on Farms

As forests have been cleared or become inaccessible, farmers have begun growing some native or endemic trees and perennial grasses for a wide variety of subsistence and commercial products, including fuel, construction material, food, spices, medicines, fiber, fodder, and resin, as well as for environmental services. Enormous potential exists for developing perennials for staple food and feed production that can compete economically with arable grains, legumes, vegetables, and feeds on soils unsuitable for continuous cultivation, and on landscapes where key ecological functions were originally played by native forests. One approach is to improve productivity by managing trees in natural stands (and, by selective harvest, modifying natural populations). Where forest is relatively abundant, this approach has considerable potential for high-value products.

The domestication of annual crops began over 10,000 years ago during what is known as the "Neolithic revolution." The process of domesticating and increasing the productivity of trees, shrubs, and perennial grasses—what some refer to as the "woody plant revolution"—began only recently for most species (Leakey and Simons 1998; Leakey 1999b). Genetic improvement programs for selected species, especially endangered or "keystone" species, could make them much more economically attractive components in ecoagriculture systems.

How these crops are spatially arranged and managed determines their value as habitat (Tolbert and Schiller 1996).

A classic and prescient work by J. Russell Smith (1958) proposed that the United States and Canada actively seek to develop "a permanent agriculture" in erosion-prone farming areas based on native tree crops grown for food starch, protein, sugar, oils, bran substitutes, and animal fodder. He noted that forest species like keawe, carob, honey locust, and the mesquites produce livestock food. The chestnut and hickory nut produce human and animal food and could help replace corn in areas not suited to growing that crop. The Persian, eastern, and other walnuts, pecans, and hickories serve as excellent sources of food for humans, and the sugar maple and honey locust provide sweeteners. What was missing, then, was the science and "horse sense" to select highly productive, early-maturing tree varieties, viable management systems, and suitable processing technology.

New horticultural techniques of vegetative propagation developed and applied to tropical trees for timber and nontimber forest products since the 1970s have dramatically reduced the time required to produce and harvest superior varieties, making them more viable for commercial production (Leakey 1999b). Research programs are now under way throughout the United States on a number of the tree crops suggested by Smith, as well as on shade-tolerant crops that can be grown in the understory of farm forests and woodlots. In seeking alternative sustainable energy sources, researchers are developing short-rotation woody crops and perennial grass systems for biomass energy production, using native species such as switchgrass (*Panicum virgatum,* an important component of the native North American tallgrass prairie), native poplars, sweetgums, sycamore, and maples. Similar initiatives are under way in Europe, Australia, China, and other temperate and Mediterranean climates.

In Africa, tree and shrub domestication efforts are already showing success in meeting livelihood security needs of farmers and conserving biodiversity. *Ziziphys mauritania* was threatened from overexploitation in dryland regions for building fences, and it had disappeared from much of the northern fringe of its range in the Sahel. Improved cultivars of *Ziziphys* were developed, focusing on the consumption and market potential for its vitamin-rich fruit. It is now grown in farm hedges for multiple functions. Supported by local and regional markets, farmers in Cameroon already plant African plum (*Dacryodes edulis*) extensively, especially among their perennial cash crops (for example, cocoa and coffee). The development of cultivars using simple vegetative propagation techniques on superior trees is under way. These techniques can reduce the time required for these trees to fruit from about ten to fifteen years to three to five years, making them economically viable for farm production.

Six Case Studies

Described below are six examples of agricultural systems that closely mimic the ecosystem natural to the area. They do so by integrating economically useful perennial plants that also contribute to ecosystem stability and biodiversity.

EXAMPLE 31. NATIVE PERENNIAL PRAIRIE GRAINS ON U.S. FARMS

One approach to reducing the need for frequent field cultivation, and the damage it causes to soils and soil fauna and flora, is to develop perennial grain crops that do not need tillage at all. Researchers in the United States have domesticated wild perennial grains that can be grown in plant mixtures that mimic the structure and function of native prairies (Soule and Piper 1992). These "farm prairies" include combinations of four basic types of perennial plants—warm-season grasses, cool-season grasses, legumes, and sunflowers—grown so that the soil is never laid bare. Perennial forms of wheat bred specially for the dry and highly erodible soils of eastern Washington State (where abandonment of grain land due to soil degradation is widespread) now yield 70 percent as much as the annual wheat varieties. The perennial mixtures also inhibit weeds, resist pathogens, and build soil fertility. Promising species already being grown on an operational scale include the Illinois bundleflower, a nitrogen-fixing legume whose seed is about 38 percent protein; leymus, a mammoth wild rye; eastern gamagrass, a bunchgrass that is related to corn but is three times as rich in protein; and Maximilian sunflower, an abundant oil producer (Pimm 2000).

The perennial mixtures can provide many benefits. By reducing seedbed preparation and cultivation, application of synthetic chemicals, and irrigation, farmers save money on energy and materials. Because the soil is not regularly disturbed, and because of the perennial vegetative cover, these mixed grain fields harbor a high proportion of the native wildlife and many times higher biodiversity than conventional monocultural grain fields (Piper 1999). More widespread adoption of these systems still requires additional research, and the adaptation of machinery for more efficient harvest of mixed grains. But this offers a potentially revolutionary opportunity for low-ecological-impact grain farming, particularly in erosion-prone environments.

EXAMPLE 32. CREATING WILDLIFE HABITAT THROUGH FARMLAND FALLOWS IN AFRICA

Improved tree fallows use species or management practices that accelerate the processes of soil nutrient and organic matter recovery after cultivation that was traditionally achieved through longer fallows of natural vegetation. Improved fallows have the greatest potential in areas where farmers have some experience with the struggle to maintain soil fertility but still have sufficient land to rely on fallow rotations. They are also promising in densely populated areas with two rainy seasons, where farmers can fallow for one season per year. In the mid-1990s, 3,000 farmers in eastern Zambia began to use improved, two-year shrub fallows with *Sesbania sesban* and *Tephrosia vogelii,* while in western Kenya several thousand farmers began using one-season fallows of *Crotolaria grahamiana, C. ochroleuca,* and *C. tephrosia.* Farmers using improved fallows in Zambia increased annual net farm income from corn, their most important enterprise, 2.8 times (Franzel, Phiri, and Kwesiga 2002). In western Kenya, a yield increase of 21 percent from a one-season fallow gave better economic returns than continuous cropping (Swinkels et al. 2002).

Farmers also value improved fallows because they prefer to invest labor rather than scarce cash resources, and they prefer to reduce risk by using home-produced substitutes for expensive purchased fertilizer. Moreover, the benefits of fallows are spread over two to three years, while nitrogen fertilizer provides benefits for only a single year. In the event of a crop failure, the farmer loses only the investment in tree-planting, which is one quarter the cost of an equivalent amount of fertilizer. In addition to providing nutrients (through leaf litter, roots, and, for some species, nitrogen fixation), tree fallows improve the soil's structure and organic matter content, enhancing its ability to retain moisture during drought years (Franzel and Scherr 2002).

Improved fallows not only increase farm productivity, food security, and sustainability but also serve as valuable habitat for wild biodiversity. Shrub and tree canopies provide nesting areas, shade, protection, and other services for birds and small mammals. Short-term fallows can control noxious weeds and pests (Gallagher, Fernandes, and McCallie 1999), reducing the need for chemical control. Increased amounts of organic matter and reduced cultivation favor soil microorganisms. In Nigeria, populations of soil microarthropods (tiny spiders and insects that play an important role in soil fertility maintenance by regulating decomposition and nutrient turnover) were found to be more than a third higher under improved fallows than under natural fallows, and much higher than under continuous annual cropping (Adejuyigbe, Tian, and Adeoye 1999). The more diverse landscape mosaic supports a wider range of wild species than does the more homogeneous landscape of continuous annual cropping.

EXAMPLE 33. INSECT AND BIRD BIODIVERSITY IN SHADE-GROWN COFFEE

Several million hectares of agricultural land in humid tropical uplands are planted in coffee (*Coffea arabica*). Traditionally, coffee bushes were grown under a canopy of native forest trees, which were important habitats for wild biodiversity. Many of the migratory songbirds that breed in North America spend part of their winters in coffee plantations in Central and South America. Coffee plantations that are designed to mimic natural systems may have even higher numbers of species than natural forests (Perfecto et al. 1996). Central American coffee agroforests are second only to moist tropical forests for species diversity, especially for forest "generalists" that can use a variety of forest habitats (Leakey 1999b). Species richness for bats and terrestrial mammals in shade coffee plantations in Veracruz, Mexico, was significantly higher than in adjacent agricultural fields, though lower than in forests (Lacher et al. 1999). Shade-grown coffee plantations harbor far more species of invertebrates such as beetles and ants than do monoculture stands of unshaded coffee.

But many coffee plantations are now intensifying production with higher-yielding sun-tolerant coffee varieties that do not require the overhead tree canopy, thereby reducing the area of habitat available to the winter migrants. Sun-grown coffee is kept shorter and planted at high densities. Time to first harvest is shorter, but plantation life span is less, use of agricultural chemicals is greater, and risk of erosion is higher. High input costs make sun-grown coffee nearly 50 percent more expensive than shade-grown coffee, even without considering environmental costs, though yield is also much higher. Sun-grown coffee has been widely promoted and adopted. But researchers concerned with agricultural sustainability and biodiversity have developed technical and business innovations to improve the profitability of shade-grown coffee systems. One approach is to establish native shade tree species in coffee plantations that also have high economic value as timber. For example, researchers in Central America have developed a system for growing *Cordia alliodora* in coffee at densities of 100 trees per hectare. The system achieves very high growth rates, produces comparable coffee yields, and raises farm income substantially (Beer, Ibrahim, and Schlonvoigt 2000). Other researchers have actively promoted marketing of environmentally friendly coffee, which provides a financial premium to farmers. The Rainforest Alliance has launched an ECO-OK certification program for coffee, and Conservation International has a similar Sustainable Coffee Initiative with sites in Mexico, Guatemala, Colombia, and Peru (Lacher et al. 1999). The potential market for shade-grown coffee exceeds U.S.$500 million globally at current retail prices (Giovannucci 2001).

EXAMPLE 34. SAVING AN AFRICAN MEDICINE TREE

Western science discovered in the mid-1960s that the bark of the tree *Prunus africana* was an effective treatment for some prostate disorders. Burgeoning demand threatened the slow-growing species—found only in the moist highlands of Africa—with extinction. In Cameroon, bark production rose from 200 tons in 1980 to a peak of 3,100 tons in 1991, from which it has since declined due to scarcity. Approximately 80 percent of the mature trees on Mount Kilim, in Cameroon's western highlands, were killed off by unscrupulous harvesters. It is estimated that if destructive harvesting continues unabated, many populations of the tree will be extinct in the wild within five to ten years. In 1995, the tree was added to Appendix II of the Convention on International Trade in Endangered Species (CITES), which requires strict regulation of trade under a licensing regime. A related biodiversity concern is that the tree's fruits are a major food source both for a large number of rare and threatened endemic birds of the mountain forests of Cameroon (including Bannerman's turaco, *Tauraco bannermani,* and Cameroon mountain greenbul, *Andropagus montanus*) and for a threatened primate (Preuss's gueron, *Cercopithicus preussii*) (Leakey 1997).

Since attempts to manage the tree sustainably in the wild seemed doomed, researchers began a concerted effort to domesticate *Prunus africana,* so that it could be profitably and sustainably grown and harvested on farms. The tree grows well in the open and is therefore adapted to intercropping; farmers can grow the tree in diverse niches, even on small farms. Long maturation and seed problems, however, made domestication difficult. New techniques of molecular analysis developed for agricultural research have accelerated progress in selecting for improved traits (for example, faster growth, more reliable seed production, seed viability) and grafting to reduce time to maturity in seed orchards. Wild germplasm has been collected from many stands in Cameroon and Kenya, and seed orchards have been established. High-quality tree germ plasm is being grown by farmers in six villages in Kenya. The farmers are being taught nondestructive methods for stripping the bark, and researchers are helping them to identify the most profitable marketing opportunities, thus achieving more secure incomes as well as biodiversity conservation (ICRAF 1999). It is hoped that farm-grown trees will supply most of the raw material for economic use of the product, relieving pressures on the natural populations and protecting the species that depend on it. Planted trees may also serve as supplemental food sources for wild species living in or around farms near protected areas or wildlife corridors.

EXAMPLE 35. SPECIES-RICH AGROFORESTS IN INDONESIA

Agroforests located in rainforest ecosystems are complex, multistrata mixtures of high-value commercial perennial crops with other subsistence, commercial, and wild species that closely mimic the structure of the humid rain forest. They cover millions of hectares in Southeast Asia, Sri Lanka, Latin America, and West Africa, most notably for the production of tea, coffee, rubber, and cocoa, but also for many minor products.

About 3 million hectares of agroforests are found today in Indonesia (examples are shown in Figure 7.3). They were developed through local initiative over the last century in two ways. Farmers modified the nearby primary forest to encourage greater production of marketable products or else introduced valuable trees into cropped land or early stages of the traditional fallow. Farmers with agroforests also typically produce staple foods in irrigated rice fields or swidden farming. These agroforests currently account for between 50 and 80 percent of total agricultural income of villagers in some rubber-producing provinces, such as Jambi (Sumatra). They also provide food and nutritional security and diversity; raw materials for construction, tools, and furniture; fuelwood; and capital assets. For example, the rubber agroforests near Gunung Palung National Park produce seeds from *Shorea stenoptera,* durian fruits, rubber, and timber, especially from the Bornean ironwood (*Eusideroxylon zwageri*), forming a mosaic of rubber gardens, fruit gardens, and dry rice fallows. A study of this system (Lawrence, Leighton, and Peart 1995) found that even though total abundance of durian and *Shorea* was greater in the primary forest, villagers gathered these products only from more accessible managed forests near the village. Agroforests are sustainable, profitable to farmers, and economically important. Just the rubber from agroforests (a quarter of the world's natural rubber) is valued at U.S.$1.9 billion (De Foresta and Michón 1994).

As to biodiversity benefits, rubber agroforests contain 250 to 300 other plant species. Study plots in the complex, multistrata damar agroforests of Sumatra dominated by resin-producing dammar trees suggest that a very high proportion of the regional pool of resident tropical forest birds, mammals, soil flora and fauna, and local plants occurs in these agroforests (Leakey 1999a). Orangutans, an endangered species, have been sighted in some agroforests, as

Figure 7.3. Cinnamon agroforest-rice mosaic in Sumatra, Indonesia.
Source: R. R. B. Leakey.

have footprints of the rare Sumatran rhino. Given the decreasing availability of land, there is considerable potential (already being exploited by some farmers) for establishing such agroforests to rehabilitate degraded *Imperata* grasslands that occur over large areas of deforested land in the region. The forest-like appearance of some agroforests led to claims that they were natural forests and therefore belonged to the state, leading to conflicts with the local people and damage to the forests. Recent legislative changes in Indonesia now recognize the managed nature of agroforests and grant farmers land rights over them (Tomich et al. 2001).

EXAMPLE 36. RAISING THE ECONOMIC VALUE OF
TROPICAL FOREST FALLOWS IN LATIN AMERICA

Shifting cultivation continues as the economic mainstay of rural communities in less densely populated regions of many developing countries. However, the imperative to evolve more permanent forms of land use has grown with rapid population growth, official delineation of remnant wildlands as protected areas, and state policies that encourage sedentary agriculture and discourage the use of fallows and fire. One strategy used by shifting cultivators is to enhance the value of the forest fallows, introducing or increasing the productivity of existing perennial species of economic value. Harvested products from these "enriched fallows" provide a stream of products and income to supplement crop production and discourage forest conversion (Cairns and Garrity 1999). By mimicking the structure and ecological function of the natural forests, shifting cultivators retain many forest species. Enrichment strategies can be used in both secondary forest fallows (forests that regenerate after initial clearing for crops and which farmers expect to clear again), and permanent secondary forests maintained for products or environmental services or because land is unsuitable for farming.

CIFOR researchers found that such secondary forests already play an important economic role in agricultural systems of colonists in the rainforests of Brazil, Nicaragua, and Peru. Products from these forests include firewood, which is used for processing cassava flour and provides two-thirds of total household cash income on Brazilian farms. Other products—roundwood, palms for home construction, game animals, fruit, and medicines—also make important contributions to livelihoods. The market value of the products—mainly fruit and charcoal—is typically 10 to 20 percent of total income. In Nicaragua, permanent secondary forests are maintained for production of the medicinal plant ipecac (*Cephaelis ipecacuana*), as well as for supplying firewood, protecting water courses, and providing fencing materials for pastures (Smith et al. 2001).

Strategic selection and management of commercial species can greatly enhance the profitability and viability of enhanced fallow systems. Capacity to resprout after fire or cutting, as shown by the trees *Tabebuia* spp. and *Caycophyllum spruceanum* in Peru, facilitates silviculture. Fire-resistant palm species that provide useful products and persist during the agricultural phase are also valuable. These include the palms *Scheelea basleriana* and *S. tessmannii* in Peru, *S. gracipes* in Bolivia, and the timber tree *Guazuma crinita* in Peru. Synergies between forest products and agricultural incomes can be exploited, as in cases of naturally regenerated timber species (such as *Schizolobium amazonicum* or *Hampea popayanensis*) that provide shade for coffee. Short production cycles (capable of producing a marketable product within the fallow period) are an obvious advantage. *Cordia* spp. and *Calycophyllum spruceanum* can produce building poles within four years of establishment. Tolerance of shade in plants other than trees is important. For example, *Cephaelis ipecacuana* (ipecac) is a shade-tolerant understory shrub cultivated under forest cover in Nicaragua. Management of trees for timber is appropriate where markets exist. Site conditions must favor a high density of fast-growing pioneer tree species, or else provide good conditions for regeneration of valuable timber species remaining from the primary forests. Farmers can use silvicultural practices to favor commercial species, such as direct seeding during the crop phase, thinning, and weeding. Enrichment planting and management techniques can enhance germination, survival, and growth of planted trees (Smith et al. 2001).

Potential Application

Of the six examples above, one is from a temperate grassland ecosystem (United States), one from a tropical woodland savanna (Zambia), and the others from humid tropical forest ecosystems. In the African fallow system, the perennial species provide principally a service role; in the other five cases—the perennial grains, coffee, African medicine tree, agroforests, and enriched forest fallows—perennial species are the principal source of income in the fields where they are present. The perennial tree crops are all much higher in value than the annual crops grown by the same farmer. In all of these examples except the indigenous Indonesia agroforests, scientific research played a critical role in systems development, and the systems are still evolving.

All but the prairie grasses are being used by a large number of farmers, yet extension systems for their further dissemination and adaptation are still quite inadequate. While the short-fallow system is fairly simple and inexpensive to adopt, the other systems involve substantial new management skills as well as relatively long time periods for commercial products to be ready for market. All of the systems can, however, be adopted incrementally, as there are no significant economies of scale. As with all of the ecoagriculture systems, wildlife benefits are greatest where farmers have adopted these measures over large areas with intentional ecosystem planning. Unlike many other practices, however, there are likely to be major benefits for many wildlife species from adoption even on limited areas, as such farms can serve as "islands" of protection because habitat qualities are so high.

From the perspectives of biodiversity and other ecosystem functions, this type of agricultural system is especially attractive. However, significant investment in research will be necessary to raise the productivity of previously neglected perennial crop species and to make them economically competitive with annual cereal and legume crops that have benefited from centuries, if not millennia, of crop improvement, well-developed market and processing infrastructure, and often from public subsidies. In many cases, new investment will also be required in extension, education, market development, and policy adaptation for new perennials to be adopted on a widespread basis. New and lower cost research methods now make this challenge much more feasible.

Chapter 8

Coexisting with Wild Biodiversity in Ecoagriculture Systems

Chapter 4 described farmers' wariness of having wildlife in and around their farms, originating from the potential threats posed by wildlife to their harvests, their livestock, or their families. Farmers' actions stemming from such concerns have been partly responsible for the depletion of biodiversity in many agroecosystems. In ecoagriculture, farmers are asked to accept much higher and more diverse populations of wild plant and animal species. While most wild species either provide benefits to farming and rural populations or have neutral impacts, it is important to note that with some species peaceful coexistence requires careful management or even separation. This chapter briefly highlights some approaches that have been successfully used to manage interactions between people and wild biodiversity in agricultural areas.

Most Wild Species Have Beneficial or Neutral Effects

The vast majority of wild species have either beneficial or neutral effects on farm production and human settlements, especially through forming parts of well-functioning ecosystems (see Box 2.2). Wild predators help to control agricultural pests (USDA 1994), nitrogen-fixing bacteria help to fertilize irrigated rice fields, soil microorganisms help to maintain soil fertility and cycle nutrients, and many insects serve as pollinators (especially of fruit trees); numerous other services in support of ecosystems could also be listed (Daily 1997).

Thus the role of biodiversity in maintaining ecosystem services is essential for supporting sustainable forms of agricultural development, and wild biodiversity also provides considerable direct benefits. These benefits include provision of meat, timber, energy, fur, fruit, medicinal plants, pets, grazing, and so forth (Heywood 1999). One can, of course, argue that all species also have an ethical right to continued existence, a view endorsed by the authors. The question is often one of balance and trade-offs. Table 8.1 summarizes the main types of benefits and costs of wild species to farmers.

It also appears that many of the most destructive wild species are not native to the area where they cause the most problems to humans. These include many disease organisms, weeds, insects (for example, Mediterranean fruit flies in California or Australia), and vertebrates (such as rats). Such pests are considered invasive alien species and are subject to vigorous efforts at prevention, eradication, or control (McNeely et al. 2001). They are not considered part of native wild biodiversity.

Approaches to Wildlife Management

Many traditional farmers recognize the multiple benefits of wild biodiversity and have developed cultural means of conservation (e.g., McNeely and Wachtel 1988). But modern pressures of growing human impacts on wild habitats often undermine traditional relationships, requiring new approaches to conflict resolution. Ecological research over the past few decades has shown that strategic interventions can often significantly reduce the number of conflicts with resident or visiting wild species.

Using research to identify potential problems is an important part of ecosystem planning, and an essential element of the ecosystem management process is monitoring of farm-wildlife interactions to enable corrective measures to be taken. Especially promising ways to enhance agriculture-wildlife coexistence are strategies whereby local farming populations benefit directly from the presence of wildlife in their landscapes, as through sharing of ecotourism revenues, direct harvesting of wild products, public assistance with wildlife control measures, and payment for biodiversity services provided (Kiss 1990).

In the cases where wild biodiversity does cause problems to agricultural operations or farm households, these must be addressed. Measures that have been implemented successfully in various parts of the world include community involvement in decision-making; improved management of livestock; improved management of agricultural lands; excluding, deterring, or scaring wildlife; managing pest populations to restore ecological balance; and paying compensation. These measures are examined below.

Table 8.1. Some Potential Costs and Benefits of Wild Biodiversity to Farmers

	Costs	Benefits
Mammals	Feed on crops (bats, monkeys, pigs, rodents, deer, elephants, rhinos, other herbivores)	Disperse seeds (bats, elephants, other herbivores)
	Prey on livestock (foxes, lynx, wolves, leopards, bears, other carnivores)	Pollinate trees, crops (bats, squirrels)
	Transmit diseases (rodents, herbivores)	Reduce number of crop-raiding herbivores (large predators)
	Prey on fish (seals, dolphins)	Attract tourists (large mammals)
	Destroy structures (elephants, other large mammals)	Provide source of meat, skins Provide genetic material (wild pigs, cattle)
Birds	Feed on fruits, grains, grasses (parrots, weaverbirds, geese, etc.)	Pollinate trees, crops
		Prey on livestock (eagles on lambs, hawks on chickens, etc.)
		Prey on insect pests
		Attract tourists
		Provide source of meat, eggs, feathers, pets
		Prey on rodents (owls, hawks, etc.)
Reptiles	Prey on fish (e.g., crocodiles)	Prey on rodents (snakes)
		Prey on insect pests (lizards)
Soil Organisms	Cause of soil-borne diseases (e.g., *Phythopthora* causes late blight in potato, *Fusarium* associated with Phylloxera in grape)	Biodegradation (cycle nutrients)
		Improve soil structure and water absorption
		Help sequester carbon
		Contribute to soil fertility
Insects	Feed on crops	Pollinate trees, crops
	Transmit diseases	Prey on insect pests (insect predators and parasitoids)
		Cycle nutrients
Trees, Shrubs, Grasses, Annuals	Compete with crops for sunlight, water, nutrients; invade crop fields and pastures	Enhance crop production through wind protection, shade, litter fall, soil conservation
	Provide hosts for crop or livestock diseases or pests	Provide useful foods, fodder, fuel, construction materials, etc.

Community Involvement in Decision-Making

Human-wild species conflicts often arise when local communities are not involved in wildlife management plans and their specific needs not considered. Above all, people want to protect their crops and livestock from damage by wild species. Local needs can be identified by communities themselves, as part of developing and implementing their own conservation initiatives. This would typically include participatory appraisal of problems and opportunities, collaborative management (an agreement that specifies the respective role, responsibilities, and rights of the different stakeholders in management), and participatory monitoring and evaluation (Borrini-Feyerabend 1997). One example to limit the conflict between people and predators occurred in the Mountain Areas Conservation Project (MACP) in northern Pakistan, where project staff helped local people themselves write the wildlife conservation plan for their region. Locals are now very concerned with development and conservation in their region. As one result, they no longer persecute endangered snow leopards (*Panthera uncia*), even when these predators kill their livestock (Campbell 1998).

Modified Management of Livestock

Some modifications in livestock husbandry can reduce attacks by predators. For example, lambing and kidding put newborns and mothers at high risk; moving the process to sheds provides protection from predators. Other such modifications include putting bells on sheep; fencing (species-specific requirements); proper disposal of livestock carcasses so that predators do not acquire a taste for livestock; keeping, rather than selling, experienced herd lead animals that can teach appropriately cautious behavior to younger animals; using guard animals (such as with donkeys in sheep and goat flocks); permitting wild prey species to intermingle with livestock; and using repellents and frightening devices (Nowell and Jackson 1996; USDA 1994). The corralling of sheep and goat flocks during the night in Africa helps to reduce attacks by lions, since lions attack mostly at night to avoid encounters with humans (Mazzola, Graipel, and Dunstone 2002). The use of guard dogs is an effective method to reduce losses due to predators, and they remain important parts of most pastoral systems. In the Abruzzo region in Italy the use of sheepdogs allows wolves and people to live harmoniously (Klaffke 1999). Guard dogs are also being used successfully in some areas of Europe where flocks are subject to lynx attacks (Stahl et al. 2001; Vandel and Stahl 1998).

It is also important to monitor and manage infectious and noninfectious diseases that may spread between humans, domestic livestock, and wildlife. When designing ecoagriculture systems, it is advisable to screen for health problems and for the presence of toxic agents. When screening reveals such agents, it is important to intervene actively to maintain domestic livestock health (to prevent transmission to wildlife) and control wildlife diseases where these pose a threat to conservation or human or livestock health (Deem, Karesh, and Weisman 2001).

Modified Management of Agricultural Lands

Some wild pests, such as insects, can be controlled by managing habitat to encourage their predators. For example, farmers concerned about insect pests can encourage the presence and activity of insect-eating birds. Possible management measures include providing suitable cover (fencerows, hedgerows, shelterbelts, woodlots, grassed waterways, and native or scrubby vegetation along roadside edges and field margins). Farmers can alter field size and shape (for example, smaller or long narrow fields support more hedge around their edges and consequently more birds); reduce pesticide use; reduce the frequency or change the timing of mechanical operations (to avoid destruction of bird eggs); or farm organically. They can provide nest boxes around farmsteads, for instance for barn swallows and purple martins (Kirk et al. 1996).

These techniques are also friendly for beneficial insects. For example, a diversified agricultural landscape increases the parasitism rate of various troublesome insects such as the pollen beetle (*Meligethes aeneus*). The distribution of diverse insect pests is affected by the edge habitat of the field. In some cases, plants disliked by a particular pest can be planted around fields where the pest occurs. For example, plum curculio overwinters outside the crop under the litter cover of the woody surroundings and is therefore more abundant during the growing season on the edge of the field. Conifers are unfavorable to the wintering survival of plum curculio, so an orchard surrounded by such trees will be less damaged. Various plants, such as buckwheat, mustard, anise, and garlic, have repulsive properties for specific insects. To enhance the effect of beneficial insects, it is important to provide alternative food resources and refuges for all stages (usually adults and larvae feed on and use different sources or hosts). An appropriate management of the field landscape should consider the ecology of the species concerned.

Excluding, Deterring, and Scaring

Various techniques to prevent wildlife from entering croplands are used by farmers, with variable success. Asian and African farmers use different measures to scare, deter, or exclude elephants, rhinoceros and wild pigs from crops. For example, electrical fencing reduces elephant damage at the community level. Techniques to prevent or at least reduce bird damage on crops include netting the crop (which can be very expensive and require extensive labor), scaring techniques such as scarecrows, and broadcasting bird distress calls. However, some birds can habituate rapidly to these techniques. Box 8.1 describes techniques that have been used to control crop damage by wild geese. Near large wildlife reserves, digging trenches has proven effective in discouraging elephants, but only where they are well designed so that elephants do not manage to cross (Nepal and Weber 1992; Sukumar 1986). (See, for

---

Box 8.1. Dealing with Excess: The Problem of Geese in Europe and North America

The protection of geese by laws that have reduced hunting in Europe and North America has led to a substantial increase of the populations since 1960. Farmers now face a conflict with these wintering populations. The majority of the geese concerned are Pink-footed (*Anser brachyrhynchus*), Greylag (*A. anser*), White-fronted (*A. albifrons*), and Brent (*Branta bernicla*) in Europe, and Snow (*A. caerulescens*) in North America. In Europe, the reduction of the traditional wintering habitat of the geese by human activity, such as wetland drainage, has increased use of croplands and in some cases a shift to these lands. Conflicts with farmers are simultaneously rising in many localities where geese cause damage on pastures, cereals, and high-value cash crops. The damage to cereals is due mainly to Greylag geese, costing U.S.$250,000 to $300,000 in one year in the Netherlands; the Brent goose causes localized losses from 6 to 10 percent of crop yield in England. In North America, the main causes for the population increase are the development of agriculture near coastal salt marshes (the traditional goose habitat) and the resulting loss of many coastal marshes. Post-harvest plants and leftover grain and cereal offered by the agricultural fields provide food for geese throughout the year, inducing a population increase.

Solutions for the conflict with European farmers include the creation of refuges for the wintering geese and scaring them away from croplands. Refuges can be made more attractive by mowing and fertilizing with nitrogen in late autumn, when not situated in a nitrate-sensitive area. Legislation could be relaxed to allow shooting in particular contexts (on agricultural land, and only for the Pink-footed and Greylag Geese, which are the most numerous; this measure could be cancelled in case of significant population decline). Compensation, even if difficult to evaluate, is being instituted in many countries. In North America, managers are considering commercial harvesting of snow geese for human consumption, and trapping and culling geese on migration and wintering sites. The example of geese demonstrates the complexity of wildlife management in relation to agriculture, the need to maintain flexibility and creativity, and the importance of monitoring and research.

Sources: Ben-Ari 1998; van Eerden 1990.

example, the Internet Center for Wildlife Damage Management at http://www.ianr.unl.edu/wildlife/solutions/handbook.) Techniques for excluding or scaring smaller animals include odor repellants (e.g., the scent of a predator), scarecrows, ultrasound devices, and nets or screens.

In developing areas that overlap with elephant habitat, procedures sensitive to the conservation status of elephants should be instituted. These passive measures to facilitate or block elephant movements include buffer zones, areas of inhospitable land between the protected area and croplands that have low attraction for the elephant (including canals, the banks of reservoirs, and human-made structures.). Elephant habitat can be enriched by providing diverse vegetation, water sources, foraging areas, or minerals to draw them away from crops and to help establish migratory routes that avoid isolation of elephant groups (Seidensticker 1984).

Restoring Ecological Balance

While ideally nature should be allowed to take her course, many landscapes shared by wild and agricultural species call for more intensive management. Grazing regimes may need to be modified to permit the control of some types of weeds or bird and insect pests, by establishing plants that provide alternative food and water sources. Many pests can be controlled by managing pest-predator populations. The recycling of resources and judicious land use can reduce the negative impacts of some wild biodiversity while increasing positive impacts of other wild biodiversity. Where large predators are absent from an area, some species of deer and other mammals may become too numerous for the ecosystem to support, forcing them to feed on fields and gardens to avoid starvation. Ecological balance can be restored in problem regions by culling programs.

Where large predators are present in agricultural landscapes, they can be controlled in a variety of ways. Selective removal of animals appears to be effective in the case of problem tigers (Dinerstein et al. 1999). However, a study in the French Jura region found that removing lynx selectively had no impact on reducing the attacks on sheep (Stahl et al. 2001). While efforts to eradicate wild cats have failed, control of specific individuals has been more successful. Large predators that are preying on livestock can be targeted by fitting a collar containing a capsule of powerful poison under the throat of a domestic animal, where a predator is likely to bite. Another approach now being tested is injecting the carcasses of livestock killed by leopards with the nauseating substance lithium chloride; this has proven successful in preventing further livestock killing in at least one case (Nowell and Jackson 1996). Over the last fifty years, popular and scientific attitudes toward predators have changed as the ecological benefits of conserving them have become more fully understood.

Paying Compensation

Where conflicts are unavoidable, mechanisms must be put in place to compensate farmers fairly for their losses. While in the past farmers were sometimes paid to destroy wildlife, such as owls, hawks, wolves, and other predators, today they are more likely to be compensated for conserving biodiversity. In Europe, compensation to farmers for livestock losses by wolves and lynx reduces the human–wildlife conflict (Klaffke 1999; Vandel and Stahl 1998). In India, compensation is paid to families of people killed by elephants (Sukumar 1986). On the other hand, in Africa, where monetary compensation schemes for elephant damage do exist, they are not always useful in resolving conflicts due to inadequate funds, difficulty in quantifying damages, or lack of change in farmer behaviors that increase the risk of conflict (Taylor 2000).

An ideal compensation mechanism should have several features. It should have a management scheme acceptable to the local community, involve a direct financial incentive, be managed by a committee including local representatives, and include close liaison between local people and government agricultural and wildlife agencies. Such schemes may have some drawbacks though. Compensation systems may be expensive, slow to administer, open to considerable abuse or blatant corruption, and cause disputes, or members of the management team may use their position for political advantage, leading to loss of faith in the system. In some cases, compensation may reduce the incentive for self-defense by farmers (and therefore could even aggravate the scale of the problem). Farmers may accept compensation but continue to kill wildlife. An alternative approach that might be considered is private or public insurance for losses of crops or livestock. Other policy tools regarding payments for biological services are discussed in Chapter 10.

Conclusion

Many approaches to management of wild biodiversity are available to farmers, and these will be adopted when it is in the interest of the farmer to do so. Most often, a combination of approaches will be required. Wild biodiversity is dynamic, so research is essential to devise and document the efficacy of control methods for specific species and agroecosystems, and to design new and more effective ones.

PART III

Policy Responses

It seems clear that ecoagriculture systems could make a considerable contribution to the goals of increasing food supplies, reducing poverty, and protecting wild biodiversity. But this potential is sharply constrained by current policy and institutional frameworks that have been shaped historically to segregate conservation and production land uses, and to emphasize short-term productivity goals for agricultural systems. Policy and institutional innovations are needed to make the transition to ecoagriculture feasible on a large scale.

Chapter 9 discusses the major policy changes needed. Price and trade policies need to be modified to reduce distorted incentives for land and input use. Legal frameworks need to be adjusted by addressing ecoagriculture in international conventions, with regulatory reform to enable more flexible types of land management, and changes in property regimes to protect local rights and increase options for farming and conservation to be undertaken jointly.

Chapter 10 discusses innovations in the marketplace that could provide greater economic incentives for farmers to practice ecoagriculture. Major approaches include technical innovation to raise profitability, new markets for products from ecoagriculture systems, and payments to farmers for biodiversity conservation.

Adoption of ecoagriculture calls for evolution of key environmental and agricultural institutions. Chapter 11 describes ways to incorporate ecoagriculture into ecosystem planning and management at a regional scale. It discusses how to integrate ecoagriculture into support institutions such as research, monitoring, extension, and support to farmers' organizations. New ways to finance investments in ecoagriculture are explored, through government, development banks and agencies, conservation organizations, civil society, and the private sector.

Finally, Chapter 12 assesses the extent to which ecoagriculture could be adopted and points to existing problem areas where policy-makers are most likely to find it an attractive solution. It closes with a call for global mobilization of leadership, research, and investment to promote ecoagriculture.

Chapter 9

Policies to Promote Ecoagriculture

Effective policy support is essential to making the transition to ecoagriculture on a scale that would produce a meaningful impact on biodiversity. This chapter presents some of the most promising policy innovations being developed and tested around the world, especially those relevant to developing countries.

Reframing Policy Goals

For the reasons explained in Part I, most developing countries need to continue placing strong emphasis on expanding agricultural supplies through domestic production. Governments need to aggressively pursue productivity increases in ways that directly benefit low-income producers, especially in the marginal lands where two-thirds of the rural poor reside. In such areas, sustainability concerns will necessarily be prominent, requiring widespread local and public investment in land, soil, and water improvements. At the same time, it is also critically important to protect and enhance ecosystem services produced in agricultural areas, such as reliable water supplies, waste cycling, and agricultural support services—especially in more densely populated regions. To the extent such countries wish to conserve their national heritage of wild biodiversity and contribute to conserving globally important resources, their efforts must focus in and around agricultural regions.

To achieve these multiple goals—agricultural supply, poverty reduction, ecosystem services, and wild biodiversity conservation—national policy goals must be reframed. Agricultural, environmental, and rural economic development strategies must be integrated so as to jointly support these goals. Policies must help to "green the middle," not just "green the edge" of farming, ranching, forestry, and fishery activities.

Price and Trade Policies

Agricultural price and trade policies have often worked to the detriment of wild biodiversity and natural habitats, as these considerations were overlooked in developing such policies. Subsidies and development strategies were designed to promote production, thus encouraging land conversion, while ignoring or at least undervaluing environmental goods and services. Such policies influence the land-use incentives for farmers and must be reformed if large-scale biodiversity conservation is to take place in agricultural regions. Shifting relative prices is necessary, though seldom sufficient, to promote conservation practice.

Agricultural Prices and Subsidies

Prices for agricultural products and inputs indirectly affect biodiversity in agricultural lands in various ways. A price decline for farmers' products commonly reduces incentives for good land husbandry, while sustained high prices (though not short-term highs) generally result in better land husbandry. For decades, most developing countries in Latin America, Africa, and Asia pursued agricultural product pricing policies that reduced farmgate commodity prices overall, either to collect greater agricultural tax revenue or reduce urban food costs to keep industrial wages low. Various subsidies and tax incentives encouraged land-clearing for agriculture and livestock production.

The imposition of export taxes on high-value tree-crop products has also been a disincentive for farmers to adopt what are often more sustainable perennial-based systems. An example of this is described in Box 9.1. Forest product prices paid to local producers have often been controlled by governments and kept artificially low to encourage national industries or provide revenue to forest agencies. Economic liberalization processes during the 1980s and 1990s reduced direct government intervention in pricing of forest as well as agricultural products, including the partial delinking of farm subsidies to commodity markets in many developed countries (OECD 1998). The impacts of such reduced intervention on biodiversity are not yet clear.

Myers and Kent (2001) estimate that subsidies promoting economically and environmentally damaging resource exploitation in agriculture, forestry, mining, and other activities cost nearly U.S.$1.95 trillion per year globally. Subsidies for agricultural chemicals in many countries have led to overapplication of chemicals and the abandonment of more sustainable alternatives. Reducing such subsidies can often indirectly benefit biodiversity in agricul-

---

Box 9.1. Leveling the Playing Field for Biodiversity-friendly Production Systems—The Case of Rattan in East Kalimantan, Indonesia

Rattan, a climbing palm cultivated in East Kalimantan, Indonesia, as part of the traditional forest fallow–based agricultural system, has long played an important role in the local and national economy. Rattan gardens are essentially secondary forests with a higher than normal population of high-value rattan, medicinal plants, fruit, and other food species. Rattan has provided an important source of cash income for farmers, and the raw material has been a major source of internationally traded rattan and, more recently, a strong domestic handicrafts industry. The system is well-adapted to the local ecological conditions. Biodiversity in rattan gardens is much higher than in surrounding areas under agricultural cultivation or in plantations with crops such as oil palm, though it is somewhat lower than in nearby high forest. However, farmers have reduced rattan cultivation over the past two decades, largely because of policy-induced changes that put rattan at a disadvantage relative to other agricultural products. Government policies designed to encourage the domestic processing industry and a strong manufacturing association sharply depressed demand and prices. Roads, industrial plantations, mining, and other new economic activities have both displaced existing rattan gardens and offered attractive alternatives that have led some rattan farmers to shift to new activities. But recent policy reforms are beginning to level the playing field. Researchers have confirmed that because rattan is cultivated, and a new "crop" can be brought into production fairly quickly (seven to ten years to first harvest), high production can be encouraged without fear of overexploitation. Thus, cultivated rattan should be treated more like an agricultural commodity than like a wild forest product. Researchers also recommend that rattan be exempt from the high export tax and from all fees charged on forest products. Taxes have already been cut, and domestic industry control has been loosened (Belcher 2000).

---

tural regions, even without more direct policy action (see Box 9.2). Provision of water for irrigation free or nearly free of charge to farmers has encouraged unsustainable and wasteful water use in many irrigated areas, depriving wild habitats of their share and leading in some places to soil salinization. Charging a price for water closer to its real economic value promotes more efficient consumption (Rosegrant and Schleyer 1995). Policies intended to assist farmers in North America and Europe are beginning to shift from commodity and input price supports—which distort commodity markets as well as having negative environmental impacts—to subsidies and payments for environmental benefits, which will be discussed in Chapter 10 (USDA 2001).

Box 9.2. Benefits of Reducing Pesticide Subsidies in Indonesia

Between 1977 and 1987, Indonesia's rice crops were threatened by three outbreaks of brown planthopper despite heavy application of pesticides provided at
subsidized prices. A new approach clearly was needed, so in 1986 a presidential
decree banned fifty-six brands of rice pesticides and established a national integrated pest management (IPM) program. Subsidies for remaining pesticides were
abolished. Between 1987 and 1997, no national outbreaks of brown plant hop-
per occurred. A survey of 2,000 Farmer Field School graduates carried out in
Indonesia in 1993 found that rice yields had increased by an average of 0.5 ton
per hectare, and that the number of pesticide applications had fallen from 2.9 to
1.1 per season. The cost of pesticide application had been reduced by more than
half. Eliminating pesticide subsidies saved the Indonesian government more than
U.S.$1 billion in ten years. Rice fields under IPM are being recolonized by plant
and animal species previously suppressed by pesticide use (Fliert 1993, reported
in Reijntjes, Minderhoud-Jones, and Laban 1999).

Trade Policies

Trade policies influence demand for land and land conversion, quality standards for produce, and market demand for agricultural products grown in a
biodiversity-friendly way. For example, promotion of livestock exports in
Latin America in the 1960s and 1970s, using policy instruments such as land
tax reduction and feed subsidies, encouraged large-scale deforestation for mar-
ginally profitable ranching activities. When these subsidies were removed,
much of this land reverted to forest (Kaimowitz 1996). But trade opportunities can also benefit biodiversity. For example, overseas demand for new products from perennial cropping systems can encourage the conversion of steeply
sloping lands from annual crops to a more sustainable and biologically richer
perennial cover. This allows countries and regions with mountainous topography to make profitable use of steep lands without harming biodiversity.

   It is difficult to generalize about the impacts of agricultural trade on wild
biodiversity, as those impacts so often reflect local ecological conditions, farming practices, and institutional arrangements. Thus the development of overseas markets for high-value vegetables can, in some areas, accelerate intensification in vegetable plots while allowing extensive lands to revert to natural
habitats. In other areas trade sharply threatens wild biodiversity, as vegetables
are planted along unprotected streambeds or in forests newly cleared for growing vegetables. Export opportunities may make it economically feasible for
farmers to implement conservation farming techniques or, alternatively, force
farmers to overuse pesticides in order to produce blemish-free products
(Thrupp, Bergeron, and Waters 1995). Opportunities to export tea and coffee

have enabled large numbers of producers on small farms in the Kenyan highlands to intensify their production sustainably, while large operations using monoculture and chemical-intensive farming for banana exports in Honduras have harmed the environment. To achieve specific biodiversity goals requires carefully tailored intervention.

Trade policies may be beneficial in cases where good land husbandry or biodiversity conservation can be recognized as legitimate commercial attributes under trade rules, but such cases are not yet common. Conflicts are already arising between international trade rules and global environmental rules, such as the Convention on Biological Diversity and the Convention on International Trade in Endangered Species (CITES), in relation to environmental certification of forest and agricultural products (Caldwell 2001). These will need to be worked out to promote ecoagriculture, especially to ensure that legitimate products produced through ecoagriculture can be clearly labeled and marketed internationally, and to ensure that payments to farmers for biodiversity services are not treated as agricultural subsidies, which are restricted under trade agreements.

Legal Frameworks and Regulations

Environmental regulations have played a central role in promoting the use of more ecologically "friendly" agricultural systems. They have been most effective in developing countries with strong political legitimacy and enforcement mechanisms, mainly for habitat protection through zoning and pollution control. A number of international conventions negotiated in the past twenty years provide a broad framework for national regulatory systems. Direct regulation, which requires strong public institutions and reliable scientific backstopping, has sometimes proven to be an unwieldy and expensive policy instrument. Decentralization of regulatory authority to local actors and greater flexibility of regulatory instruments are increasingly used to reduce costs and enhance local "buy-in."

International Conventions

Several international conventions have been developed to address concerns in relation to biodiversity. The most central is the U.N. 1992 Convention on Biological Diversity (CBD). The CBD has begun to focus more attention on biodiversity in agricultural regions. In decision IV/6 the 1998 Conference of Parties to the CBD suggested: "Governments, funding agencies, the private sector, and nongovernmental organizations should join efforts to identify and promote sustainable agricultural practices, integrated landscape management of mosaics of agriculture and natural areas, as well as appropriate farming systems that will reduce possible negative impacts of agricultural practices on biological diversity and enhance the ecological functions provided by biological diversity to agriculture." In implementing this decision, the 1998 Conference identified a series of measures that can be taken to improve the relationship between agriculture and wild biodiversity. Their recommendations include:

- Identify key components of biodiversity in agricultural production systems; monitor and evaluate the effects of different agricultural practices on wild biodiversity; and modify practices to attain the desired levels of biodiversity.
- Redirect various support measures provided to agriculture that run counter to the conservation of wild biodiversity.
- Implement targeted incentive measures that have positive impacts on wild biodiversity, and assess the impact of new activities in order to minimize adverse impacts of agriculture on wild biodiversity.
- Encourage the development of technologies and practices that increase productivity and reduce degradation; and reclaim, rehabilitate, restore, and enhance wild biodiversity, especially that which is relevant to agricultural development.
- Empower indigenous and local communities and build capacity for in situ conservation and sustainable use.
- Develop measures and/or legislation to encourage appropriate use of agricultural chemicals and to discourage excessive dependence on them.
- Study the positive and negative impacts of intensification and extensification of agriculture on wild biodiversity.

Other international conventions have important implications for biodiversity conservation in agriculture. The 1982 United Nations Convention on the Law of the Sea governs commercial fishing in international waters. The 1973 Convention on International Trade in Endangered Species (CITES) seeks to regulate markets for products from endangered species. The 1951 International Plant Protection Convention (amended in 1997), housed at FAO, and the various regional conventions established under the IPPC, were set up to secure common action to prevent the spread and introduction of pests of plants and plant products, and to create appropriate measures for their control. The Ramsar Convention on Wetlands of International Importance, established in 1971, provides guidance and support to countries to reduce wetland conversion and to re-establish wetlands previously converted to agriculture. The World Heritage Convention Concerning the Protection of the World Cultural and Natural Heritage of 1972 has set up a system of World Heritage sites, including highly valued, rare-habitat areas. The Convention on the Conservation of Migratory Species of Wild Animals of 1979 helps to protect the habitats of numerous migratory species, especially birds.

The U.N. Convention to Combat Desertification and the Framework

Convention on Climate Change were devised to meet other environmental objectives, but they stipulate the desirability of achieving biodiversity conservation in their implementation. For example, options in the 1997 Kyoto Protocol to the U.N. Framework Convention on Climate Change addressed land use, land-use change, and forestry. The 1997 Protocol enables countries to meet a small part of their commitments to carbon-emission reduction by investing in agroforestry and forestry land uses that sequester carbon. Kyoto-driven afforestation and reforestation in biologically degraded regions could contribute significantly to the restoration and enhancement of habitats for wild species.

Redefining Protected Areas

The international community has evolved in its approach to designating specific lands for biodiversity conservation as a tool to achieve the goals of the CBD. According to the IUCN (1994), a protected area is defined to be an area of land or water dedicated to the protection and maintenance of biological diversity, and of natural and associated cultural resources, and managed through legal or other effective means. But recognizing the need for a range of approaches to protected areas, from strict protection to sustainable use, the IUCN revised its system of protected area categories based on objectives for management. The new categories are defined in Box 9.3. Nature reserves, national parks, and natural monuments, by definition, need to be protected against resource harvesting on a commercial scale. However, ecoagriculture may be permitted, or even encouraged, in other categories of protected areas—such as species management areas, protected landscapes, and managed-resource protected areas.

National Legal Frameworks for Biodiversity Conservation

Environmental strategies and conventions for global conservation that emerged in the mid-1970s were a stimulus for many nations to prepare new environmental laws, including national action plans that included habitat protection in agricultural regions. The institutions responsible for implementing the international conventions provided technical and legal support to help countries design and implement those plans (Hannam 2001). In East Africa, for example, in response to the CBD, countries including Ethiopia, Malawi, Tanzania, and Zimbabwe formulated new policies for biodiversity conservation in the late 1990s (Place and Waruhiu 2000).

These legal frameworks still often suffer from lack of coordination between biodiversity goals and other environmental law. Kenya, for example, has more than seventy pieces of legislation relating to natural resource management, but they are often contradictory. During the 1990s, many countries moved toward more integrated environmental law and policy, with biodiversity conservation goals part of soil, vegetation, and watershed management strategies, all linked by overarching environmental planning and assessment legislation (Hannam

---

Box 9.3. Categories of Protected Areas in the IUCN System

I. *Strict Nature Reserve/Wilderness Area.* Areas of land and/or sea possessing outstanding or representative ecosystems, and outstanding or representative geological or physiological features and/or species, available primarily for scientific research and/or environmental monitoring; or large areas of unmodified or slightly modified land and/or sea, retaining their natural character and influence, without permanent or significant habitation, which are protected and managed so as to preserve their natural condition.

II. *National Park: Protected Areas Managed Mainly for Ecosystem Conservation and Recreation.* Natural areas of land and/or sea, designated to (a) protect the ecological integrity of one or more ecosystems for this and future generations, (b) exclude exploitation or occupation inimical to the purposes designated for the area, and (c) provide a foundation for spiritual, scientific, educational, recreational, and visitor opportunities, all of which must be environmentally and culturally compatible.

III. *Natural Monument: Protected Areas Managed Mainly for Conservation of Specific Features.* Areas containing one or more specific natural or natural/cultural features that are of outstanding or unique value because of their inherent rarity, representative aesthetic qualities, or cultural significance.

IV. *Habitat/Species Management Area: Protected Areas Managed Mainly for Conservation through Management Intervention.* Areas of land and/or sea subject to active intervention for management purposes so as to ensure the maintenance of habitats and/or to meet the requirements of specific species.

V. *Protected Landscape/Seascape: Protected Areas Managed Mainly for Landscape/Seascape Conservation and Recreation.* Areas of land, with coast and sea as appropriate, where the interaction of people and nature over time has produced an area of distinct character with significant aesthetic, cultural, and/or ecological value, and often with high biological diversity. Safeguarding the integrity of this traditional interaction is vital to the protection, maintenance, and evolution of such an area.

VI. *Managed-Resource Protected Area: Protected Areas Managed Mainly for the Sustainable Use of Natural Ecosystems.* Areas containing predominantly unmodified natural systems, managed to ensure long-term protection and maintenance of biological diversity, while providing at the same time a sustainable flow of natural products and services to meet community needs.

Source: IUCN 1994.

2001). These integrated strategies pose their own challenges, because unless carefully designed, environmental services for human use (soil erosion control, watershed management) may pose trade-offs with wild biodiversity conservation goals (Daily 2000).

Decentralization of Biodiversity Policy

Historical reviews demonstrate that for most countries centralized, top-down biodiversity conservation is seldom effective, except where large budgets are available for enforcement and the society concerned is willing to accept a rather undemocratic conservation process. Formal authority over natural resource management has been devolved to local levels in a number of developing countries over the past decade, including Bolivia, Zimbabwe, and the Philippines. This reform should make it easier for local people to play a role in designing and managing protected areas. But even where the enabling policy framework exists, devolution processes pose some immediate problems, given the weaknesses of local authorities. In many countries, local administrators and elected officials have little training in biodiversity and natural resource management, and limited resources at their disposal. Some local communities who have been customary resource managers may be disempowered (Place and Waruhiu 2000). Stronger local governments can potentially help to protect farmers' rights in natural resource policy, but the effects on biodiversity will depend on local interest and commitment.

Successful decentralization requires local participation and capacity building, as well as enabling policies, laws, and institutions. Incentives need to be created that allow local communities to keep income generated by sustainable use of reserves and other biodiversity assets, with conditional subsidies where needed. Regulations need to be appropriately enforced, especially against powerful local, corporate, or special political interests. Stakeholder forums and ecoregional managers need to be given decision-making authority (Lutz and Caldecott 1996).

Regulation of Agricultural Use and Management Practices

Direct government regulation at national, provincial, and municipal levels is widely used to protect biodiversity from damage from agricultural practices. Most such rules have protected economically important types of biodiversity such as valued fisheries, often by regulating pollutants. For example, rules may limit the quantity of livestock wastes disposed of in or near waterways or require waste processing. Rules may be designed to protect human communities as well; farmers upstream from a public water system may be required to practice organic farming (as in some of the watersheds above Tegucigalpa,

Honduras). In some countries, pollution controls are voluntary so long as a decline in pollutant levels can be confirmed. In other places, all farmers in a region may be regulated because of the cost of site targeting and the lack of methods to monitor pollutant flow. Site-specific (not blanket) regulations are usually needed for effective biodiversity management. For example, limits on the allowable catch of a particular fish species in a particular season, or maximum stocking densities, need to be specified. Because invasive alien species continue to be such a major threat to wild biodiversity, well-functioning national systems for controlling plant and animal movements across borders are essential, along with systems of quarantine and monitoring of released alien species (Wittenberg and Cock 2001). Since little is known about how genetically modified species interact with ecosystems, they need to be subjected to the same monitoring as invasive alien species.

Hand-in-hand with such regulations, it is essential to encourage producers and agro-industry to develop technology that enables producers to comply with regulations at a reasonable cost. Otherwise, regulatory efforts will result in political backlash, serious damage to the agricultural economy, or simply widespread noncompliance and expensive enforcement. As methods for identifying ecological hotspots and monitoring resource quality improve through geographic information systems, it should be possible to reduce costs for regulatory approaches that target producers who contribute most to pollution problems. Pollutant limits could be established for whole communities or watersheds, allowing flexibility for the community to determine the easiest and least-expensive options for compliance.

In designing habitat protection plans for agricultural regions, it is common to establish zoning rules or easements governing the use or management of critical areas, such as riparian zones, wetlands, or wildlife migration routes. Australia has instituted new national regulations to control agricultural land-clearing in environmentally sensitive areas, including protection of listed ecological communities and listings of native vegetation loss considered to be of "national environmental significance" or part of "a key threatening process," thus triggering legal action (Glanznig and Kennedy 2001). For ease of monitoring and enforcement, and often due to lack of information on the sensitivity of wild species to levels and types of human intervention, many of these rules strictly exclude local people from managing or using protected areas in any way. Regulations originally intended to protect native vegetation (e.g., harvest restrictions and permit requirements) sometimes have had the perverse effect of preventing farmers from practicing agroforestry and farm forestry and even discouraging natural regeneration. Many countries are exploring ways to revise these regulations (Current, Lutz, and Scherr 1995). As shown in the case studies for Strategy 1 in Chapter 6, protected areas can often be managed to permit economic use by farmers and still protect biodiversity. As another example, in South Africa it was widely believed by the public, environmental NGOs, and the national Forestry Commission that the woodcarving industry, based on wood illegally harvested from state or private forests, was resulting in deforestation. Detailed research by CIFOR demonstrated, however, that not only was woodcarving a crucial source of livelihood for many households, but also the amount of wood consumed annually was only a minor percentage of the waste produced by the logging industry. As a result, the government is easing up on woodcarving restrictions (Campbell et al. 2002).

It may be difficult for small farms and farm organizations to comply with overly strict environmental regulations. More flexible systems are being tested, though with mixed success (Heid 1998). Some countries have moved significantly away from regulation as the primary environmental policy instrument. The Netherlands, for example, now focuses almost exclusively on building farmer capacity to understand and manage biodiversity, emphasizing education and self-regulation as a long-term change strategy (Bekkers and Verschuren 1996).

One quasi-market mechanism recently developed to provide more flexibility in meeting biodiversity protection goals is tradeable development rights (TDRs). TDRs grant a limited set of rights for commercial development in areas designated for conservation. These can be sold to public or private sector conservation interests or exchanged for development rights on land outside the restricted use areas. The sale of TDRs provides the financial means to compensate for restricting property rights. Only the development rights are sold or exchanged, not the land itself, so communities or owners can continue nondegrading activities. The exchange value of TDRs should reflect a balance between the buyer's willingness to pay for the public good and the seller's estimation of his or her forgone development benefits. With government regulation, the price of TDRs is determined by the value of local development opportunities. In the state of Oregon (USA), TDRs are being used to protect salmon habitat on farms and towns along major rivers. A TDR system is being considered for forest protection in the state of Minas Gerais in the Brazilian Amazon. There, a longstanding requirement that 20 percent of each private farm or property be maintained under natural forest cover may be modified to allow landholders, under certain conditions, to meet this requirement through combinations of properties, rather than on a property-by-property basis (Chomitz 1999).

Property Rights for Biodiversity Protection

Clear property rights help to conserve wild biodiversity because they strengthen the legitimacy of conservation areas and actions. Yet such laws need

to take into account the needs and rights of people living in and near the conservation areas in order to be effective. Legislators should pay particular attention to recognizing local farmers' rights in relation to the designation and
management of protected areas, protecting indigenous rights in biodiversity-
rich areas, integrating biodiversity considerations into water-rights regulations,
and resolving rights over wild genetic resources.

Recognizing Farmers' Rights in Protected Area Designation and Management

Most protected areas for biodiversity have been established on public lands or
under eminent domain by national government agencies. In many cases, local
farmers had actively used these lands or claimed them under customary rights.
Losses suffered by local people were particularly significant in shifting cultivation systems where fallow lands were presumed by the authorities to be
"unused," and for common lands where extractive activities were important to
local livelihoods, especially for the poor. Recognition of this issue has led to
the incorporation of safeguards to protect local land and usufruct rights for
sustainable agriculture, hunting, and forestry (e.g., McNeely 1999). Many of
these rights are now reflected in various international environmental conventions, including the CBD. Mechanisms for protecting against and compensating for losses due to protected-area designation are now also part of national
legislation in many countries.

Indigenous Land Rights for Biodiversity Conservation

A high proportion of remaining wild biodiversity is found in areas of traditional settlement where indigenous resource management systems are still
functioning. For example, 80 percent of the remaining natural forest in Mexico—and that with the greatest biodiversity—is on lands controlled by indigenous people (Scherr, White, and Kaimowitz 2002). The CBD recognizes that
indigenous communities embodying traditional lifestyles can be relevant to
conservation and sustainable use of biodiversity (articles 8j, 10c, 17.2). However, in many developing countries, as a result of colonial rule, indigenous
claims to natural resources have been denied during nationalization of natural
resources at independence, or the establishment of protected areas. In the
process, traditional rules regulating resource access have lost their legitimacy,
invariably leading to overexploitation of resources. Even where land tenure for
agriculture is secure (through titling or usufruct rights to individuals or communities), indigenous people have often lost many of their rights to manage
natural resources.

Recent initiatives have been successful in establishing indigenous people's rights to manage protected areas in order to conserve both biodiversity and compatible agricultural systems. Nearly one-fourth of the forest estate in the most forested developing countries is now owned (14 percent) or de facto controlled (8 percent) by indigenous and rural communities, as a result of recent government recognition and devolution of local claims. A majority of Latin America's natural forest is now under indigenous control (White and Martin 2002). In Nicaragua, the Miskito people have formed their own NGO to manage the Miskito Coast Protected Area, overseen by a commission including government, regional, NGO, and community representatives (Barzetti 1993). In the Philippines, a local NGO established by the Ikalahan Tribe is managing the 14,730-hectare Kalahan reserve in Luesan. The NGO is implementing an integrated program of community forest management and the extraction of nontimber forest products leading to production of jams and jellies from forest fruits, extraction of essential oils, collection and cultivation of flowers and mushrooms, and manufacture of furniture. As early as 1975, the South Pacific Conference on National Parks and Reserves recommended that governments "provide machinery to enable the indigenous people involved to bring their land under protection as national parks or reserves without relinquishing ownership of land, or those rights in it which would not be in conflict with the purposes for which the land was reserved." India's Joint Forest Management policy was discussed in Chapter 6.

In Papua New Guinea, the government has established wildlife management areas where local communities co-manage resources. Management committees have instituted measures such as establishing royalties for the taking of game and fish by outsiders, hunting technique restrictions, prohibiting collection of crocodile eggs, fishing technique restrictions, and restrictions on logging (Eaton 1985). Australia's Uluru National Park, 132,566 hectares large and designated as a World Heritage site under the World Heritage Convention, contains the renowned Ayres Rock and is jointly managed by the Anangu Aboriginal traditional owners and the Australian National Parks and Wildlife Service using a combination of traditional knowledge and modern techniques.

Water Rights for Biodiversity Protection

Complex sets of ground, surface, and irrigation water rights in agricultural areas govern access to water by farmers for irrigation and livestock, by industrialists for processing needs, and by settlements and cities for domestic needs. Only recently has water been legally reserved in some parts of the world to preserve wildlife habitat. State law in California, for example, now prohibits water transfers that would have an unreasonable impact on fish, wildlife, or other instream uses. The U.S. Endangered Species Act of 1973 prohibits water transfers that could harm or harass endangered species or cause a significant loss of their habitat. In Mexico, the water law of 1992 mandates that the quality of water required in the discharge be specified in the granting of water rights, and the responsible national agency can restrict water use in the event of damage to ecosystems, overexploitation of aquifers, and other environmental impacts. Environmental protection objectives have also been incorporated successfully into both market-based and administrative mechanisms for water allocation (Rosegrant and Schleyer 1995). Increasingly, processes being developed for negotiating water rights among diverse stakeholders in a watershed or irrigation district include conservation groups or other negotiators representing the interests of biodiversity conservation (Meinzen-Dick and Bruns 2000). Ten of the primary international actors in the fields of water resource management, conservation, and health have established a Global Dialogue on Water, Food, and Environment to examine future water needs and policies for nature and food production at a global scale (IWMI 2002).

Property Rights for Genetic Resources

The rising dominance of private companies, rather than public-sector research institutions, in research on the genetic improvement of agricultural species, and the promising commercial prospects for GMOs in agriculture and other sectors, has ushered in a period of intense debate and conflict about "property rights" for genetic resources. Who "owns" a gene? Who should benefit from the commercial application of that gene? Will patenting of genetic improvements restrict farmers and local people from using and distributing the native plants or indigenously developed varieties that were the original source of the gene? Should farmers be compensated financially for past or current in situ conservation of genetic material from valuable domesticated or wild plants and their wild relatives? The legal frameworks that are ultimately established internationally and nationally to govern these rights will have a profound effect on farmer, agribusiness, environmentalist, and research incentives to maintain, control, and access biodiversity.

With entry into force of the CBD in 1993, bioprospecting and transfer of benefits arising from the use of genetic resources have become much more complicated. Today's bioprospectors must meet the CBD's article 15 requirements for prior informed consent, access on mutually agreed terms, and the fair and equitable sharing of benefits. They must also address issues of intellectual property rights and technology transfer; obtain appropriate permits to collect materials, enter land, and export and import materials; satisfy phytosanitary standards (to control plant pests and diseases); satisfy CITES requirements for protecting endangered species; and ultimately meet regulatory requirements for product safety standards. Thus bioprospecting depends for its success on the shared and realistic expectations of the partners and their ability to meet one another's needs.

The Philippines has already introduced restrictive legislation governing access to genetic resources, while access and benefit-sharing measures have been concluded or are under development in Australia, Fiji, India, Indonesia, Malaysia, Thailand, and elsewhere (ten Kate and Laird 1999). Meanwhile, indigenous groups in Latin America have proposed a moratorium on bio-prospecting deals until they are able to strengthen their knowledge and nego-tiating capacity, and some have proposed to organize an indigenous cartel to enhance local bargaining power over wild species' genes and local knowledge of wild species (Voegel 2000).

In 1994, the Future Harvest international research centers formalized their status as trustees, rather than owners, of the ex situ germplasm collections they hold (mostly of domestic crop species and some wild relatives) by signing legally binding agreements with the United Nations Food and Agriculture Organization. The germplasm collections must remain in the public domain, and genetic resources should remain available without restriction to all users. A new standard Material Transfer Agreement binds recipients of germplasm held or developed by the Future Harvest centers to the terms of the FAO agreements (IPGRI 2000).

Conclusion

Government policy, oversight, and regulation, whether at international, national, provincial, or municipal levels, play a valuable role in protecting bio-diversity in agricultural regions. Though progress has been made in the past decade, more is needed, with a much stronger focus on providing policy sup-port for the development of ecoagriculture.

Chapter 10

Market Incentives for Ecoagriculture

Many approaches have been developed to provide incentives to encourage farmers, foresters, and fishers to conserve biodiversity, or at least to avoid activities that lead to the loss of biodiversity. Often, such incentives are based on traditional behaviors, such as taboos and other social pressures, that balance public and private interests, or else on regulatory penalties. As described in Chapter 9, reshaping policies to encourage biodiversity is necessary for ecosystem conservation. But these are unlikely to be sufficient. Since the profitability of agriculture, no matter how marginal, ultimately drives habitat conversion and degradation, only the profitability of conservation can protect biodiversity. In a world increasingly dominated by market transactions, ways must be found to provide market incentives for ecoagriculture.

To promote ecoagriculture, systems and practices should be designed so that they are not only profitable for farmers, but *more* profitable than conventional practices. Investment needs to be targeted to produce the research and management breakthroughs that enable farmers to raise output and/or reduce their costs. Growing consumer interest in biodiversity means that there are also opportunities to market products that are produced in ecoagriculture systems. There are also emerging opportunities for farmers to receive payments for the biodiversity and other ecosystem services they provide, either by the direct beneficiaries or through public agencies or supporters of conservation. This chapter discusses these diverse incentives for ecoagriculture adoption.

Promote Profitable Ecoagriculture Systems

Ecoagriculture designers and practitioners can learn from decades of experience in promoting soil conservation and agroforestry systems in developing countries. One of the major lessons is that farmer adoption is far more rapid and more sustainable when conservation outcomes are embedded in practices that *also* increase productivity and profitability (Enters 2001; Current, Lutz, and Scherr 1995). Financial payments generally should be used only to finance the costs of transition and to encourage early adopters to experiment with the practice—to demonstrate its profitability and encourage its adaptation to local needs—not to subsidize continuing a practice that is not economically viable (Scherr and Franzel 2002). The latter may be theoretically justifiable in terms of a social cost-benefit analysis, but it is difficult to implement in practice without long-term financing.

In North America and Europe, the remarkably high level of farm subsidies (only gradually being redirected toward conservation) is a historical relict that reflects the small size of the farming population and the economic and political capacity of high-income countries to pay for such subsidies. Such a public subsidy approach is difficult to defend in high-income countries, and highly inappropriate for low-income countries that have far more pressing uses for public funds. Instead, public funding for ecoagriculture will be more effective if directed toward research, ecosystem planning and monitoring, technical assistance, and support to local farming and conservation organizations to adapt ecoagriculture to their local conditions. Indeed, emphasis should rather be on removing subsidies for agricultural systems and practices that threaten biodiversity.

While good data on farm income impacts were not available for all of the case studies described in Chapters 6 and 7, many of them demonstrated the economic potentials of ecoagriculture. In seven cases, economic benefits were neutral or negative. Among the three money-losers for farmers, the Chesapeake Bay pollution controls (#20) and biodiversity set-asides in the United Kingdom (#9) were the result of regulation, and the perennial grains case in the United States (#31) is still experimental technology. Four cases had neutral effects on farm income: the gene sanctuary in Turkey (#5), the wetlands reserves in the United States (#8), the Amazonian saki reserves on private farms (#7), and the Australian Landcare reserve (#3). The remote sensing technology case (#23) had no direct impact on farm income.

But all the other twenty-eight cases had clear positive benefits for local livelihoods. In three cases, farmers benefited from pest control technologies by avoiding large income losses: the cassava mealybug biocontrol (#14), the biocide against grasshoppers (#18), and the cattle vaccine against rinderpest (#30). For the Costa Rican orange plantation (#2) and agricultural intensification in Honduras (#13), farmers themselves reported that they benefited economically. In the protected areas of Nepal (#1) and Tanzania (#4), significant cash and subsistence benefits to the community were reported, and in the cassava case, food security was much enhanced.

In fifteen cases net incomes increased, based on documentation that gross income increased significantly while costs declined or remained stable. The five more rigorous studies show significant increases in income: a tripling of wage income for upland farmers in the Philippine irrigation case (#12); a 100 percent income increase among dairy farmers in Brazil's Atlantic Forest (#15); a 30 to 80 percent increase in total farm income for dairy grazing in the United States (#21); and a near-tripling of net farm income for corn from improved fallows in Zambia (#32). In the final example, preservation of traditional practices preserves an important source of income. Cash income from enriched tropical forest fallows in Latin America (#36) represented 10 to 20 percent (in some cases up to two-thirds) of total cash income. Certainly these represent sufficient financial incentives to encourage large-scale adoption.

Markets for Products from Ecoagriculture

The power of the market need not harm biodiversity. It can also be harnessed to protect biodiversity through commodity markets for wild products, certified biodiversity-friendly products, and agroecotourism.

"Green" Markets: Certification for Biodiversity

Another way to use markets to support biodiversity is to provide a premium for agricultural commodities that are grown as part of ecoagriculture systems. The most important instrument that has been designed to achieve this has been producer certification. The global trade in certified organic agriculture was worth over U.S.$21 billion worldwide in 2000. In the United States, the organic and natural foods industry has grown 20 percent each year during the past decade and now earns $10 billion in yearly sales, with the potential to double to $20 billion in the next four years (Clay 2002). In Austria, the European country where organic products have become most important, 10 percent of the food consumed is now organic. The World Organic Commodity Exchange (www.wocx.net) represents more than 2,500 organic products, including textiles, furniture, cosmetics, wine, vegetables, fruits, pet food, baby food, ice cream, and water. One might wonder about "organic water," but the indication of public interest in such products is high and growing, often mainly in response to human health concerns but increasingly because of environmental concerns as well.

Consumers in many high-income countries are interested in supporting better habitats for migratory birds in agricultural lands. Working with collaborators in the Sustainable Agriculture Network, the Rainforest Alliance, an international conservation organization, has established a certification program for coffee plantations that maintain forest cover, limit agrochemical applications, and control soil erosion. Rainforest Alliance–certified coffee, bananas, and oranges from Guatemala are now widely available in the United States, and vendors who sell them emphasize their environmental advantages over standard products. Chiquita Brands International has invested heavily in certification, reducing its chemical use by 80 percent in order to maintain access to a U.S.$350 million banana market in Europe and reduce insurance costs (Perfecto et al. 1996). The Nature Conservancy has developed "conservation beef" for premium beef markets and is marketing it on the Web. The designation certifies that the beef was produced according to high conservation values and without use of hormones. An example of a special certification program to protect an endangered fish species is described in Box 10.1.

A number of organizations around the world have also begun to certify that forests are being managed and harvested in a sustainable manner. In 1993, the Forest Stewardship Council (FSC) was formed by several conservation organizations and retailers. Other certification programs have been recently established in developing countries. By the year 2000, about 82 million hectares had been certified, including 24 million hectares by the FSC (Rametsteiner and Simula 2001). Certified forests now account for about 10 percent of the total land under timber concession in Latin America, 5.2 percent in Africa, and 1.3 percent in the Asia-Pacific region. The FSC is beginning to certify nontimber forest products such as Brazil nuts, the resin base for chewing gum, and cork (WWF 2000). In addition to environmental certification

---

Box 10.1. Salmon-Safe Farming in the United States

The Salmon-Safe Agricultural Products program in the U.S. Pacific Northwest awards a label to farmers who protect salmon habitat. The label was created in 1997 to provide an economic incentive for farmers to switch to crops and agricultural practices that help reverse the decline of wild salmon stocks throughout the region. To earn the label, farmers must minimize or avoid chemical use, keep livestock out of streams, plant trees or conserve existing vegetation in riparian zones, and use cover crops or plant forest on steep hillsides to minimize erosion. By the end of 1999, more than 4,000 hectares of sensitive farmlands had been independently certified and enrolled in the program. Apples, peaches, pears, wine, juices, and dairy products from these farms are now being marketed with the Salmon-Safe label in over 200 natural food stores and supermarkets in the Pacific Northwest (Johnson 2000).

criteria, FSC and other certification schemes require protection for local forest communities and users. Evidence suggests that certification commonly spurs efficiency gains in forest businesses such that certification pays for itself, at least over time (Clay 2002). It is projected that 200 million hectares of forest will be certified by 2005.

Major merchandisers, such as the Swedish multinational furniture retailer IKEA, are agreeing to use only certified timber. Both manufacturers and consumers in many developed countries have indicated a preference for certified timber products and even the willingness to pay a small price premium (Pearce, Putz, and Vanclay 1999). Over 600 member companies have joined forest and trade networks around the world, including Home Depot (North America) and B&Q (United Kingdom). Such certification is an indication that consumer choice can affect the way that resources are managed, providing a sort of feedback between resource management and the consumers.

Certification of biodiversity impacts may become a consideration in financial markets, as "green" mutual funds seek agricultural industries that contribute actively to sustainable development (Daily and Walker 2000). Large companies traded on stock exchanges around the world are judged by potential investors according to a variety of criteria. Increasingly, some of those criteria relate to environmental sustainability; a growing number of mutual funds exclusively invest in environment-friendly companies. These companies can achieve a competitive advantage by marketing their products as sustainably produced and packaged, and by advertising their environmental responsibility in managing corporate land, water, and forest resources. With further efforts to educate and animate both investors and the public, company performance as stewards of biodiversity might also be rewarded.

Market Development for Products from Wild Species

While hundreds of species have been domesticated, they still represent a small fraction of the potentially useful plants and, to a lesser extent, animal species. Domestication was historically limited by technical difficulties in breeding useful and economically desirable characteristics that would make domestic production worthwhile, and by the relative abundance and ease of harvesting wild populations. Both of these limitations have been substantially overcome—one by modern science and the other by the dramatic decline in wild habitat and populations. There is thus a huge potential for domestication and economic trade in wild species that were heretofore ignored (Leakey 1999b). Such developments could help to conserve genetic diversity, reduce pressure on remaining wild populations, and offer opportunities for farmers to diversify and increase their incomes in more environmentally sustainable ways. An example of one corporation's attempts to seek out nontraditional forest products for industrial use is described in Box 10.2. Buffalo and "beefalo" burgers in the United States demonstrate domestication and hybridization, respectively, of a wild species. Also, jicama and starfruit (carambola), unknown to most U.S. consumers a decade or two ago, now appear regularly in supermarkets.

The development of markets for new products, however, is a multifaceted challenge. Potential consumers need to learn about the product and its value. If the product is a raw material to be processed, then processors need to learn about processing requirements (for example, sawmills may need to modify their equipment for new timber species). Quality standards must be developed

---

Box 10.2. Private Auto Company Promotes Farm Biodiversity

In the early 1990s, Daimler-Benz set up its Environment Department to examine the durability and quality of the company's products, the life cycle of the products, the opportunities for recycling parts, and the possibility of using natural materials in vehicle production. One outcome was the decision that the company should support basic and applied research on the use of renewable, natural materials in automobile manufacturing. Daimler-Benz established the "Poverty and Environment in Amazonia" (POEMA) program, in cooperation with the Federal University of Pará in Brazil and a rural cooperative called PRONAMAZON. Among other development activities, the project examined the suitability of a range of products from Amazonian plants for industrial uses. The result was the creation of *flexiform,* an alternative to fiberglass for use in interior paneling of vehicles. This material is now used in C-Class Mercedes-Benz cars, as are several other natural product components, such as coconut fiber headrests fabricated by PRONAMAZON. Among the commercial benefits from flexiform are its lower weight (20 percent lower than conventional materials), its shock resistance, and its resistance to splintering. In addition, the manufacturing process has lower energy requirements and the material causes less wear-and-tear on manufacturing tools.

Raw materials being grown for Daimler-Benz by local farmers include coconut fibers, jute, sisal, curana (*Ananas erectifolius*), ramie (*Boehmeria nivea*), castor oil, rubber, cashew oil, andiroba (*Carapa guianensis*), and indigo (*Indigofera arrecta*). These are grown together with a wide range of food crops and other trees producing important subsistence products within multilayer agroforestry mixtures. This biodiversity-friendly system provides a monthly income of U.S.$353 by the twelfth year after establishment, a 450 percent increase over a typical local wage. Following the merger of Daimler-Benz and Chrysler Corporation, the company announced plans to repeat the Brazilian experience close to another factory in South Africa. Meanwhile, in Brazil, the POEMA program has expanded to become BOLSA AMAZONICA and has extended its activities to other parts of Amazonia. *Source:* Mitschein and Miranda 1998.

so that buyers and sellers can agree on a standard price. Pricing information needs to be widely known so that producers and consumers can plan knowledgeably for the future. Research and development are typically needed to reduce marketing costs and promote more diverse and more efficient production systems. Reliable transportation systems are needed, especially for perishable products. Seed, planting stock, or breeding stock of known provenance needs to be readily available to producers. The private sector will and should handle the bulk of these roles. The public sector and civil society can play a valuable catalytic role in disseminating information, promoting environmentally friendly consumerism, bringing together key market actors, removing subsidies for competing products, reducing regulatory hurdles for producers and intermediaries, and supporting the needed research (Scherr and Dewees 1994). In Southeast Asia, markets for locally grown tree products and benefits to local people from market activities have been improved through "green" marketing, niche specialization, marketing services offered by extension programs, and market price information systems (Raintree and Francisco 1994).

Agroecotourism

Interventions to increase wild biodiversity in and around farm fields can enhance the aesthetic, cultural, and environmental value of an agricultural landscape. Tourism can be promoted to exploit these resources as part of efforts to encourage tourist visits to already-established protected areas or to extend tourism to farm landscapes. Locally grown and processed products based on sustainable agricultural systems and local wild products can be marketed to tourists, and local people can also earn income by providing lodging, board, and guide services. Livelihood benefits from such tourism in developing countries are greatest where resources have very high biodiversity value, the transportation infrastructure is good, and local people are directly involved in providing goods and services to tourists (Honey 1999). The Nepal and Tanzania case studies in Chapter 6 (Examples 1 and 4) are good examples.

European protected areas often incorporate agricultural lands. Switzerland, a country heavily dependent on tourist revenues, subsidizes farmers in mountain areas to maintain traditional, mixed agriculture-wildland landscapes, which are perceived as being of enhanced value to tourists. Some Italian protected areas have ecotourism activities linked explicitly to organic agriculture, with local farmers serving as guides and locally grown organic products being sold to tourists. In Mt. Etna Park (59,000 hectares), pastures and agriculture occupy about 30 percent of the land (Compagnoni 2000; Spampinato 2000). In Brandenberg, Germany, 6,980 hectares of organic farms have been established within the Schorfheide-Chorin Biosphere Reserve (Voegel 2000). Protected area management here includes developing and testing management methods for buffer zone agriculture that takes better care of native wildlife, in order to attract tourists with both wildlife and traditional agriculture. In Slovenia's Triblavski Natural Park and Sneznik Regional Park, organic farming is being evaluated as a possible tool for maintaining the traditional rural landscape and protecting and enhancing wild biodiversity (Slabe 2000). Some large landholders—notably in Latin America and Eastern and Southern Africa—allocate at least part of their property for conservation purposes, with the goal of incorporating tourism.

Payments for Biodiversity Services

In some cases, the potential income and other values from ecoagriculture may be insufficient to motivate local people to adopt such systems. But the value of protected habitat to other users in the region or downstream (for biodiversity or other environmental services), or to the global community, may indeed be much greater than its agricultural use. Possible examples include the value of reduced sedimentation for downstream fisheries, reduced water pollution for downstream populations' drinking water, increased carbon sequestration for reducing global warming, or rare species for concerned conservationists.

Ecoagriculture can draw on new approaches to compensate farmers and farming communities for these biodiversity and natural habitat values that are being developed all around the world (Daily and Ellison 2002; Pagiola, Bishop, and Landell-Mills, forthcoming; Powell, White, and Landell-Mills 2001; USDA 2001). Guiding the economic development process toward paths that are compatible with ecosystem protection is basically sound, because farmers often are the principal agents of ecosystem change or are in the best position to protect the ecosystem.

Offering Tax Advantages for Maintaining Biodiversity

Tax policy has been widely used in developed countries to promote biodiversity conservation. Differential land taxes are sometimes applied, with biodiversity-conserving uses taxed at the lowest rate or exempt from taxes. In sensitive agricultural landscapes important for wild biodiversity protection, policymakers may also provide income tax deductions or tax credits for farmers who establish and manage biodiversity-rich systems. These systems can be defined in terms of protecting natural habitats in and around the farm (for example, riparian zones, native forest remnants, native grasslands), or managing biodiversity-friendly production systems (for example, avoiding use of pesticides known to harm endangered species, or use of shade-grown coffee ensuring bird habitat). Periodic recertification is needed to confirm eligibility.

Paying Farmers to Maintain Protected Areas

Farmers and other rural landholders who manage their land and resources to provide environmental services—like biodiversity—produce environmental services of real value to their local, national, and/or global communities. The concept of payments for environmental services implies that farmers should be compensated directly for providing these services. Numerous approaches have been used and are being developed further to reward farmers financially for biodiversity conservation services.

Some conservation programs purchase permanent conservation easements from farmers, covering all future rights to develop the land for farming or other purposes, although the property itself remains in the name of the owner. The farmer is compensated for the loss of the future economic benefits that might have come from that piece of land, through a lump sump payment or, more often, payment of an annuity. Some agreements specify landowner responsibility to conserve and manage wildlife.

Such resource conservation agreements are probably the most widely used instrument to conserve biodiversity in farming areas in the United States and Europe. They can be a cost-effective alternative to public or private purchase of land for establishing protected areas, or paying for permanent conservation easements. For example, an evaluation of public costs of managing over 350,000 hectares of existing public lands in subtropical southwest Florida considered to be critical to the survival of the Florida panther (*Felis concolor*) found that the cost of conservation agreements on private lands was roughly equivalent to current conservation expenditures for the public lands reviewed; it was one-half to one-quarter as expensive as the estimated costs of purchasing privately owned lands or paying for permanent conservation easements to conserve privately owned panther habitats in the region (Main, Roka, and Noss 1999).

In the European Union, over 7 million hectares of land were diverted from cereal and oilseed production under short-term set-aside schemes in 1995–96, including more than 15 percent of agricultural land in the United Kingdom, Germany, France, Italy, and Spain (European Commission 1999). Fallow land in Europe increased from around 1 million hectares in the early 1980s to over 4 million hectares by the mid-1990s, providing additional opportunities for wildlife. Farmers in the United Kingdom are being encouraged to maintain and restore grass margins, conservation areas, and uncropped field margins to preserve biodiversity. In the United States, the Conservation Reserve Program includes nearly 15 million hectares of cropland, most of which have been seeded to grass. But this reserve program also contains 800,000 hectares of land managed for special wildlife practices, over 100,000 hectares of wetlands, and 1 million hectares of land planted to trees, all of which benefit wild biodiversity (USDA 1997; OECD 1997a). Thanks to this program, tree corridors established between formerly isolated woodlots have reduced the dispersal barriers to wildlife and lowered the degree of landscape fragmentation. A similar grassland reserve program is being established that would compensate farmers and ranchers for easements of grasslands. To date, conservation easements have been used most widely in the developed countries, but they are also increasingly being used in developing-country projects supported by private conservation organizations or conservation trust funds. Under some circumstances, direct payments may offer a lower-cost and more targeted alternative to projects promoting more sustainable production systems (Ferraro and Simpson 2000). Several such instruments have been used for biodiversity protection in the state of Minas Gerais, Brazil (Bernardes 1999). A landmark agreement was recently reached between six environmental organizations and a small community in Mexico, as a result of which the community will receive U.S.$250,000 over fifteen years to preserve the nesting habitat of the western thick-billed parrot (*Rhynchopsitta pachyrhyncha*) on a 2,400-hectare old-growth forest near the Cebadillas village in northern Chihuahua. An "International Habitat Reserve Program" has been proposed as an international counterpart to the set-aside programs of the OECD.

Such measures do not automatically enhance biodiversity. Many of these programs were originally set up as instruments to reduce crop production or transfer income to farmers. Thus, attention was not always paid to ensuring spatial configurations at the landscape scale that would genuinely enhance biodiversity, nor to ensuring that agreed management changes were made. Kleijn et al. (2001) evaluated the contribution of such schemes to the protection of biodiversity in intensively used Dutch agricultural landscapes; they found no positive effects on plant and bird species diversity. The Netherlands has been implementing management agreements designed to conserve biodiversity on farms since 1981, often obliging farmers to postpone agricultural activities on individual fields until a set date that will allow certain species of birds to safely hatch their chicks. Other management agreements are designed to conserve species-rich vegetation in grasslands, restricting the use of fertilizer or postponing the first mowing or grazing date. They found that some of the management activities may have had perverse effects. For example, postponing the first mowing or grazing date may have adversely affected the abundance of soil fauna that certain bird species use for food, leading to an "ecological trap" where the cues that individual birds use to select their nesting habitat (for example, food availability) are decoupled from the main factor that determines their reproductive success (delayed mowing/grazing). They conclude that the primary concern of farmers is necessarily to secure their income, so nature conservation will be of secondary importance to them, especially in the context of a farming system that is driven by economic pressures to increase its intensity. These results clearly indicate that more effective approaches are needed, requiring that such joint agriculture–environment schemes are accompanied by a scientifically sound evaluation plan and are carefully designed to be ecologically appropriate.

To date, all these types of programs typically determine biodiversity value solely by the current type of land use. More specific indicators are needed to reduce conservation and transaction costs, especially as a wider range of buyers becomes interested in "purchasing" biodiversity services. State Forests of New South Wales, which is responsible for managing Australia's largest planted forest estate, has pioneered the development of environmental management services and products to facilitate private investment in forestry projects. This organization is exploring various options to assign a market value to forests planted and managed primarily for biodiversity services, in particular a "biodiversity credit" that could be sold to interested conservation groups and private companies wishing to enhance their "green" image. One approach to calculating the value of a biodiversity credit is shown in Box 10.3.

Paying Farmers for Ecoagriculture Production

Farm conservation payment programs are beginning to move toward more targeted strategies that reward biodiversity enhancement in agricultural systems (strategies 4, 5, and 6 in Chapter 7). For example, in the United States, conservation and sustainable farming groups are lobbying for the Conservation Security Act, which would reward farmers for good practices on working lands on the basis of a "Farm Results Index." The Nature Conservancy has developed a program that allows farmers with small-scale forests to put trees (not the land) in a Conservancy-managed "forest bank" in return for an annual payment funded from sustainable management of the bank's timber resources (R. Curtis, The Nature Conservancy, personal communication, February 28, 2001). In Europe, most countries have "agri-environment" programs that compensate farmers for biodiversity-enhancing agricultural practices. For example, in the United Kingdom government subsidies to organic agriculture amounted to U.S.$36 million in 1999, designed to convert over 60,000 hectares to organic production by 1,000 farmers. A further U.S.$20 million has been earmarked for this purpose over the next six years (OECD 2000).

Some payments also support maintenance or expansion of land under agricultural production where such uses are threatened (for example, by market forces or urban development). Such systems are found primarily in Europe, although efforts to maintain peri-urban agriculture are growing in the United States and Canada. In Switzerland "ecological compensation areas," including extensive and low-intensive pasture, floral meadows, and other farmland

---

Box 10.3. Calculating a Biodiversity Credit for Native or
Planted Forest

Can a "tradeable unit of biodiversity" be defined? One possible approach might
include the following variables:

B = Bios, the unit being transacted (annual units of biodiversity conservation)
A = Area (hectares)
E = Ecosystem value (could be scored from 0 to 10, with exotic species = 0,
    to unique, rare, or endangered ecosystem types reestablished or protected
    = 10)
L = Locational modifier (isolated forest without connectivity = 0.5; forest con-
    nected to existing functional habitat = 1.0; forest creating new connectiv-
    ity between areas of critical habitat = 1.5)
S = Species population additions (could estimate habitat supply or likely pop-
    ulation increases for rare, threatened, or endangered species of plants and
    animals, then multiply that number times 1 for rare, 3 for threatened, and
    10 for endangered species)

The equation for the calculation would then be:
$$B = (A \times E \times L) \div S$$
    This would create a very high value for a piece of intact rainforest in Brazil
and a low value for an exotic pine monoculture, with a range of values between.
The unit is based on an ability to sell annual rights or to contract over the long
term. The owner of the ecosystem would be able to sell as many years of conservation as desired, but the owner would need to register a conservation right
over the property and maintain a pool of biodiversity credits that would always
have to exceed the number of credits in the market. Any disturbance to the
ecosystem would be controlled by a biodiversity credit accounting standard
(Brand 2000b).

---

reserved for ecological purposes, have now expanded to cover almost 8 percent of agricultural land; these areas have had a positive effect on species diversity, especially for beetles and other insects. In Austria around 40,000 hectares
of meadows of high ecological value were signed up for similar protection,
while the United Kingdom has targeted ecologically valuable types of semi-
improved grassland for maintenance through the Environmentally Sensitive
Area and the Countryside Stewardship schemes. Canada's Permanent Cover
Program has helped the prairie bird population recover by increasing grasslands and forage production (OECD 1998). One of the advantages of environmental payments—unlike traditional payments to leave land out of production—is that they are a form of transfer to farmers that is likely to be
approved by the World Trade Organization.

Paying Farmers for Protecting Wildlife

"Conservation contracting" involves paying farmers directly for specific conservation outcomes, rather than land uses. This approach reduces the set of critical parameters that practitioners must effect to achieve conservation goals, permitting more precise targeting and more rapid adaptation over time, and strengthening the links between individual well-being, individual actions, and habitat conservation. This can create a significant local stake in ecosystem protection. In the Netherlands, for example, Musters et al. (2001) propose that farmers be paid for clutches of wild birds bred on their land so that they are paid for what they produce rather than compensating them for losses in income due to restricting their farming practices. In small-scale experiments, breeding success was significantly higher on farms where farmers were paid for meadow-bird clutches than on farms where farmers were not paid. The approach seems to be interesting to farmers, and it is less expensive than compensation for income losses. The level of compensation is about U.S.$35 per clutch as compared to costs of U.S.$85 to $340 per clutch when compensating for income losses. Some NGOs in the United States are paying farmers who can provide evidence of wild wolf dens on their farms, and the Safe Harbor government program exempts farmers from many regulations if endangered species are found to inhabit their land.

Paying for Bioprospecting Rights

Some outside groups have a strong economic interest in ensuring that wild biodiversity is conserved. Pharmaceutical companies, for example, have been willing to pay local people, governments, or companies who own or control large tracts of land under native vegetation. These payments may provide compensation for maintaining the resource intact and/or pay royalties related to important information or genetic material obtained from the resource. Agencies or private companies with an interest in in situ conservation of wild relatives of domesticated crops or livestock may be willing to pay local communities to conserve the natural environment within which such species would thrive and evolve, as described in the Turkey gene sanctuary example (Example 5 in Chapter 6).

Paying Farmers for Other Environmental Services Compatible with Biodiversity

Payments to farmers for establishing vegetation and managing their farms to provide carbon sequestration, water, salinity control, or other environmental services could potentially be combined with payments for biodiversity conservation to encourage farmer investment in ecoagriculture.

Watershed services. Various countries are experimenting with payments to land managers in upper regions of watersheds to protect water quality and guarantee water levels to downstream users for urban consumption or irrigation (Daily and Ellison 2002). Experience from various parts of the world demonstrates that good, natural vegetative cover, which keeps watershed ecosystems healthy and enables them to produce a steady and reliable supply of water, may also provide good biodiversity protection. For example, 7,600 hectares of cloud forest in the La Tigra National Park in Honduras provide the capital city of Tegucigalpa with 40 percent of its drinking water at a cost of about 5 percent of its second-largest source because the latter has to undergo expensive purification treatment. Examples of payments to farmers include self-organized private deals between France's Perrier-Vitel water bottling company and upstream farmers to encourage reforestatation, trading schemes for agricultural pollution emissions, and payments by New York City to landowners in the Croton and Catskills watersheds to maintain water quality (Johnson, White, and Perrot-Maître 2001).

Carbon sequestration services. The 1997 Kyoto Protocol of the United Nations Framework Convention on Climate Change allows companies to pay farmers and forest owners for carbon sequestration (planting trees to absorb carbon dioxide) to offset industrial emissions. At this time, eligible land uses include afforestation and reforestation, which may involve agroforestry practices, community or private tree plantations, forest rehabilitation, and ecosystem restoration (Smith and Scherr 2002). Pilot projects and private-sector initiatives are already under way in many parts of the world; many involve farming and forest communities. Such projects typically involve an energy firm, such as Wisconsin Electric Power Company or American Electric Power, and an intermediary NGO, such as The Nature Conservancy or CARE. Financial instruments are being developed that would allow credits for these payments to be traded in secondary and futures markets and thus be included in investment portfolios (Wilson, Moura Costa, and Stuart 1999).

Conclusion

In the context of globalization and government policy interventions in agriculture and biodiversity, ideal ecoagriculture market interventions should be relatively simple, be carefully designed for the local context, and have a reasonable probability of success. They need to be relevant at the scale of ecosystems, of sufficient magnitude to actually affect behavior (yet not so attractive that they draw in people from areas not benefiting from such measures), and capable of reducing the social and political conflicts over resource allocation that so often threaten biodiversity.

Chapter 11

Institutions to Support Ecoagriculture

For farming communities and businesses to make the transition to ecoagriculture, good technologies, good laws, and supportive markets must be promoted and supported by strong institutions, as described in the previous chapters. In addition, institutions must be created or adapted for ecosystem management, research, monitoring, extension, and financing.

Ecosystem Planning and Management at a Bioregional Scale

Policies, legal frameworks, and some types of policy instruments to promote biodiversity need to be developed at the national or state level. Many countries have established institutional umbrellas or mechanisms that promote comprehensive multi-sector discussion on biodiversity issues, such as Kenya's National Environmental Secretariat and Ethiopia's Environment Protection Authority/ Institute for Biodiversity Conservation and Research. Beyond the local and national levels, regional and global ecosystem planning initiatives such as the Mekong River Basin Commission for wetlands conservation can also contribute to the management of shared biodiversity resources.

However, because so much ecosystem management by definition must be undertaken within a defined geographic area, policy design and governance must be tailored to local conditions, with local input. Experience in numerous developing countries suggests that conservation enforcement services have the best morale and are best accepted locally when they work on behalf of a level of society higher than that of the community, but not as remote as that

231

of the nation—the bioregional level, such as the Guanacaste Conservation Area in Costa Rica, where wardens work on behalf of a regional committee of local stakeholders (Caldecott and Lutz 1996). Other institutions capable of providing leadership at the bioregional level include river basin development authorities, forest agencies, and state government planning and environmental agencies. The trend toward decentralization of authority in government agencies could have positive implications for integrating agriculture, forestry, and biodiversity, because it would make activities less compartmentalized and increase accountability to local stakeholders (Place and Waruhiu 2000).

The old models of watershed, river basin, and protected area planning imposed theoretically "optimal" solutions that failed to incorporate ideas from actual land managers and therefore were often ignored in practice. Such models left little scope for local experimentation with alternative solutions to achieve environmental goals.

Ecosystem management calls for the emergence of new types of land-use planning institutions and tools to coordinate public and private investment, regulate zoning, and monitor changes in the condition of biodiversity (McNeely 1999). New approaches provide more flexibility for ongoing adaptation of program designs, and more opportunities for partnerships with NGOs, public agencies, and the private sector (Barborak 1995; Ebregt and de Greve 2000; Margoluis et al. 2000).

While decision-making needs to draw on the expertise of technical and policy specialists to estimate the likely outcomes of different options, final policy design should reflect a negotiated outcome among different farmer groups, environmental organizations, and other resource-user groups (Buck et al. 2001; MacKinnon et al. 1984). Planning on an ecosystem scale requires that the interests of different stakeholders be understood and reconciled. This requires first that conflicts and commonalities of interest be clearly identified, a process that may take a few years. Then a communication process must be created that allows differences to be negotiated and shared understanding created. Policies and products can then be designed to build stakeholder commitment; these should be implemented in a way that allows continuing communication and shared understanding (Miller, Chang, and Johnson 2001). Low-cost methods for participatory natural resource planning have been tested successfully in low-income rural areas (e.g., Queblatin, Catacutan, and Garrity 2001).

Abrupt transitions are more likely to cause conflict; gradual change is often easier for diverse groups to accept. Organizations promoting ecoagriculture should recognize that changes in agricultural systems must be strategically designed for gradual implementation. For example, Denmark, which aims to become a pesticide-free country within five years, has developed a phased, methodical transition to alternative technologies (Gemmill 2002). A phased approach allows time for fine-tuning the procedures, and for positive demonstration projects to build community support.

Without a genuine "buy-in" from stakeholders, implementation of ecoagriculture policy objectives and strategies will not be effective. New techniques of interactive landscape planning can be invaluable to foster such cooperative processes (Rhoades and Stallings 2001). Recent reviews by The Nature Conservancy of nine community-based land-use planning efforts in conservation areas in the United States and Latin America (Chung 1999), and by the Biodiversity Support Program of their projects in various tropical countries (Salafsky, Margoluis, and Redford 2001), found a wide range of successful institutional models, from indigenous organizations to coalitions of local and national NGOs to public agencies.

A Central Role for Local Farmers in Ecoagriculture Development

> No nation will have lasting conservation on private lands until landowners are excited about the land and understand that environmentally sound land use is not a limit on personal freedom but rather a positive exercise of skill and insights (OECD 1997a).

It has become clear that to achieve real results on the ground, it is essential for local farmers—including the poor—to play a central role in planning and implementing legislative frameworks, property rights, payment systems, market innovations, and ecosystem planning organizations. Agricultural ecologists have learned to respect the wisdom inherent in much traditional practice, and conservationists increasingly have come to appreciate that throughout history wild biodiversity has been conserved as part of the agroecosystems that farmers used to support themselves. The "rule of indigenous environments" is that "where indigenous peoples have a homeland, biologically rich environments are still found" (Nietschmann 1992). Indeed, many of the lands most critical for biodiversity conservation in the world today are inhabited by indigenous or traditional peoples. They are valuable partners in developing ecoagriculture systems.

Many innovative approaches have been developed by conservationists in recent years to systematically engage local farmers in ecosystem management planning. For example, conservation biologists and local people can jointly develop guidelines for extraction of economically important products from protected areas (see Box 11.1).

Participatory monitoring can help to engage the interest and support of local populations, including schools and civic groups, in biodiversity conservation. Simple indicators widely understood and easily collected, evaluated, and accessed by all participants are usually the most cost-effective. Examples of these are described in the later section on monitoring.

Box 11.1. People's Protected Areas in Madhya Pradesh, India

India's protected area network has been largely targeted at faunal conservation, with its main strategy being to reduce human pressure on forest habitats. A new concept of People's Protected Area (PPA) has emerged that also aims to achieve biodiversity conservation, but through strategies that facilitate poor people's access to physical, material, human, social, and environmental assets from protected areas. PPA considers these public lands the "Poor People's Pool of Assets." PPAs are consistent with IUCN Category VI (see Box 9.3). The network of PPAs focuses mainly on biodiversity-rich buffer zones, fringe areas, and corridors of national parks and wildlife sanctuaries. It aims at converting open-access resources into a community-controlled natural resource, increasing both local people's income from minor forest product collection and processing, and biodiversity protection.

The legal framework was developed from existing legislation that confers ownership rights to minor forest products on village-level institutions. The state government of Madhya Pradesh has specified additional rights. The program is being supported by a public agency, the Madhya Pradesh State Minor Forest Produce (Trading and Development) Cooperative Federation Limited, an organization of approximately 5 million forest produce gatherers comprising 1,947 primary cooperatives and 84 district unions. At present the state has over 30,000 collection centers, and the annual turnover of minor forest product trade runs to more than 5 billion rupees (over U.S.$104 million). Research, marketing, and investment activities are being organized to support growth in sustainable harvest and income from minor forest products. So far, PPA projects are being implemented in parts of several districts, with the intention of developing a statewide PPA network (MFP 2000).

A variety of strategies can be used to link biodiversity conservation with local interests. For example, conservation may be linked to community health initiatives using a barter strategy that trades provision of health services for local behavior that sustains conservation. Or health services may provide an entry point to gain trust, goodwill, and community decision-making capacity that can then be turned to conservation activities. A "bridge" strategy helps local people to see the connection between their own health needs and conservation, for example by addressing a water pollution–related health problem to begin an environmental education campaign that shows the relationships among sanitation, aquatic habitat quality, and health. A symbiotic strategy develops project interventions based on known common ground between health needs and conservation goals, such as forest conservation to maintain water quality (Margoluis et al. 2001).

Indigenous and local communities can be empowered to build capacity for in situ conservation and sustainable use based on their traditional approaches. For example, subsistence Yup'ik hunters in the remote wetlands of the Yukon-Kuskokwim Delta of Alaska have developed new waterfowl conservation practices and attitudes crucial to the survival of several species of Pacific migratory birds. This development resulted not from enforcement of official regulations, but by creating the minimum necessary conditions for voluntary conservation to emerge as a cultural practice. These conditions include (1) a social mechanism that maintains the rules within the community and prevents outsiders from cheating; (2) the perception by hunters that they can influence the availability of game; (3) a vested interest in the continued availability of the resource; and (4) the availability of sufficient overall resources to meet basic needs (Zavaleta 1999).

Local initiatives can be encouraged and rewarded through public recognition. For example, Land for Wildlife is a voluntary property registration program begun in 1998 in Queensland, Australia, to recognize and support landholders who provide habitat for wildlife on their property in conjunction with other land uses. By February 2001, 1,300 properties had been registered, with a total area of terrestrial habitat of 50,000 hectares, of which 13 percent has been identified as either endangered or "of concern" (Millar 2001).

While community involvement in ecoagriculture development is essential, it is no panacea. Local communities do not always have peaceful relations with neighboring communities; thus building support networks requiring inter-village cooperation is not always easy. Many biodiversity problems may be of fairly recent origin as a result of expanding populations, immigration, and resulting increases in consumption; for these, traditional solutions may not be effective. The fact that local communities are often well adapted to their local environmental conditions does not automatically mean that they are going to make wise decisions. Deciding how to invest scarce resources in assets that mature over several decades (such as forest trees) or are highly mobile (such as migratory species of waterfowl) is a sophisticated task. Any community faces a challenging set of problems when it tries to govern and manage complex multispecies and multiproduct resource systems whose benefits mature at varying rates and are under pressure from competing groups of humans at every step (Ostrom 1998).

The best approach to this complex of problems appears to be greater long-term commitment from resource management agencies to help communities improve their resource management programs, enforce agreed-upon regulations, and assist with research and monitoring (Wood, McDaniel, and Warner 1995). While people-oriented conservation approaches can appear more challenging than more authoritarian production approaches, experience suggests there is no substitute, in the long term, for effective collaboration with local people (Wilshusen et al. 2002).

Local Farmer Organizations for Landscape Management

Because effective biodiversity conservation requires site-specific management, sustained interest and action by local farmers is critical. Some of the most exciting institutional innovations are voluntary organizations developed to enable small groups of local farmers to jointly evaluate, plan, and implement landscape improvements of mutual benefit. Often governments or civic organizations play a strategic role in providing information, training, professional services, or small grants to catalyze group activity. Successful locally based initiatives are exemplified by the Landcare movement begun in Australia, the conservation catchment programs of Kenya, and local forest user groups in Nepal; these are described below.

Landcare is a voluntary, community-driven conservation phenomenon begun in Australia in the mid-1980s; it now includes over 4,500 community groups across the country—nearly a third of the farming community. Landcare is a partnership among communities, government (the National Landcare Program), NGOs, and corporations to address the degradation of Australia's soil, water, and biodiversity. The foundation of the program is voluntary groups of land users. The basic institutional rationales for working with such groups include the following:

• An understanding by the land user of the physical and biological processes and interactions involved in sustainable land management is essential;
• Land degradation can only be solved by land users;
• Project/program relevance is maximized if land users develop and implement them; that is, participatory decision making by those who "own the problem" is fundamental to developing effective solutions;
• Groups of landholders with a common problem will develop and implement more effective projects than individual land users alone, or individual land users working with a government agency; and
• Attitudes usually change slowly, though group dynamics can accelerate development of new approaches and systems across a community (Campbell 1994).

Many Landcare groups have actively pursued wild biodiversity conservation goals, as illustrated by Example 3 in Chapter 6, while many more have positively affected wild biodiversity as a result of sustainable agricultural practices. Landcare has expanded through federations of local groups to become an important influence on public land policy in Australia. The Landcare model has now spread to other countries, particularly in Southeast Asia.

In Kenya, with so much of the country's prime agricultural land located on hill slopes, the government has for decades been concerned with managing soil erosion. However, efforts to promote adoption of soil conservation practices and structures at farm and field levels have had mixed results, and these have generally failed to address environmental problems affecting entire watershed areas. In the late 1980s, a new "catchment approach" was introduced, concentrating resources and efforts in a 200- to 500-hectare watershed, with dozens or hundreds of small farms. In this approach, local people take the lead in identifying opportunities and priorities for soil, water, and biodiversity conservation in their catchment area. The public extension program provides short-term technical support for their initiatives, including many that improve the quality of wildlife habitat. These include fodder production for stall-fed livestock (to reduce grazing pressure), riverbank protection, rehabilitation of degraded areas, and planting of both exotic and native tree species. Farmers dig ditches to catch rainfall runoff, so it can filter slowly into the soil, and harvest rainwater to irrigate crops.

Since adoption of the catchment approach, the number of farms assisted by extension agents has increased from 50,000 to 100,000 per year, without a corresponding increase in cost to the government. Biodiversity benefits have resulted from the rehabilitation of 70 hectares per year of degraded area, protection of 300 kilometers of riverbank, control of 200 kilometers of gullies, and raising and planting of 1.3 million tree seedlings. Studies show that incomes and land values of participating farmers have increased (Kiara 2001).

In Nepal, the organization of local forest user groups (described in Example 1 in Chapter 6) has enabled local villagers to manage village forests. Users have become much more actively involved in forest management and have a strong feeling of ownership. Improved forest management practices have been developed, involving women and the poor in activities such as thinning, pruning, fire control, forest protection, and harvesting. The species composition of flora and fauna, the area of tree crown cover, the microhabitat of invertebrates, mosses, fungi, and lichens, and the overall habitat have all improved, indicating an overall improvement of the forest ecosystem. The resulting community forests are providing ecological stability as well as income and making the forest user groups more sensitive to the conservation objectives of their forests. In at least some forests, wildlife populations have increased along with species numbers (Beek, Rai, and Schuler 1997).

Support Institutions for Ecoagriculture

As argued in Chapters 5 through 8, the successful development and dissemination of ecoagriculture depend most on technological advances that make it possible for agricultural systems to become more productive, while also providing good habitat for wildlife. Thus new or adapted institutions are needed to provide the ecoagriculture research, monitoring, extension, and education to develop and explain new techniques.

Agricultural, Ecological, and Wildlife Research

Ecoagriculture requires the integration of research input from agricultural, ecological and wildlife sciences, as well as from the people managing land and other resources. This calls for reorganization of research efforts at international, national, and local scales.

The sixteen Future Harvest research centers collaborated in developing more than a third of the ecoagriculture examples presented in this book. But only 10 percent of their current budget is allocated to the theme of biodiversity, and most of that is found in the forestry, agroforestry, and fisheries research institutes (CGIAR 2000). With committed leadership and new funding, however, all of the centers could quickly expand ecoagriculture research within the context of existing and proposed systemwide initiatives on agricultural biodiversity, sustainable mountain development, hillside development, livestock production, soil-water-nutrient management, integrated pest management, water management, and rainforests and the poor.

One successful example of international research collaboration relevant to ecoagriculture is the systemwide initiative of Future Harvest research centers to develop alternatives to shifting cultivation (ICRAF 1999). Another example is the Millennium Ecosystem Assessment—a consortium of the World Resources Institute, the United Nations Development Programme (UNDP), the United Nations Environment Programme (UNEP), the World Bank, and many national and international partners (World Resources Institute 2000). DIVERSITAS, an international program for biodiversity science (co-sponsored by the International Council for Science, the Scientific Committee on Problems of the Environment, the International Union of Microbiological Societies, the International Union of Biological Sciences, and the United Nations Economic, Social, and Cultural Organization) is proposing to develop a new network on "greening agriculture" (DIVERSITAS 2001).

Strong field collaboration between researchers and farmers at the local level is essential. For example, since 1993 the Project on People, Land Management, and Environmental Change (PLEC), of the United Nations University, has managed a collaborative effort between scientists and small-scale farmers in twelve developing countries. Their studies integrate locally developed knowledge of soil, climate, biological resources, and other physical factors with scientific assessments of the quality of these factors in relation to crop production. PLEC is seeking to devise a set of sustainable agricultural technologies that maintains biological diversity both on farm fields and in the surrounding lands, and also improves production and income (Stocking 2001).

Regional and local programs of ecoagriculture development can usefully apply "adaptive management" approaches that incorporate research into conservation action. These initiatives integrate design, management, and monitoring to systematically test initial assumptions about technologies and management practices, in order to adapt the technologies to fit local conditions. This approach has been widely applied in tropical forest management and conservation programs (Buck 2001; Salafsky, Margoluis, and Redford 2001) and is the de facto approach used in developing many other innovations, such as natural vegetative strips and intensive dairy grazing (Examples 26 and 21 in Chapter 7).

There are also important opportunities for private sector research. Large-scale private agribusinesses are likely to undertake research to integrate aspects of ecoagriculture into their own farming systems where such aspects are more profitable, to meet regulatory requirements, and/or to take advantage of ecoagriculture market opportunities. Private companies can also develop new types of agricultural services that would better protect biodiversity, and they could contract with farmers to provide such services. Companies might, for example, sell pest control services to farmers rather than simply selling them products.

The challenge of promoting wild biodiversity in agricultural landscapes also requires investment in applied policy research. Rigorous investigation of the experiences of ecoagricultural policy innovations would help scientists and policy-makers identify their strengths and weaknesses, evaluate cost-effectiveness, and document benefits and trade-offs. Such research calls for inter-disciplinary collaboration among agricultural and ecological scientists, economists, social scientists, local people, and land managers at farm and landscape scales. Currently, policy research is limited by inadequate funding, short research periods, and narrow disciplinary focus. Longer-term monitoring of policy innovations will bring the greatest payoff in improved understanding and design.

Unfortunately, just as the challenge of integrating wild biodiversity into agricultural systems is becoming recognized as an important part of the agenda, resources for public agricultural research around the world have declined (Conway 1997). Biodiversity issues remain poorly integrated into national agricultural research institutions, and collaboration between these institutions and environmental research and conservation organizations is minimal. This decline in public agricultural research and extension investments reflects, in part, the growth in private-sector research, particularly in areas like biotechnology, whose benefits can be easily captured by the innovator. Some private-sector firms are indeed contributing to more sustainable and biodiversity-friendly agriculture, and in many cases the public sector could contract private research firms to undertake needed studies. But the answers to most of the critical research questions will offer limited opportunities for patenting or sales. Rather, they contribute to public understanding of ecological and agricultural processes and foster new forms of management, whose benefits accrue to all farmers, the public, and wildlife. Research to address such questions should therefore be considered "public goods research" and supported by public funds.

Monitoring Wildlife in Agricultural Areas

If serious, large-scale programs are to be developed to support ecoagriculture, much more effective monitoring is necessary. The *Pilot Analysis of Global Ecosystems: Agroecosystems* (Wood, Sebastian, and Scherr 2000) concluded that very little of the necessary data to assess the interactions between agriculture and biodiversity is now being monitored. Currently published national data on land use show only net land-use changes and thus understate the true scale of agricultural conversion. Higher-resolution spatial data collected over time are needed. The size of many protected areas is known, but their precise geographic boundaries are not easily determined. Improved road-network, land-use, and land-cover data would help to assess the level of habitat fragmentation in agricultural landscapes. Improved data on production systems diversity could be used as a proxy for the size and quality of wildlife habitat within agricultural areas. Little data has been collected systematically on the abundance of wild flora and fauna in and around agricultural production areas, and on the impacts of specific crop combinations and management changes on wildlife populations (Wood, Sebastian, and Scherr 2000).

Some of these challenges can be addressed by national agencies, including improvements in the use of remote sensing, collecting information on site location and geographic characteristics for agricultural census and survey data so that these can be linked to environmental data, and long-term monitoring of the environmental features of a panel of farms. But much of the necessary information to guide farmers' and local communities' efforts requires locally based monitoring. It is important to be able to document the impact of interventions, not just in terms of areas where practices have changed, but actual changes in wildlife species diversity and numbers.

For example, the Dunelands Farmers Foundation in The Netherlands, a group of 220 farmers working in and around a dune area in the process of becoming a national park, has developed its own wildlife gauge. An instrument to measure nature conservation value at the individual farm level, it provides an objective score of the density and quality of wildlife on a particular farm, quantifies the efforts of the individual farmer on behalf of wildlife, and stimulates greater awareness among farmers about nature and wildlife on their farms. The gauge may be linked to financial rewards through the Subsidy Ruling for On-farm Nature Conservation of the Dutch government.

The Land Stewardship Council, a private, nonprofit, membership-based organization to support sustainable agriculture in the midwestern United States, has developed a monitoring "toolkit" for farmers to use with neighbors in assessing wildlife and other environmental variables on their farms. The toolkit was developed by a twenty-six-member interdisciplinary team from universities, public agencies, and farmers; it can be downloaded from www.landstewardshipproject.org (G. Boody, Land Stewardship Project, personal communication, February 20, 2001). The Biodiversity Support Program (a collaborative project of the WWF, The Nature Conservancy, and World Resources Institute) developed a threat reduction assessment method that is intended to be a low-cost practical alternative to monitoring systems based on intensive biological data collection (Margoluis and Salafsky 2000). Participatory research projects have also been successfully organized in developing countries to collect farm information for developing biodiversity-related innovations (L. Sperling, International Center for Tropical Agriculture [CIAT], personal communication, March 2001) and for water quality and aquatic biodiversity (Deutsch and Orprecio 2001).

Efforts are under way, through the Global Invasive Species Programme (GISP) (a consortium coordinated by the Scientific Committee on Problems of the Environment [SCOPE]), to develop an easily accessible invasive-species-monitoring data acquisition system that links data monitoring with research results, information on identification, and information on ecology and control of invasive alien species. Under the Convention on Biological Diversity, a new International Pollinators Initiative was created in 1999 to monitor pollinator decline and its impact on pollinator services, and to identify restoration requirements to sustain pollinator diversity in agriculture and related systems. The FAO has established a European Network on Pollination, and a Pollination Work Group has been set up recently in Kenya. Other such efforts have also been established elsewhere, indicating a healthy interest in this topic.

Extension and Farmer Support

To meet the challenges posed in this book will require changes in the institutions providing technical assistance and organizational support to farmers. They will have to embrace the synergies and opportunities of agriculture-biodiversity linkages and collaboration. Extension professionals need to become information brokers, providing information to farmers and local communities on ways to design farms and landscapes to enhance both agricultural and biodiversity values. Another key role will be that of the facilitator in multistakeholder, multiobjective landscape planning. The extension agent will need to play a third role providing support to strengthen local farmer organizations.

Individuals playing these diverse roles will need training in biodiversity and habitat protection, agricultural production, and group facilitation. Ideally, diverse public, civic, and private organizations—including farmer federations, commercial advisory services, and technical assistance entities sponsored by agribusiness—will evolve to provide these support services. Jurisdictional confusion and competition among public agriculture, forestry, and conservation extension programs must be resolved in order to enable more effective service in the areas of shifting cultivation, agroforestry, and integrated conserva-
~~~~~~~~~~ment, all of which could have significant biodiversity

benefits. Programs for professional certification in ecoagriculture may be developed.

Decision-making processes at all levels of management need to be supported by applied research that yields a constant flow of information on the productive and ecological functioning of agricultural systems. Applied researchers must develop the kinds of information that will help answer ecosystem-specific questions posed in planning exercises about trade-offs, thresholds, and so on. Successful models of this approach may be seen in ICRAF's on-farm agroforestry research networks of eastern and southern Africa (Franzel and Scherr 2002), as well as in some of the agriculture and natural resource management projects of the International Fund for Agricultural Development (IFAD). In East Anglia and the West Midlands of the United Kingdom, the Arable Stewardship Scheme, a three-year pilot program begun in 1998, has been helping farmers re-create and enhance wildlife habitat in arable areas by testing suitable methods in commercially practical ways (RSPB 2001).

Many grassroots organizations are already working with farmers to help them meet their information needs in fields such as marketing, decentralized energy resources, vocational training, genetic resources, approaches to crop protection, seed exchange, organic farming, water management, agroforestry, and so forth. These information dissemination processes are ideally suited to feed information about wild biodiversity to the farmers who are most in need of such information.

For example, in India People's Biodiversity Registers (PBR) have been established to promote folk ecological knowledge and wisdom by devising more formal means for their maintenance and by creating new contexts for their continued practice. The PBR Program is based on fifty-two study localities in eight Indian states, primarily in rural and forest-dwelling communities that cover all of the climatic zones of the country (Gadgil et al. 1998). By recording local information with full acknowledgment of the sources, the program serves as a means of sharing any benefits that may flow from further economic utilization of such information. The comprehensive data collection engaged some 350 researchers and 200 assistants from village communities (Chatre et al. 1998). Two of the PBRs, one from a village in Rajasthan and one from a village in Orissa, recorded examples of spontaneous establishment of regulated-use regimes that led to resource recovery. In both cases, revegetation and water harvesting increased fodder and crop production. Following the initial success of the PBRs, a large number of people from all over India have expressed an interest in undertaking PBR exercises in their own area, with similar interest expressed from Brazil and South Africa. The hope is that PBRs will evolve into a useful tool supporting a process of community-based management of living resources contributing to the conservation of wild biodiversity and rewarding folk knowledge (Gadgil et al. 1998).

Education

Building the intellectual capacity, especially in the developing countries, to work on biodiversity and agriculture issues is a top priority. Since the 1970s, great strides have been made in promoting environmental education in primary schools around the world and in integrating environmental science into high school curricula. Environmental NGOs have also worked to develop informal educational materials and media to reach those not in school, and even to translate some of this material into local languages. However, relatively little progress has been made in providing environmental curricula as part of farmer and agricultural science education, particularly relating environmental services directly to agriculture or teaching environmentally sensitive land husbandry.

The challenge for higher education is even more acute. The developing world—home to approximately 80 percent of the world's population—can claim just 6 percent of the world's scientists and technicians. Of these, a relatively small proportion now work in fields directly relevant to biodiversity conservation, and a disproportionate number of these are concentrated in Brazil, Mexico, India, and East Asia (Cracraft 1999). Numerous programs have been established over the past decade in fields relevant to ecoagriculture, such as agroecology, ecological economics, wildlife sciences, agroforestry, and ecosystem restoration, but most of these are still modest in size and resources and very few focus explicitly on developing ecoagriculture systems as defined here.

Financing Investment in Ecoagriculture

The lack of investment in measures to conserve wild biodiversity exists in part because biodiversity is basically a public good, being provided to everyone, rather like education, defense, and law and order. Because many of the economic benefits of biodiversity are available to all, any one individual, community, or commercial firm has fewer incentives to conserve this resource than would society as a whole. Market economies typically underprovide public goods like biodiversity, because the full social benefits of these goods are beyond appropriation by markets. To address this gap, various instruments have been developed in recent years to finance biodiversity investment that are well suited also to ecoagriculture. These instruments include government funds, international development agency grants and loans, conservation trust funds, civic contributions, and private-sector investments (Bayon, Lovink, and Veening 2000).

Government Investment

Governments will continue to play a leading role in financing and influencing the financing of biodiversity conservation. In the 1990s, many developing countries cut public spending on agriculture, as international donors and lenders pressed for smaller government. As a result, during 1990–96 agriculture expenditures grew less than 3 percent annually in low-income countries (excluding India and China) and 2 percent in Africa, not enough to keep up with population growth (Pinstrup-Andersen and Cohen 1998). Increasing agricultural-sector investment in general can provide greater scope for integrating biodiversity-enhancing designs into agricultural programs. Linking political support from environmentalists with political support from the farming and agricultural communities has the potential to reverse the longstanding marginalization of agriculture and to encourage agriculture to move toward greater sustainability (Bridges et al. 2001).

But specific government investments in biodiversity conservation in agricultural areas are also needed, principally for research, extension, education, monitoring, ecosystem and protected area management, and payments to farmers for highly valued biodiversity services. In many cases, rather than directly finance all these activities, government can play a facilitating role in mobilizing investment resources from other sources outside the agriculture sector. As many biodiversity benefits are public goods, it is appropriate to finance ecoagriculture from general revenues or civic organizations. But other benefits are potentially private goods, the value of which is commercially marketable (for example, genetic resources, tourism resources, pollination services, and water resources). These benefits could be financed, at least in part, through fees for users who receive the benefits (Shah 1995).

Special taxes could be imposed as well, with revenues targeted explicitly for biodiversity conservation. For example, an "ecological" value-added tax was introduced in four states of Brazil in 1991, following state legislation to reallocate the tax according to environmental criteria. The ecological tax revenues are distributed to municipalities according to the extent to which they restrict land uses in favor of conservation and water protection. The mechanism explicitly recognizes the need to compensate municipalities for forgone income, and it makes payments linked to environmental performance indicators (Bernardes 1999; Richards 1999).

Rural Financial Institutions

Farmers should be able to access the rural banking system for profitable ecoagriculture investments and operations to the same extent as investments for conventional farming. Credit programs can also be designed especially for ecoagriculture, with public programs to target, guarantee, or subsidize bank loans. Private payments for the environmental services of particular agroecosystems could be channeled into ecoagriculture loan programs for farmers in that agroecosystem. Such a program, called Pro-Ambiente, is now being developed to promote sustainable agriculture in the Amazon rainforest (Nepsted 2001).

International Development Banks and Agencies

Some financing for ecoagriculture may be available from international development banks and bilateral aid agencies. Bilateral assistance from many developed countries already provides considerable resources for biodiversity conservation in developing countries, and some of those resources could be allocated to ecoagriculture. The Biodiversity Support Program mentioned earlier is an example. This collaborative program among three conservation organizations (which included some ecoagriculture activities) was funded by USAID for twelve years (Margoluis and Salafsky 2000). However, overseas development aid to agriculture plummeted almost 50 percent in real terms between 1986 and 1996 (Pinstrup-Andersen and Cohen 1998). Reversing that trend is a priority, and ecoagriculture may provide an attractive focus to combine donor interests in poverty reduction and conservation.

Meanwhile, governments of developing countries are eligible for loans and grants for biodiversity conservation from many international development finance institutions. Since 1992 the Global Environment Facility (GEF), managed by the World Bank, UNDP, and UNEP, has funded biodiversity protection programs for resources of global importance. The GEF is now spending over U.S.$100 million per year on biodiversity conservation of all types. The World Bank, GEF, and Conservation International recently designated U.S.$150 million for a Critical Ecosystem Partnership Fund to help local groups protect the world's biodiversity hotspots. The GEF's new Operational Program 13 specifically targets agricultural biodiversity in productive landscapes (GEF 2000). By early 2001, more than seventy projects of the World Bank worth U.S.$23 billion had some (usually minor) biodiversity components in "sustainable use in agro-landscapes." Of the Bank's own resources committed to these projects (U.S.$7.5 billion), $1.5 billion was targeted to biodiversity. The World Bank also has a few large-scale projects focused on biodiversity, such as the Mesoamerican Biological Corridor (described in Box 5.2), which is jointly funded with UNDP (Miller, Chang, and Johnson 2001). The Inter-American Development Bank has been active in biodiversity conservation in Latin America, for example by co-financing the EcoEnterprise Fund discussed in Chapter 7. International public institutions including the IFAD have also begun to integrate biodiversity objectives into development planning in agricultural regions.

Conservation Trust Funds

Considerable resources are available from national and international sources to cover the one-time cost of purchasing land or permanent farm easements for protected areas. A chronic challenge, however, is to finance the continuing maintenance of those areas, whether in large protected areas or in the interstitial lands around farms. Fire control, invasive pest control, monitoring of endangered species, and protection from encroachment must often be locally funded. Yet national and particularly municipal and state governments usually have no revenue source for such activities. One approach to fund public or civic management is the establishment of conservation trust funds that collect ear-marked revenues and disburse them for environmental and conservation pur-poses. They usually operate on the basis of a capital endowment fund that gen-erates interest for financing environmental activities. Many Latin American countries now have some kind of trust fund, exemplified in the well-developed system found in Colombia. Indonesia has built up a reserve of more than U.S.$700 million in its National Reforestation Fund from a 32 percent share of forest fees. At the international level an "International Rainforest Fund" was proposed by UNEP based on a charge proportionate to the gross national product of each country.

Trust funds in twenty tropical countries and almost all the transitional economies of Eastern Europe have been funded by debt swaps, GEF contri-butions, and other multilateral and bilateral aid funds. Innovative domestic funding methods have sometimes also been used to supplement external financing, such as a tourist tax in Belize and a tax on airline tickets in Algeria (Richards 1999). It has been suggested that the Highly Indebted Poor Coun-tries Initiative (established by the International Monetary Fund and the World Bank in 1996) be used to reduce the unsustainable debt burden and allocate savings to conservation trust funds. For example, in Madagascar, 5 to 10 per-cent of the Initiative's proceeds have gone to support environmental programs (Guerin-McManus and Hill 2001).

Civil Society

The environmental movement began as a citizens' initiative, so it is not sur-prising that many of the most important activities supporting biodiversity in agriculture today are carried out by NGOs. International NGOs such as the World Wide Fund for Nature (WWF) plus thousands of national NGOs in developing and developed countries directly channel and leverage both finan-cial and in-kind contributions from the general public to support biodiversity. The Nature Conservancy leverages private investments and donations of over U.S.$400 million per year to fund opportunities to keep land in conservation trusts. A large part of their work on ecosystem protection in the United States is now with farmers and ranchers, a reflection of their focus on ecosystem man-agement (B. Boggs, The Nature Conservancy, personal communication, Febru-ary 22, 2001). In many developing countries, civic groups finance protected areas and provide support services to innovative farmers. Governments may, as in the United States, encourage such investments through tax incentives.

Farmers' organizations may also mobilize to raise resources needed for bio-diversity investments, particularly where these directly benefit farming, as in the reestablishment of habitat for pollinating insects or windbreaks for pro-tecting valued crops and livestock. Governments, trust funds, and civic organ-izations in many parts of the world are helping to catalyze farmer action to enhance biodiversity through innovative cost-sharing programs of various types. For example, in North Dakota (USA), a monetary legal award of U.S.$13 million for environmental damage to local wetlands granted in the mid-1980s was used to fund the quasi-public North Dakota Wetlands Trust. Its board is composed of three members appointed by the state governor from among irrigation, energy, and real estate interests, and three members nomi-nated by conservation organizations such as the National Audubon Society and Ducks Unlimited. The group funds research and pilot projects on wetlands protection (B. Rusmore, Institute for Conservation Leadership, personal com-munication, February 20, 2001).

Private Sector

A little-recognized movement promoting environmental responsibility has begun in some segments of the private sector. Indeed, participation by private-sector interests could offer truly global-scale potential for biodiversity protec-tion, by providing options for financing far beyond the reach of other actors. For example, a growing number of venture capital firms are promoting invest-ment in biodiversity, and often targeting agricultural enterprises. One of these, the EcoEnterprise Fund, is sponsored by an international conservation NGO collaborating with the business sector and obtaining co-financing from an international development bank. A2R, a partnership between a Latin Ameri-can business firm and an American bank, is investing in various agribusinesses in the Amazon Basin to improve industrial and business performance while ensuring sustainable, biodiversity-friendly production from rural suppliers. Direct industrial supply arrangements that integrate biodiversity conservation, such as the Daimler-Benz POEMA project (described in Box 10.2) and the sustainable forestry sourcing programs of some furniture makers (described in Chapter 10), are becoming more common.

Agribusiness is likely to increase direct investment in ecoagriculture for a variety of reasons: to avoid or comply with environmental regulations; respond to ecosystem management plans; protect the environmental services support-ing long-term land and forest investments; or respond to concerns about the sustainability of production from outgrowers (private producers contracting to supply an agricultural processing company). Future growth can also be expected in private consulting and contracting services that specialize in eco-agriculture for landscape and whole farm management. Many of these firms

will strive to remain state-of-the-art and will be positioned to translate scientific findings into practical ecoagriculture systems for private producers, government agencies, and conservation programs.

Meanwhile, a suite of frameworks and tools has emerged to start making
sustainability an operational concept in the corporate environment. The Natural Step program, an overarching framework developed in 1989 in Sweden,
is preeminent in Europe and now attracts major U.S. corporations. This program analyzes the "ecological footprint"—or ecological impact—of each
company and recommends strategies, industrial methods, and business practices to reduce that footprint. A scheme called the Global Reporting Initiative
is under development to incorporate a wide array of environmental measurements into annual business reporting that can be used to encourage corporate
environmental accountability. Efforts to address climate change serve as a useful corporate model for pursuing sustainability issues. Twenty-one corporations (largely Fortune 500 companies, with U.S.$550 billion in collective
annual revenue) have joined the Pew Center on Global Climate Change to
research alternative courses of action. Member corporations are taking voluntary steps now to reduce carbon emissions (Daily and Walker 2000). A similar,
worldwide mobilization could be spurred by corporations with agricultural
and agribusiness interests to conserve biodiversity in agricultural regions.

Payments for Environmental Services

As discussed in Chapter 10, payments for environmental services—particularly
for carbon sequestration (to mitigate climate change) and for watershed or
biodiversity services—could provide a new source of financing for ecoagriculture. Since the performance of ecoagriculture in providing these services
would have to be monitored and certified in order to determine payment,
such schemes could become an important source of local data to improve
analysis of ecoagriculture systems.

Conclusion

Mainstreaming the practice of ecoagriculture requires that the approach be
embraced and supported by institutions responsible for economic, land-use,
and conservation planning—by farmer organizations; by public and civic institutions of agricultural research, extension, and education; and by financial
institutions responsible for public, private, and civic investment in agricultural
production and in conservation. Well-organized initiatives are needed to introduce and strengthen ecoagriculture activities in such institutions.

Chapter 12

Bringing Ecoagriculture into the Mainstream

As we enter this new century, the world faces two major land-use challenges: how to stem the tide of biodiversity loss, which is at its highest rate ever; and how to feed an additional 75 to 85 million people each year, most of them in countries suffering the greatest biodiversity loss. These challenges will only be met if those individuals and institutions promoting biodiversity and those promoting agricultural development work together.

Governments, scientists, and conservation organizations have begun to respond to the looming loss of biodiversity. In the decade or so since the publication of Wilson and Peter's groundbreaking *Biodiversity* (1988), considerable progress has been made in promoting the conservation of the world's variety of genes, species, and ecosystems. The concept of biodiversity has led to comprehensive new approaches to conservation (WRI, IUCN, and UNEP 1992; McNeely et al. 1990). The Convention on Biological Diversity has been ratified by more than 180 countries. But most of those efforts have been directed to conserve habitat and species in remote regions of low population density. In more populated agricultural regions, protected areas are being fenced off on maps, but they have little hope of functioning effectively to protect biodiversity unless meaningful strategies to modify agriculture in the surrounding environment can be implemented.

Previous chapters have described numerous examples of agricultural innovations that have successfully protected endangered wild animal and plant species by encouraging farmers to protect or restore natural habitat, by reducing negative impacts of agriculture on neighboring wildlife habitat, and by increasing the value of farmland itself as wildlife habitat. Concrete on-the-ground improvements from ecoagriculture have already been accomplished; some of these are described in the thirty-six case studies presented in Chapters 6 and 7. Interestingly, many of these positive biodiversity impacts were achieved serendipitously; the main concerns of innovators (at least initially) were to improve agricultural productivity or sustainability. It might be expected, then, that intentional pursuit of such complementarity will lead to even greater impacts, and more quickly.

This final chapter considers the role of ecoagriculture in overall conservation strategies and assesses the potential scale of adoption. Some promising opportunities for integrating biodiversity conservation, agricultural production, and poverty reduction initiatives are highlighted. Three critical elements needed to get the ecoagriculture agenda moving on the scale required to achieve biodiversity goals are emphasized: visionary leadership to achieve collaboration between agriculturalists and environmentalists; expanded research to devise the needed technological and institutional innovations; and financial investment in ecoagriculture.

Ecoagriculture's Role in Conservation Strategies

Some conservationists have expressed concern that pursuing ecoagriculture strategies could undermine initiatives to establish more protected areas, or be used to justify the further expansion of agriculture into undeveloped areas. Such concerns are unwarranted so long as ecoagriculture is promoted within an ecosystem approach. Indeed, ecoagriculture is both supportive of and complementary to other conservation strategies.

Large protected areas have a critical role to play in global conservation strategies; they provide a haven for wild plants, animals, and microorganisms that are sensitive to human intervention, that require large contiguous habitats for their survival, or that may pose a danger to human safety. For this reason it is important to ensure that all the major habitats of the world are represented in a system of reserves. Several of the ecoagriculture strategies explored in this book seek *explicitly* to enhance both the extent and the viability of protected areas, by increasing their benefits to neighboring farming populations and by increasing the productivity of agricultural land, which can free up new lands for protection.

But in a world where the current generation's grandchildren may number 9 billion, such protected areas are not enough. Even for those species that can thrive only in protected areas, the habitat quality and environmental impacts of surrounding landscapes will be an important determinant of conservation success. In the much larger areas dominated by human settlements and agricultural land use, it will be critical to design a matrix of land uses that can host rich species diversity by strategically integrating a network of reserves with compatible agricultural systems containing small patches of varied habitats.

Conservationists are accustomed to think of areas of production agriculture, livestock husbandry, forestry, and aquaculture as areas bereft of wild biodiversity (some have characterized them as "zones of sacrifice"), whose extent must be minimized. But this view ignores the livelihood needs, not to mention the cultural values, of much of the world's population. The world needs a new vision that sees productive agricultural landscapes as potential opportunities for biodiversity conservation—not for all wild species, but for many. Certainly the essence of agricultural production is to create habitats conducive to the growth of selected valuable domesticated species on a sustainable basis that is economically viable. Some crops, such as annual grains, are more sensitive than others to weeds and shade. However, many agricultural practices that threaten biodiversity contribute little to productivity, and these are amenable to significant improvements. Many traditional and modern agricultural practices and systems are already compatible with high biodiversity. And agriculture itself benefits from many elements of wild biodiversity, as do the ecosystem functions of landscapes that indirectly support wildlife. New research has uncovered a surprising level of biodiversity even in many intensive agricultural systems. Indeed, much of today's "wild" has grown up from abandoned agriculture fields, grazing lands, and managed forests, and their species mix has been fundamentally shaped by human management.

While large wild animals like rhinos and lions are unlikely ever to find a niche within intensive agricultural systems, it is feasible to carve a place for them in some extensive farming and pastoral systems. The scope for greater coexistence between agriculture and myriad smaller species certainly seems promising. At a minimum, it should be possible to maintain wild plant and animal populations currently present in agricultural regions, even as production grows. As farmers gain experience and skill in ecoagriculture, it should be possible to increase wild biodiversity significantly in many types of farming systems. The best way to achieve this is to empower the people who live on these lands with the resources to enhance their biological value for production, as well as ecosystem function and biodiversity. Ecoagriculture offers a strategy to heighten the ecological "literacy" of rural communities, which in turn could strengthen the communities' commitment to broader conservation strategies, including protected areas. In many cases, of course, progress in wildlife conservation will depend upon complementary progress in other aspects of ecosystem management—waste management, controlling industrial pollution and urban sprawl, and infrastructure development.

Potential Scale of Adoption and Impact

The thirty-six examples in Chapters 6 and 7 illustrate that wild biodiversity *can* be preserved in agricultural regions if farmers find ways to gain economically from environmentally beneficial changes in farming systems. The examples show that local farmers and rural communities can be natural allies with environmentalists in promoting wild biodiversity, and that technical innovations in agriculture can build on that alliance. Table 12.1 summarizes the key features of the thirty-six examples. Some of the benefits achieved by these case studies include:

- *Biodiversity:* All had positive impacts on biodiversity, either indirectly (Examples 11–15) or directly.
- *Farm income:* Of the thirty-six cases, twenty-eight demonstrated clearly positive economic benefits, in a few cases doubling or tripling farm income. Five cases had a neutral impact on incomes, and in three cases (one experimental system and two responses to regulation), incomes were less than in conventional agricultural practice.
- *Poverty:* In twenty-five of the examples, the principal beneficiaries of ecoagriculture innovations were poor, small-scale farmers in developing countries. Poor farmers benefited from reduced production risks, increased assets, and more secure livelihoods. Another eleven cases benefited middle- to high-income producers in developing (four cases) and more developed countries (seven cases). Ecoagriculture systems were adapted to farmers' financial circumstances: wealthier producers used more capital-intensive approaches, while poorer farmers used more labor-intensive approaches.
- *Food or fiber supply:* More than two-thirds of the examples improved food or fiber supply; the others had a neutral effect, except for two examples from developed countries with high surplus production.
- *Scale of adoption:* Of the thirty-six ecoagriculture examples, at least eight are already being practiced on more than 1 million hectares and another nine on more than 100,000 hectares. The other nineteen are being practiced on small areas, although the potential scale of application beyond the study sites is much larger for most of these.

Thus, even with minimal research, extension, and policy support, millions of hectares of agricultural land have begun the transition to ecoagriculture. All over the world, local communities, environmental organizations, farmer groups, and researchers are taking action in their own agricultural areas to enhance wild biodiversity. But to have a meaningful impact on the global state of wild habitat conservation and species protection requires replicating such efforts on an unprecedented scale.

Table 12.1. Scale and Impacts of Ecoagriculture—36 Examples

Strategy	Site	#	Case	Type of Farming System	Climate Zone	Scale of Adoption (Hectares)	Supply Impacts	Sustainability Impacts	Biodiversity Impacts	Livelihood Impacts
1	Nepal	1	Buffer zones to protect rhinos and tigers	Community forest mgmt.	Subhumid tropics	100,000	*/−	?	++	+ (poor)
1	Costa Rica	2	Orange farm cooperates with protected area	Perennial crop	Humid tropics	50,000	+	+	++	+ (nonpoor)
1	Australia	3	LandCare groups plan for biodiversity goals	Annual, grazing, marginal lands	Semi-arid subtropics	100,000 +	*	+	++	* (nonpoor)
1	Tanzania	4	Livestock co-managed with wildlife in reserve	Livestock herding	Semi-arid tropics	800,000+	+	+	+	+ (poor)
1	Turkey	5	Agricultural gene sanctuary	Agropastoral, silvipastoral	Temperate	50,000 +	+	+	++	* (various)
1	Philippines	6	Marine reserves help fish and fishermen	Aquaculture	Humid tropics	100,000	+	+	++	+ (poor)
2	Brazil	7	Landowners protect Amazonian saki	Grazing	Humid tropics	10–20,000	*	*	++	* (nonpoor)
2	U.S.	8	Protecting wetlands on farms	Various	Various	100,000 +	*	+	++	* (nonpoor)

(continues)

Table 12.1. *Continued*

Strategy	Site	#	Case	Type of Farming System	Climate Zone	Scale of Adoption (Hectares)	Supply Impacts	Sustainability Impacts	Biodiversity Impacts	Livelihood Impacts
2	U.K.	9	Biodiversity set-asides on farms	Annual crops, quality land	Temperate	100,000 +	−	+	++	− (nonpoor)
2	Costa Rica	10	Farmland corridors for wild animals	Perennial crops, dairy grazing	Humid tropics	50,000 +	+	+	++	+ (nonpoor)
3	Zambia	11	Introduction of cassava	Intensive crops, marginal lands	Subhumid tropics	1,000,000 +	+	+	+	+ (poor)
3	Philippines	12	Irrigated rice reduces fallow agriculture	Intensive crops marginal lands	Humid tropics	100,000 +	+	+	++	++ (poor)
3	Honduras	13	Improved vegetable, coffee technology	Intensive crops, marginal lands	Subhumid tropics	10,000 +	+	−/★/+	++	+ (poor)
3	West Africa	14	Biocontrol of cassava mealybug	Intensive crops, marginal lands	Subhumid, semi-arid	1,000,000 +	+	+	+	+ (poor)
3	Brazil	15	Dairy technology releases pasture for protected area	Dairy cattle grazing	Humid subtropics	Less than 10,000	+	+	++	++ (poor)
3	Mexico	16	Carbon farming	Extensive cropping	Humid tropics	100,000 +	+	+	++	+ (poor)

	Region	#	Description	Farming system	Climate	Population				
4	China	17	Pest control through intercropping in Yunnan	Intensive irrigated	Subtropics	100,000 +	+	+	+	+ (poor)
4	W. Africa	18	Save storks and songbirds with a natural biocide	Agropastoral	Semi-arid tropics	10,000,000 +	++	+	+	+ (poor)
4	Vietnam	19	Clean water through IPM for rice	Irrigated annual crops	Humid tropics	100,000 +	+	+	+	+ (poor)
4	U.S.	20	Restoring fisheries in the Chesapeake Bay	Intensive crops, livestock	Temperate	100,000 +	★	+	+	− (nonpoor)
4	U.S.	21	Intensive dairy grazing conserves water quality	Intensive livestock	Temperate	100,000 +	★	+	+	++(nonpoor)
4	Costa Rica	22	Organic cocoa production in buffer zone	Intensive perennial crop	Humid tropics	10,000 +	+	+	+	+ (poor)
4	Kenya	23	Saving native fish in Lake Victoria	Intensive crops, marginal lands	Humid tropics	100,000 +	+	+	+	? (poor)
5	U.S.	24	Managing flooded rice for wildlife habitat	Irrigated annual crops	Temperate, subtropics	100,000 +	★	+	++	+ (nonpoor)
5	Southern Africa	25	Dambo irrigation in wetlands	Irrigated annual crops	Semi-arid tropics?	50,000 +	+	★	+	+ (poor)

(continues)

Table 12.1. *Continued*

Strategy	Site	#	Case	Type of Farming System	Climate Zone	Scale of Adoption (Hectares)	Supply Impacts	Sustainability Impacts	Biodiversity Impacts	Livelihood Impacts
5	Philippines	26	Erosion control with native vegetative strips	Intensive crops, marginal lands	Humid tropics	20,000 +	+	+	+	+ (poor)
5	Global	27	Preserving soil micro-fauna by minimum tillage	Intensive annual crops	Various	10,000,000 + +	+	+	+	+ (nonpoor)
5	Costa Rica	28	Trees in pastures	Livestock grazing	Humid tropics	1,000,000 +	+	+	+	+ (various)
5	India	29	Regenerating native forests	Natural forest management	Semi-arid, subhumid	10,000,000 + +	+	+	++	+ (poor)
5	East Africa	30	Veterinary vaccine stops lethal wildlife disease	Pastoral	Arid, semi-arid tropics	10,000,000 + +	+	+	+	+ (various)
6	U.S.	31	Native perennial prairie grains	Intensive crops	Temperate	10–20,000	–	+	++	– (nonpoor)
6	Africa	32	Turning farmland fallows into wildlife habitat	Fallow-based cropping	(Sub)humid, tropics	100,000 +	+	+	++	++ (poor)
6	Latin America	33	Biodiversity in shaded coffee	Perennial crops	Humid tropics	100,000 +	*/–	+	+	+ (poor)

						Supply	Sustainability	Biodiversity	Livelihood
6	Africa	34	Saving an endangered African medicine tree	Fallow-based, marginal lands	Humid tropics	Less than 10,000	+	+	+	+ (poor)
6	Indonesia	35	Species-rich agroforests	Fallow-based cropping	Humid tropics	1,000,000 +	+	+	++	+ (poor)
6	Latin America	36	Raising economic value of forest fallows	Fallow-based cropping	Humid tropics	1,000,000 +	+	+	++	+ (poor)

For supply impacts: ++ positive/increased significantly; + modest increase; ★ neutral/no change; – negative/decline; ? uncertain.

For sustainability impacts: + more sustainable; ★ neutral/no change; – negative/decline; ? uncertain

For biodiversity impacts: ++ positive/increased extent of natural habitat; + improved quality of habitat; ★ neutral/no change; ? uncertain

For livelihood impacts: ++ positive/increased net income; + protected food security; ★ neutral/no change; – negative/decline, ? uncertain

A major new thrust is required to promote biodiversity in agricultural regions through ecoagriculture. To be sustainable, such initiatives must help farmers and farming communities fulfill their own needs while making their systems more biodiversity-friendly. Agriculture is the engine of growth for the poorer but often biodiversity-rich countries, and it is still economically important in developed countries. But this engine must be retooled to ensure that biodiversity conservation is part of it. We need to add a catalytic converter of sorts, to ensure that we achieve supply and income goals without destroying valuable biodiversity resources in the process. We must energetically, if realistically, seek the many opportunities to achieve both goals together. It is time to "mainstream" the practice of ecoagriculture.

Opportunities for Promoting Biodiversity in Agricultural Regions

Setting priorities is essential, particularly for public investments in research, monitoring, and ecosystem management at international and national levels, to promote biodiversity in agricultural regions. National dialogue on these issues has begun in North America and is further advanced in some European countries. But the greatest challenges—and greatest need for international support—are in tropical developing countries. An assessment of priorities is beyond the scope of this book. Instead, highlights are presented below for some situations where opportunities in the developing countries appear to be greatest in terms of their contribution to conserving biodiversity, raising or maintaining agricultural productivity, and/or reducing poverty.

Where Biodiversity Is the Top Priority

The environmental community has diverse opinions about how to establish geographic priorities for saving biodiversity. Different groups target areas with the highest species richness, the highest likelihood of success, the highest rate of loss of unique biota, the highest cost-effectiveness, the most competent management authority, the authority most in need of help, or the widest spread of habitat types (McNeely 1997). But regardless of the criteria, ecoagriculture could play a critical role in supporting and enhancing protected areas under pressure from agricultural conversion, pollution, or fragmentation. Ecoagriculture can protect biodiversity by:

- *Preventing land conversion.* At a time when huge tracts of tropical forest are burned in certain seasons, even species that are reasonably abundant over large territories may be seriously damaged within just a few months, as with west Borneo orangutans that suffered greatly from Indonesia's forest fires.

For this reason, some conservationists recommend concentrating efforts in geographic areas where especially diverse or distinctive biota lie in the path of increasing habitat destruction (Sisk et al. 1994). Agriculture at the forest margin is central to these efforts. Technological innovations for farm-fallow, perennial-crop, and forest management techniques can help, along with policy innovations involving transfer payments to farmers for biodiversity conservation, shared buffer zone management, and agricultural intensification in source areas for potential forest migrants.

- *Protecting hotspots.* The "hotspot" approach of Conservation International suggests that areas with the highest species richness or number of endemic species and intense development pressures should be top priorities for conservation. Protected areas in such hotspots will need to be surrounded by well-designed and well-managed agricultural landscapes that provide sufficient ecosystem networks to maintain key species, and that help maintain the environmental services required. Many hotspots are located within major agricultural regions. Traditional agricultural systems may need to be sustained in some hotspot areas where they support rich assemblages of species.

- *Protecting representative habitats.* In 1992, participants at IUCN's Fourth World Congress on National Parks and Protected Areas agreed to a nonbinding goal of protecting 10 percent of each major habitat type by the year 2000 (McNeely 1992). However, fewer than one-third of the world's countries have achieved this goal (World Conservation Monitoring Centre 2000). Greater success in negotiating new protected areas and managing them effectively may be achieved through new types of collaboration with farming communities. Initiatives to protect the highly threatened habitats of tropical Mesoamerica, for example, could benefit from ecoagriculture strategies.

- *Protecting wild relatives of domestic species.* A promising opportunity lies in the protection of habitats deemed important for the wild relatives of domesticated crop and livestock species. Such protection would help to ensure the continued availability of genetic resources for agricultural improvement, thus appealing to farmers, agriculturalists, and indigenous farming communities already engaged in in situ conservation of traditional varieties. Reserves for wild relatives of potatoes in Peru, corn in Mexico, coffee in Ethiopia, citrus in India, wild sheep and goats in the Himalayas, and oil palm in West Africa are promising examples.

Where Agricultural Productivity Is the Top Priority

The greatest potential for eliciting the strong support of farmers and agricultural policy-makers for ecoagriculture will be in those areas where it contributes to agricultural productivity. The most promising strategies for enhancing productivity include:

- *Facilitate transition to more sustainable agricultural systems.* Ecoagriculture can help to support the transition of low-productivity, environmentally degrading farming systems to more sustainable systems. This strategy may apply to extensive farming systems, enabling shifting cultivators to remain viable or to sustainably intensify production. And it may enable intensively managed, but low-output, systems to raise productivity and become more sustainable. Many of the initiatives of the Future Harvest centers for integrated natural resource management address these issues (CIFOR 2000).
- *Restore environmental services critical for agriculture.* Farmers in areas with obvious environmental problems that threaten productivity, such as salinization of water supplies and soil or loss of agricultural pollinators, may be enthusiastic about collaborating in habitat restoration, when such action also provides collateral benefits for agricultural production. For example, in some parts of the Indian Punjab, farmers experiencing salinization are planting salt-resistant tree species and restoring native vegetation, both of which have had positive results for biodiversity.
- *Enhance the value of land and forest assets.* Many of the improvements in water, soil, and vegetation quality that are required to restore the wild habitat value of landscapes also serve to increase the asset value of land. Farmers may be especially interested in ecoagriculture when they seek to rehabilitate degraded farmland to raise productivity, or when they perceive opportunities to enhance the economic value of their land, water, or vegetation assets through payments for environmental services. Habitat restoration could be integrated into these economically motivated land improvements at a lower cost than through stand-alone efforts. Costa Rica's programs supporting native timber plantations and agroforestry, for example, appeal to farmers because of the high economic returns of timber harvests. In some cases payments for water or planting trees to offset emissions of carbon dioxide may provide incentives to manage the landscape in ways that significantly enhance habitat quality.
- *Develop new or higher-value agricultural products.* The increase in national and international trade, together with urban migration, offers new opportunities to market commodities traditionally grown in the wild and consumed locally. By improving product quality, production management (in wild stands or on farm), and marketing systems, these products can become significant income-earners, as illustrated by the tropical fruits produced in Indonesia's agroforests (Heywood 1999). Expanding international and national markets for environmentally certified products can encourage ecoagriculture. Regional policy-makers appreciate the positive economic impacts of new farm-income opportunities.

Where Poverty Reduction Is the Top Priority

Given the importance of natural resources to the livelihood security of poor people throughout the developing world, actions to enhance biodiversity in agricultural regions have the potential either to greatly benefit poor rural people or to harm them. Conservation goals need to be realigned to accommodate the social development priorities of local communities. Ecoagriculture will have the greatest appeal to farmers, policy-makers, and institutions concerned with poverty reduction when it contributes to the following goals:

- *Improve natural assets of the poor.* Areas with high levels of rural poverty will benefit most from biodiversity interventions that improve habitats in ways that enhance the productive value of poor people's land, forest, and water assets. For example, ecoagriculture may emphasize good land husbandry, the establishment of native species that also provide economic benefits in landscape niches accessible to the poor, or habitat restoration that protects community water resources. These actions might not achieve the same level of biodiversity conservation found in protected areas, but they could ensure that conservation achievements are sustained over time, given the self-interest of local people.

- *Protect wild species important to poor people's health and livelihoods.* Opportunities abound for protecting wild species that are of particular importance to the livelihood of poor people. Farmers who have traditionally earned income or obtained food from wild species now under threat may be very interested in ecoagriculture programs that increase both wild species populations and the potential for sustainable harvests of these species.

- *Ensure the provision of environmental services critical to poor people's livelihoods.* Adequate provision of clean water, control of damaging landslides, and other environmental services are essential to the survival and well-being of the rural poor. Poor farming communities may actively seek partnerships with conservationists to promote ecoagriculture as a means of overcoming threats to such services. In the Philippines, for example, poor local communities have joined with the municipal government to reestablish riverine vegetation along major waterways with the use of native species (Catacutan and Mercado 2000).

- *Supplement incomes with transfer payments.* Emerging transfer-payment schemes and markets for environmental services that benefit the poor have promising potential, so long as strategies are consciously pursued to address the needs of the poor (Smith and Scherr 2002). The poor may benefit not only through biodiversity payments, but also through strategic interventions in the design of land management strategies for other services, such as improving water quality or offsetting carbon dioxide emissions. The supplemental income from these payments, earned from providing biodiversity protection (or other services), may increase household consumption and make it economically feasible for the poor to manage their agricultural lands in more sustainable ways.

Challenges for Achieving Ecoagriculture

Working Together—Environmentalists, Agriculturalists, and Advocates for the Poor

Whether the opportunities and potentials for developing ecoagriculture discussed above are realized will depend on mobilizing resources to meet three challenges: leadership for new partnerships, research, and investment.

The first major challenge in practicing ecoagriculture on a large scale is to develop visionary leadership among conservationists, agriculturalists, and advocates for the poor. These leaders must work effectively together to communicate the vision of ecoagriculture as a mainstream development strategy. They need to help the public reconceptualize agriculture's role within global ecosystems and motivate action to reflect that new thinking.

While most biodiversity conservation advocates still remain focused on establishing and managing protected areas or addressing the needs of particular species, the widespread transition to the ecosystem management approach has brought new awareness of the importance of agricultural land use in ecosystem function. This transition has forced groups unaccustomed to working together, or even formerly in conflict, to find ways of collaborating. Environmentalists, agricultural specialists, and social scientists have different conceptual frameworks and research methods. Professional "cultures" sometimes can be incompatible, or at least difficult to understand by other specialists. Language, culture, and class issues constrain effective collaboration of trained professionals with farmers and resource managers. But many have overcome such obstacles, slowly building the trust and understanding needed for real collaboration on complex, long-term goals (Ruitenbeek and Cartier 2001). Key factors common to successful initiatives include a practical problem-solving focus, professional facilitation support to ensure clear and constructive communication, good faith efforts to seek "win-win" solutions, and willingness to experiment and learn from both successes and failures. Proactive attention to team building and establishment of agreed norms for collaboration seem to be essential for these partnerships to be sustainable.

Important steps have already been taken by many individuals and groups in the international environmental and agricultural communities. Some conservation organizations, such as The Nature Conservancy, have shifted much of their effort to working with farmers and ranchers in areas of high biodiversity value. The World Wildlife Fund has begun a new initiative to promote use of "best management practices" for conservation by agricultural commodity producers operating in areas of threatened biodiversity. International agency funding for conservation and protected area activities increasingly includes components that support ecoagriculture in project areas.

Scientists from the agricultural and conservation communities are beginning to collaborate more closely. For example, in July 2001 the Society for Tropical Veterinary Medicine (an organization of professionals concerned primarily with livestock diseases of the tropics and the production of healthier livestock) and the Wildlife Disease Association (a fifty-year-old organization of scientists concerned primarily with the study of management of diseases of wildlife populations) met together for the first time. They jointly issued the "Pilanesberg Resolution" pointing out the interrelatedness of wildlife and livestock health; the resolution also highlighted an array of factors that can affect the success of development and/or conservation efforts and recommended specific actions to enhance the sustainability of such programs (Karesh et al. 2002).

Many organizations with a mandate for poverty reduction have begun to invest more heavily in ecoagriculture. NGOs including OXFAM and CARE; international agencies including IFAD; and bilateral aid programs such as the United Kingdom's Department for International Development (DFID), USAID, and the German Agency for Technical Cooperation (GTZ) have focused on building natural assets of the rural poor, including biodiversity, through improved natural resource management. The Equator Initiative—a partnership of the UNDP with the IUCN, Brasil Connects, the Government of Canada, the IDRC, The Nature Conservancy, the Television Trust for the Environment, and the UN Foundation—is working to catalyze and reward community efforts in the tropics that both reduce poverty and enhance biodiversity (www.equatorinitiative.org).

Most public agricultural institutions have not pursued the agenda of wild biodiversity conservation aggressively, preoccupied as they are with tackling urgent demands to address the many conventional production and sustainability challenges that still face the agricultural sector, in a time of sharply reduced public funding. While wild biodiversity is not yet a high priority for most farmers at the moment, some sustainable farming organizations have undertaken important initiatives. The Wild Farm Alliance was recently organized in the United States explicitly to promote family farms as wildlife habitat. Farmer networks in India and the Philippines, with thousands of members, have made commitments to promote ecoagriculture. Some private-sector firms have begun to explore options for profitable management of agriculture for biodiversity conservation. For example, the Iisaak Regional Forest Management Plan in British Columbia, Canada, resulted from several years of negotiation between private forest companies, indigenous peoples' groups, environmental activists, local unions, and local governments. The goal was to ensure biodiversity protection while also logging sustainably (Iisaak Forest Resources 2000). Growing environmental concerns in both developed and developing countries are creating a more fertile ground for such developments.

Innovation—Pushing the Research Frontier

The second challenge, development of environmentally sustainable and financially profitable ecoagriculture systems that can be integrated into regional ecosystem management, is one of the most compelling scientific and technological challenges of this century. At present, the three goals of agricultural growth, poverty alleviation, and biodiversity conservation seldom complement one another, given existing production systems, landscape organization, and political economy. In many cases, science lacks even the fundamental information about ecological interactions between agricultural and wild species; about the number of individuals needed to maintain stable, sustainable wild populations for different species; or about the thresholds for maintaining habitat quality that would allow the design of better systems. As scientific understanding deepens, more general principles will be uncovered, which will aid in the design of new management systems. But still the specific solutions for most places are uniquely defined by their particular configuration of resources, uses, and users. Agricultural research needs to identify and promote types of technological change that enhance productivity while simultaneously enhancing, or at least not degrading, the resource base and the biodiversity upon which it ultimately depends.

Prolific experimentation is going on at local and subregional levels, and lessons can already be drawn from experiences such as those described in this book. The research community could be mobilized to provide strategic support. But the current scale of work in this area is a small fraction of what will be necessary to make a difference at a global, or even ecoregional, scale. Not only must researchers answer fundamental questions about the design and potential impacts of ecoagriculture, they must develop and adapt new technology and ecological management practices to the many different types of agroecosystems in which threats to wild biodiversity accompany threats to food security. Researchers also need to determine which policy incentives and institutional devices are most sustainable and helpful for enhancing biodiversity in specific settings. The problems are daunting, but human ingenuity provides the means to do far better than what is being done now in terms of both agricultural productivity and environmental protection. A global effort, based on new partnerships, is needed to mobilize research and innovation on the necessary scale.

Investment—Mobilizing Venture Capital for Ecoagriculture

The third major challenge is to mobilize the investment necessary to transform strategically selected agricultural areas to ecoagriculture. Around the world—in farm households, boardrooms, economic planning councils, and parliaments—one often hears the subtle if not explicit message that agriculture represents the "economy of the past." Some argue that the future, even of largely agrarian tropical countries, lies instead with industry and commercial services. They are mistaken. While high-tech industry, telecommunications, and other sectors will play an important complementary role, agriculture will remain a foundation for economic growth and food security in low-income countries for the foreseeable future. In fact, while the challenges for agricultural development have probably never been as pressing as today, its prospects are revolutionary. The world stands on the threshold of technological innovations in microbiology and ecology that will transform the foods we eat, establish agriculture as a major source of biofuels, medicines, and industrial raw materials, and radically alter production systems.

One of the most profound—but also most exciting—challenges confronting the world is to transform agriculture from a major threat to biodiversity to a valued contributor. To do so requires investment to:

- promote awareness among policy-makers, scientists, conservationists, agriculturalists, farmers, and the general public of the potential of ecoagriculture to increase food supply, conserve wild biodiversity, and raise farmer incomes;
- advance scientific understanding of ecoagriculture principles and strategies, to realistically assess the potentials of ecoagriculture and critically evaluate its present limitations and key information gaps; and
- accelerate the development and adoption of successful ecoagriculture methods in selected critical hotspots for global biodiversity loss and rural poverty.

With the right technical, financial, institutional, and policy support in place, the anticipated rise in food demand in the developing world over the next fifty years could serve as an engine of sustained economic growth and radical poverty reduction, especially in the vast rainfed regions of developing countries. Innovations in irrigation could release precious water resources for nonagricultural use. Ecoagriculture initiatives to protect natural habitats and promote wild biodiversity in agricultural regions have the potential to contribute actively to meeting poverty-reduction objectives. At the same time, a failure to invest and plan adequately in developing-country agriculture could have profoundly negative consequences: widespread hunger and insecurity, further environmental degradation, and a sharp acceleration in biodiversity loss.

What ecoagriculture lacks today is the kind of venture capital that made the Internet and computer miniaturization possible. Many millions of individual farmers and investors stand to gain significantly from ecoagriculture, and large benefits will accrue to the public and to wildlife from more secure provision of environmental services. But unlike the case of improved seeds or fertilizers, the returns from much of the necessary research and development cannot so easily be captured directly by the innovator; such work will be less financially attractive to the private sector. Thus, a substantial increase in public and civic investment is necessary to achieve ecoagricultural objectives. To foster the broadest possible innovation, diverse organizations must be involved—including farmer and environmental organizations, NGOs, national and international research centers, and networks to encourage collaboration.

Conclusion

The examples highlighted in this book illustrate that the potential benefits of ecoagriculture are real. Throughout history, humans have shown a tremendous capacity to adapt to changing conditions. While today's wild biodiversity is under unprecedented pressure, promising signs of innovation are coming from many parts of the world—from low-income farmers who are most directly dependent on threatened wild resources, as well as from scientifically trained agroentrepreneurs. Innovative ecoagriculture approaches can draw together the most productive elements of modern agriculture, new ecological insights, and the knowledge local people have developed from thousands of years of living among wild nature. The hope is that the innovations described in this book will help inspire the mainstreaming of ecoagriculture over the next generation, enabling both people and the rest of nature to prosper far into the future.

Glossary

Adaptive management: A continuous loop between implementing field actions to manage natural resources, monitoring the affected ecosystem and human responses, comparing the results against expectations, and adjusting future actions, with each reiteration of activity based on past experience. The adaptive management approach to protecting biological resources rests on a willingness and ability to react to new information as it becomes available.

Agricultural biodiversity (synonym: *agrobiodiversity*): The variability among living organisms associated with the cultivation of crops and rearing of animals, and the ecological complexes of which those species are part. This includes diversity within and between species, and of ecosystems.

Agriculture: The process of modifying natural ecosystems to provide more goods and services for people through the nurturing of domesticated species of plants and animals; systems often use high inputs of energy in various forms.

Agrobiodiversity: Short term for agricultural biodiversity.

Agroecosystem: An ecological and socioeconomic system, comprising domesticated plants and/or animals and the people who husband them, intended for the purpose of producing food, fiber, and other agricultural products.

Agroforestry: A land-use system that intentionally combines the production of herbaceous crops, tree crops, and animals, simultaneously or sequentially, to take fuller advantage of resources. Agroforestry encompasses a wide variety of practices, including intercropping of trees with field crops or grasses, planting trees on field boundaries or irrigation dikes, multistory and multispecies forest gardens or home gardens, and cropping systems using bush or tree fallows.

Alien species (synonym: *exotic*): A species, subspecies, or lower taxon introduced outside its normal past or present distribution; includes any parts, gametes, seeds, eggs, or propagules of such species that might survive and subsequently reproduce. Also called nonnative, nonindigenous, or foreign species.

Allele: One of the normal alternate forms of a gene, located at a specific point on a chromosome, that accounts for a particular trait.

Aquaculture: The propagation and raising of aquatic organisms under human control, as in fish ponds.

Biodiversity: Short for biological diversity.

Biodiversity prospecting: The search for economically valuable genetic and chemical resources in nature.

Biological diversity (sometimes shortened to *biodiversity*): The variability among living organisms from all sources, including terrestrial, marine, and other aquatic ecosystems and the ecological complexes of which they are part; this includes diversity within species, between species, and of ecosystems (Convention on Biological Diversity or CBD, article 2). More generally, the totality of genes, species, and ecosystems in a particular region or the world.

Biological integrity: The capacity to support and maintain an integrated, adaptive community with a biological composition and functional organization comparable to those of the natural systems of the region. Also, the measure of a system's wholeness, including presence of all appropriate elements and occurrence of all processes at appropriate rates. Unlike diversity, which can be expressed simply as the number of kinds of items, integrity refers to conditions under little or no influence from human actions; a biota with high integrity reflects natural evolutionary and biogeographic processes. This definition ignores the reality that human influence is now pervasive.

Biological resources: Genetic resources, organisms or parts thereof, populations, or any other biotic component of ecosystems with actual or potential use or value for humanity. The combination of two important properties distinguishes biological resources from nonliving resources: they are renewable if conserved, and they are destructible if not conserved. The practical target of activities aimed at conserving biodiversity, they are called "components of biological diversity" in the Convention on Biological Diversity.

Bioregion: A part of the earth's surface whose rough boundaries are determined by natural rather than human dictates. One bioregion is distinguishable from another by its flora, fauna, water, climate, soils, and land-forms, and the human settlements and cultures those attributes have nurtured.

Biosafety: The safe transfer, handling, and use of any living modified organism resulting from biotechnology. More broadly, managing the release of transgenic or other organisms into the environment and the potential they represent for causing environmental and economic damage.

Biosphere reserve: A site recognized under the Man and the Biosphere Programme of UNESCO, normally including a protected area, a surrounding buffer zone where resource use is limited, and a transition area where cooperation with local people and sustainable resource management practices are developed.

Biotechnology: Any procedure or methodology that uses biological systems, living organisms, or derivatives thereof to make or modify products or processes for specific use. Recently, some have used the term to refer especially to genetic engineering, which is only one of many applications.

Buffer zone: An area that surrounds a protected area and either serves to provide benefits to nearby human communities or to mitigate adverse effects from human activities outside the area. For example, some buffer zones are intended to protect surrounding agricultural areas from damage by wildlife.

Carnivore: An animal that eats the flesh of other animal species.

Carrying capacity: The maximum population that can be sustained indefinitely in a given area without changing the ecosystem in ways that will eventually reduce the sustainable population. This balance between population and resources is a dynamic one; it is influenced by changes in human technology as well as natural factors.

Climax vegetation: The vegetation that would naturally grow in a particular habitat without human interference, natural catastrophe, or climate change.

Community (biological): All of the groups of animals or plants living together in the same area or environment and usually interacting to some degree.

Comparative advantage: A principle of economic trade theory that a country will benefit by exporting the product in which it has a greater (or "comparative") economic advantage, and import the commodity in which its advantage is relatively less, even if the country has an absolute advantage (or disadvantage) compared to its trading partner.

Connectivity: A measure of how spatially continuous a vegetative corridor, biological network, or matrix is.

Conservation: The rational and prudent management of biological resources to achieve the greatest sustainable current benefit while maintaining the potential of the resources to meet the needs of future generations. In natural resource economics, conservation is a rate of use of a biological resource that ensures that the same or a greater quantity of that resource will be available in the future; thus conservation includes preservation, maintenance, sustainable utilization, restoration, and enhancement of the natural environment.

Corridor: A strip of a particular type of land that differs from the adjacent land on both sides. Such corridors may have important ecological functions, including conduit, barrier and habitat.

Critically endangered species: Defined by IUCN as one that has suffered a population size reduction of 80 percent or more over the past ten years or three generations, whichever is longer.

Dambo: A shallow, seasonally waterlogged depression at or near the head of a drainage network. Also called bani, vlei, marai, boli, or fadama in various parts of Africa.

Development: A process of social and economic advancement, in terms of the quality of human life. The term often implies the dominant Western worldview, involving such elements as a belief in progress, the inevitability of material growth, the solution of problems by the application of science and technology, and the assumption of human dominance over nature. Alternative philosophies are suggested by terms such as "sustainable development" or "participatory development."

Disturbance regime: A process that periodically affects a habitat, such as fire, flooding, or insect outbreak.

DNA (Deoxyribonucleic acid): The universal genetic code for all living organisms.

Domestication: The process of improving the genetic characteristics and management of wild species to make them suitable for farm production. In genetic terms, domestication is accelerated and human-influenced evolution.

Ecoagriculture: Land-use systems designed to produce both human food and ecosystem services, including habitat for wild biodiversity.

Economic externalities: Changes in human welfare due to unintended side effects, often of an environmental nature, that are not directly captured in the market transaction. Also, costs that are generated by the producer but not paid for by him or her; the effect of a project felt outside the project and not included in the valuation of that project. An example is the impact of pollution from upstream livestock operations on the water quality of downstream water consumers, where the producers do not compensate consumers for additional costs of water purification. When a beneficial or detrimental externality is quantified in monetary terms and included in market valuations, it is said to have been "internalized."

Ecoregion: A relatively large unit of land or water containing a geographically distinct assemblage of natural communities that share a large majority of their species, dynamics, and environmental conditions.

Ecosystem: A dynamic complex of plant, animal, and microorganism communities and their nonliving environment interacting as a functional unit in a specific place. Applied by some to cover only major ecosystem types or biomes, such as tropical rainforests.

Edge species: A species of plant or animal that is commonly found in the marginal zone of a biological community.

Endangered species: A species whose population has declined by 50 percent over the past ten years or three generations, whichever is longer, and where the causes of the reduction are not demonstrably reversible or not clearly understood, may not have ceased, or could recur.

Endemic: Native to, and restricted to, a particular geographical region. Highly endemic species are those with very restricted natural ranges; they are especially vulnerable to extinction if their natural habitat is eliminated or significantly disturbed.

Environmental resources: Natural systems that produce services of potential benefit to people such as clean air, clean water, attractive scenery, and so forth.

Environmental services: Beneficial functions that are performed by natural ecosystems, including hydrological services (water supply, filtration, flood control), protection of the soil, breakdown of pollutants, recycling of wastes, habitat for economically important wild species (such as fisheries), regulation of climate, and so forth.

Epiphyte: A nonparasitic plant that grows on another plant and gets its nourishment from the air, such as certain orchids, mosses, and lichens.

Establishment (of species): The process of a species in a new habitat reproducing at a level sufficient to ensure continued survival without infusion of new genetic material from outside the system.

Eutrophication: Overenrichment of a water body with nutrients, resulting in excessive growth of some organisms such as algae and depletion of dissolved oxygen, which in turn causes the death of other organisms such as fish.

Ex situ preservation: The preservation of biological resources outside their natural habitats, as in zoos, aquariums, and botanical gardens as well as in tissue cultures and seed banks. Note that this is not considered conservation as defined here; compare *in situ* conservation.

Extinction: An irreversible process whereby a species or distinct biological population forever ceases to exist. The IUCN defines a species as extinct in the wild when it is known to survive only in cultivation, in captivity, or as a naturalized population (or populations) well outside its historic range. A taxon is presumed extinct in the wild when exhaustive surveys in known and/or expected habitat, at appropriate times (diurnal, seasonal, annual), throughout its historic range have failed to record an individual.

Fallow: A crop field left uncultivated for a period of time, so as to regain its productive capacity. Fields left uncultivated for a short period are grass fallows; longer periods involve bush fallows, and still longer resting periods involve natural forest fallows. Farmers may plant or manage vegetation to enhance its fallowing function, or for economic benefits during the fallow period.

Fauna: All of the animals found in a given area; usually, the total number of animal species in a specified period, geological stratum, geographical region, ecosystem, habitat, or community.

Flora: All of the plants found in a given area.

Fragmentation: The breaking up of a habitat, ecosystem, or land-use type into smaller parcels.

Functional group: Two or more species that perform similar ecological functions and roles and may be able to replace each other to some extent.

Gene: A functional unit of heredity that controls a particular inherited characteristic. Composed of a sequence of DNA located at a specific locus or place on the chromosome, it is a stretch of DNA that tells a cell how to make a particular protein.

Genetically modified organism (GMO): An organism into which has been inserted—through genetic engineering—one or more genes from an outside source (either from the same species or from an entirely different species) that contains coding for desired characteristics, such as herbicide resistance or an antibacterial compound.

Genetic diversity: The full range of species, subspecies, and distinct biological populations of plants, animals, and microorganisms; within a species or population, the full range of genes contained by the species or population. Also refers to the amount of genetic information among and within individuals of a population, species, assemblage, or community.

Genetic drift: Variation in the genetic makeup of a species over time, often resulting from environmental change or isolation.

Genetic resources: Species, subspecies, or genetic varieties of plants, animals, and microorganisms that currently provide important goods and services or may be capable of providing them at some time in the future. Given the rapid increase in biotechnology and limitations of current knowledge, virtually all plants, animals, and microorganisms qualify as genetic resources.

Genome: The complete complement of genes of an organism. In humans, the genome contains approximately 50,000 genes, though some estimates are far lower than this.

Globalization: Worldwide economic integration of many formerly separate national economies into one global economy, mainly through free trade and free movement of capital as by multinational companies, but also by easy or uncontrolled migration.

Guild: A group of species having similar ecological resource requirements and foraging strategies, and therefore having similar roles in the community.

Habitat: The physical and biological environment on which a given species depends for its survival; the place or type of site where an organism or population naturally occurs.

Hectare: Unit of land in the metric system, equivalent to 2.471 acres.

Hedgerow: A narrow corridor of woody vegetation and associated organisms that separates open areas. Examples include hedges, fencerows, shelterbelts and windbreaks.

Herbivore: A species that eats plants.

Heterozygosity: The state of a plant or animal having one or more recessive characteristics in its genetic code and therefore not breeding true to "type." The opposite of homozygosity.

Hotspot (of global biodiversity loss): One of twenty-five terrestrial regions of the world with especially high species richness or high number of endemic species, that are highly threatened with species loss because natural habitats have already been reduced to 30 percent or less of their original land surface area. The hotspots were defined by scientists at Conservation International, an international environmental NGO.

Indigenous peoples: Social groups that have resided in a region for a long period of time and whose social and cultural identity differs from that of the dominant society in a particular region. Their identity is often strongly connected to their ancestral lands in ways that influence conservation behavior. No definition of indigenous peoples has been agreed upon internationally, but the principle of self-identification has been accepted by the International Labour Organisation, the United Nations Draft Declaration on the Rights of Indigenous Peoples, the World-Wide Fund for Nature, and the World Commission on Protected Areas. They are distinctive from other vulnerable social groups insofar as they are recognized by international law and by some states as autonomous seats of power within the state, and they exercise collective rights as groups.

In situ conservation: The protection of ecosystems and natural habitats and the maintenance or recovery of viable populations of species in their natural surroundings and, in the case of domesticated or cultivated species, in the surroundings where they have developed their distinctive properties (Convention on Biological Diversity).

Integrated conservation and development project (ICDP): A means to reconcile conservation and community interests through promoting social and economic development in communities adjacent to protected areas, using a bioregional approach that links the protected area to the surrounding lands, often through the mechanism of buffer zones (Wells et al. 1992).

Integrated natural resource management (INRM): A conscious process of incorporating multiple aspects of natural resource use into a system of sustainable management, to meet explicit production goals of farmers and other land users (e.g., profitability, risk reduction), as well as goals of the wider community (e.g., cultural values, sustainability).

Integrated pest management (IPM): The use of all appropriate techniques of controlling pests in a coordinated manner that enhances, rather than destroys, natural controls. If pesticides are part of the program, they are used sparingly and selectively so as not to interfere with natural competitors.

Intellectual property right: A right enabling an inventor (and more recently, a discoverer) to exclude imitators (or subsequent discoverers, or prior discoverers who did not file legal claim to the intellectual property right) from the market for a limited time.

Intercropping: The growing of two or more crops simultaneously on the same piece of land. Benefits arise because crops exploit different soil, water, light, and other resources, or mutually interact with one another, to raise yields or control pests and weeds.

Intensification (agricultural): Process by which additional labor, capital, or other inputs are used to increase agricultural production on a given unit of land.

Introduction: The movement, by human agency, of a species, subspecies, or lower taxon (including any part, gametes, seeds, eggs, or propagules that might survive and subsequently reproduce) outside its natural range (past or present). This movement can be either within a country or between countries.

Invasive alien species: An alien species whose establishment and spread threaten ecosystems, habitats, or species with economic or environmental harm. These are addressed under article 8(h) of the Convention on Biological Diversity.

Invertebrates: Animals lacking a backbone, such as insects.

Keystone species: A species of plant or animal that has impacts on the community or ecosystem that are disproportionately large relative to its abundance. Also called an umbrella species.

Landscape: A mosaic where a cluster of local ecosystems is repeated in similar form over a kilometers-wide area. A landscape is characterized by a particular configuration of topography, vegetation, land use, and settlement pattern that delimits some coherence of natural, historical, and cultural processes and activities.

Landscape connectivity: The extent to which different patches of habitat are linked together. The opposite of fragmentation.

Landscape diversity: The spatial variation of the various ecosystems within a landscape.

Leaching: A physical process by which water draining through the soil (as rainfall or irrigation water) carries away dissolved soil nutrients that are important for crop production.

Litter: The surface layer of a forest or crop field, in which the leaves and other organic material are in the process of decomposition.

Management (of biodiversity): The efforts of humans to select, plan, organize, and implement programs designed to achieve specified goals. Biodiversity management activities can range from protective measures to ensure that human influences are minimized to greater interventions required to maintain diversity, install facilities, control populations, or eradicate alien species.

Marginal lands: Areas that are unable to support permanent or intensive agriculture without significant investment in land or water management. Without proper management, ecologically fragile marginal lands may degrade quickly following cultivation.

Mariculture: Saltwater aquaculture.

Matrix: The background ecosystem or land-use type in a mosaic.

Microorganisms: Organisms of microscopic or ultra-microscopic size, including bacteria, blue-green algae, yeast, protistans, viroids, and viruses.

Mosaic: A pattern of patches, corridors and matrix in a landscape.

Native species (synonym: *indigenous species*): A species, subspecies, or lower taxon living within its natural range (past or present), including the area that it can reach and occupy using its own legs, wings, and wind- or water-borne or other dispersal systems, even if it is seldom found there.

Natural resources: Resources supplied by nature. These are commonly sub-divided into nonrenewable resources, such as minerals and fossil fuels, and renewable natural resources that propagate or sustain life and are naturally self-renewing when properly managed, including plants and animals as well as soil and water.

Neotropical Region: A zoogeographical unit comprising South America, the West Indies, and Central America south of the Mexican plateau. The region is often sub-divided into Antillean, Brazilian, Chilean, and Mexican sub-regions.

Network: An interconnected system of corridors.

Nongovernmental organization (NGO): A nonprofit group or association organized outside of institutionalized political structures to realize particular social objectives (such as conserving nature) or serve particular constituencies (such as local communities).

Opportunity cost: The benefit foregone by using a scarce resource for one purpose instead of for its best alternative use. For example, an opportunity cost of establishing protected areas may be the value of reduced agricultural production from the area.

Organic agriculture: A type of farming that relies on the earth's own natural resources to grow and process food. Organic practices include cultural and biological pest management; they prohibit use of synthetic chemicals in crop production and antibiotics or hormones in livestock production.

Overexploitation: Unsustainable use of a natural resource leading to the depletion or degradation of the resource and consequent loss of its availability or productivity.

Parasite: An organism living in or on another living organism and obtaining part or all of its nourishment from it without providing commensurate benefits to it. Parasites often have structural modifications adapting them to this way of life.

Participatory development: An approach to "development" that empowers individuals and communities to define and analyze their own problems, make their own decisions about directions and strategies for action, and lead in those actions. The approach is contrasted with "top-down" development processes, in which outsiders, with greater socioeconomic and political power, make the key decisions about local resource use and management.

Patch: A relatively homogeneous nonlinear area in a landscape that differs from its surroundings. (The internal microheterogeneity present is repeated in similar form throughout the area of a patch.)

Permaculture: The conscious design and maintenance of agriculturally productive ecosystems to create the diversity, stability, and resilience of natural ecosystems. This form of agriculture seeks the harmonious integration of landscape and people providing their food, energy, shelter, and other material and nonmaterial needs in a sustainable way.

Pest: Any species, strain, or biotype of a plant, animal, or pathogenic (disease-causing) agent injurious to plants, plant products, animals, or people.

Physiognomy: The apparent characteristics, outward features, or appearance of ecological communities or species.

Pollinators: Animals such as butterflies or bats that transfer pollen from the anther to the receptive area of a flower, enabling seed plants to reproduce.

Pollution: The contamination of an ecosystem, especially from human activities.

Population: A group of individual organisms living in a particular geographical space and sharing common ancestry that are much more likely to mate with one another than with individuals from another such group. When a population has observable characteristics that distinguish it from other populations, it is sometimes called a subspecies. Also, a group of organisms of one species, occupying a defined area and usually isolated to some extent from other similar groups, or geographically defined subdivisions of a species that form a group whose members differ genetically from other members of the species.

Primate: The order of mammals that includes humans, the apes, monkeys, and lemurs, characterized by increasing perfection of binocular vision, specialization of the hands and feet for grasping, and enlargement and differentiation of the brain.

Productivity (in agriculture): The relationship between the average real output of economically usable products, divided by an index of all fixed and variable inputs. "Land productivity" is defined as total output divided by the land area where outputs were produced; "labor productivity" as total output divided by total labor input.

Protected area: A geographically defined area that is designated or regulated and managed to achieve specific conservation objectives (Convention on Biological Diversity, article 2); an area of land or sea especially dedicated to the protection and maintenance of biological diversity, and of natural and associated cultural resources, and managed through legal or other effective means (1992 World Congress on National Parks and Protected Areas).

Public goods: Economic goods whose consumption by any one person does not affect their potential for consumption by others. More than one person can consume a single unit of a public good at the same time, so public goods are jointly consumable. Global public goods include the maintenance of biological diversity and avoiding anthropogenic climate change.

Remnant vegetation: Small patches of native plants that remain after conversion of landscapes to agricultural or other use.

Resource: Anything that is used directly by people. A renewable resource can renew itself (or be renewed) at a constant level, either because it recycles quite rapidly (water), or because it is alive and can propagate itself or be propagated (organisms and ecosystems). See also natural resources and environmental resources.

Ruminant: Any mammal that chews its cud, including cattle, sheep, and antelope.

Shifting cultivation: Any cyclical agricultural system that involves clearing of land—usually with the assistance of fire—followed by phases of cultivation and fallow periods. The fallow period may range from only a few years to several decades. Also called swidden agriculture.

Silvipastoral system: A land-use system combining trees with grass and other fodder species on which livestock graze. The mixture of browse, grass, and herbs often supports mixed livestock species.

Species: A group of interbreeding organisms that seldom or never interbreed with individuals in other such groups, under natural conditions; most species are made up of distinct subspecies or populations.

Species diversity: The number and frequency of species in a biological assemblage or community.

Species richness: The number of distinct species in a given site.

Stakeholder: An individual or institution having an interest (a "stake") in how a resource is managed.

Subsidy: Economic assistance granted directly or indirectly to individuals or organizations to encourage activities designed to satisfy the needs of the public or a particular group. A subsidy is discretionary and revocable, and generally conditioned upon certain rules being observed.

Sustainable agriculture: A way of producing a stable food supply in perpetuity without degrading the natural resources that support production processes.

Sustainable development: The use of natural, human, and economic resources that meets the needs and aspirations of the current generation without compromising the ability of future generations to meet their needs and aspirations.

Sustainable use: The use of biological resources in a way and at a rate that does not lead to the long-term decline of biological diversity, thereby maintaining its potential to meet the needs and aspirations of present and future generations.

Swidden: A cultivated field in an agricultural production system of shifting cultivation or rotational fallow.

Taxon (plural: *taxa*): A taxonomic group, such as a species, genus, or family, in a formal system of classification.

Trophic level: An organism's hierarchical ranking among species in the food chain.

Use: Any human activity involving an organism, ecosystem, or nonrenewable resource that benefits people. The activities range from those having a direct impact on the organisms, ecosystems, or nonrenewable resources concerned (such as fishing, farming, and mining) to those having little or no impact (such as appreciation and contemplation).

Vertebrates: Animals containing a backbone, such as mammals, birds, and fish.

Vulnerable: In reference to a species, one that has an observed, estimated, inferred, or suspected population size reduction of at least 30 percent over the past ten years or three generations, whichever is longer.

Watershed: The area drained by a river or river system, including upper catchments and valleys or floodplains.

Watershed catchment area: The upper area drained by a river basin, where the natural vegetation intercepts rainfall to help replenish underground water supplies as well as streams and rivers.

Wetlands: Transitional areas between terrestrial and aquatic ecosystems in which the water table is usually at or near the surface or the land is covered by shallow water. Wetlands can include tidal mudflats, natural ponds, marshes, potholes, wet meadows, bogs, peatlands, freshwater swamps, mangroves, lakes, rivers, and even some coral reefs.

Wild biodiversity: All nondomesticated plant and animal species.

Wildlife: Living things that are neither human nor domesticated, commonly used to refer to fauna.

References

Abramovitz, J.N. 1996. *Imperiled Waters, Impoverished Future: The Decline of Freshwater Ecosystems.* Worldwatch Institute, Washington, D.C.

Adejuyigbe, C.O., G. Tian, and G.O. Adeoye. 1999. Soil microarthropod populations under natural and planted fallows in southwestern Nigeria. *Agroforestry Systems* 47(1–3):263–272.

Agarwal, A., and N. Sunain. 1999. *Community and Household Water Management: The Key to Environmental Regeneration and Poverty Alleviation.* Poverty and Environment Issues Series No. 2. United Nations Development Programme and the European Commission, New York.

Ahmed, Mahfuzuddin, and Philip Hirsch, eds. 2000. *Common Property in the Mekong: Issues of Sustainability and Subsistence.* ICLARM Studies and Reviews 26. Penang, Malaysia: International Center for Living Aquatic Resources Management—The Fish Center and the Australian Mekong Resource Center.

Akhter, F. 2001. Agricultural biodiversity and the livelihood strategies of the very poor in rural Bangladesh. Paper presented to the International Symposium on Managing Biodiversity in Agricultural Ecosystems, 8–10 November, Montreal, Canada.

Alcorn, Janice B., ed. 1993. *Papua New Guinea Conservation Needs Assessment.* (2 vols.). Biodiversity Support Program, Washington, D.C.

Alexander, Richard B, Richard A. Smith, and Gregory E. Schwarz. 2000. Effect of stream channel size on the delivery of nitrogen to the Gulf of Mexico. *Nature* 403:758–760.

Alfaro, Milena. 2000. Environmental and social factors: Organic Community Products Eco-Enterprises Fund, a Fund of The Nature Conservancy and partners, Washington, D.C. Unpublished memo.

Altieri, M.A. 2001. Agroecology: Principles and strategies. *Overstory* 95. Available at http://www.agroforester.com/overstory.

Altieri, M.A., and Laura C. Merrick. 1987. *In-situ* conservation of crop genetic resources through maintenance of traditional farming systems. *Economic Botany* 41(1):86–96.

Altieri, Miguel A., ed. 1990. *Agroecology.* Springer-Verlag, Berlin.

Altieri, Miguel. 1995. *Agroecology: The Science of Sustainable Agriculture.* Westview Press, Boulder, Colorado.

Amaral, W., G.J. Persley, and G. Platais. 2001. Impact of biotechnology tools on the characterization and conservation of biodiversity. In G. Platais and G.J. Persley, eds., *Biodiversity and Biotechnology: Contributions to and Consequences for Agriculture and the Environment*. The World Bank, Washington, D.C.

Anderson, A.B., ed. 1990. *Alternatives to Deforestation: Steps toward Sustainable Use of the Amazon Rainforest*. Columbia University Press, New York.

Anderson, J. H., J.L. Anderson, R.R. Engel, and B.J. Rominger. 1996. Establishment of on-farm native plant vegetation areas to enhance biodiversity within intensive farming systems of the Sacramento Valley. In W. Lockeretz, ed., *op. cit.*, pp. 95–102.

Angelsen, Arild, and David Kaimowitz. 2000. When does technological change in agriculture promote deforestation? In D.R. Lee and C.B. Barrett, eds., *Agricultural Intensification, Economic Development and the Environment*. CAB International, Cambridge, U.K.

Angelsen, Arild, and David Kaimowitz, eds. 2001. *Agricultural Technologies and Tropical Deforestation*. CAB International, Wallingford, U.K.

Angermeier, P.L., and J.R. Karr. 1994. Biological integrity vs. biological diversity as policy directives: Protecting biotic resources. *BioScience* 44:690–697.

Arimura, G., R. Ozawa, T. Shemoda, T. Nishioka, W. Boland, and J. Takabayshi. 2000. Herbivore-induced volatiles elicit defence genes in lima bean leaves. *Nature* 406:512–515.

Aylward, Bruce, and Alvaro Fernández González. 1998. *Institutional arrangements for watershed management: A case study of Arenal, Costa Rica*. Collaborative Research in the Economics of Environment and Development. Working Paper No. 21. International Institute for Environment and Development, London.

Balvanera, Patricia, Gretchen C. Daily, Paul R. Ehrlich, Taylor H. Ricketts, Sallie-Anne Bailey, Salit Kark, Claire Kremen, and Henrique Pereira. 2001. Conserving bio-diversity and ecosystem services: Conflict or reinforcement? Center for Conservation Biology, Stanford University, and Instituto de Ecología, Universidad Nacional Autónoma de México. Draft.

Bamforth, S. 1999. Soil microfauna: Diversity and applications of protozoans in soil. In W. Collins and C. Qualset, eds., *Biodiversity in Agroecosystems*. CRC Press, New York, pp. 19–25.

Barao, S.M. 1996. Integrated resource management at work: A case study. In W. Lockeretz, ed., *op. cit.*, pp. 159–164.

Barber, C.V., S. Afiff, and A. Purnomo. 1995. *Tiger by the Tail? Reorienting Biodiversity Conservation and Development in Indonesia*. World Resources Institute, WALHI, and the Pelangi Institute, Washington, D.C., and Jakarta.

Barborak, James R. 1995. Institutional options for managing protected areas. In J.A. McNeely, ed., *Expanding Partnerships in Conservation*. Island Press, Washington, D.C., pp. 30–38.

Barrett, G.W., T.L. Barrett, and J.D. Peles. 1999. Managing agroecosystems as agrolandscapes: reconnecting agricultural and urban landscapes. In W.W. Collins and C. Qualset, eds. *Biodiversity in Agroecosystems*. CRC Press, Inc., Boca Raton, Fla., pp. 197–213.

Barzetti,Valerie, ed. 1993. *Parks and Progress: Protected Areas and Economic Development in Latin America and the Caribbean*. IUCN (World Conservation Union) and Inter-American Development Bank,Washington, D.C.

Batisse, Michel. 1997. Biosphere reserves:A challenge for biodiversity conservation and regional development. *Environment* 39(5):7–15, 31–33.

Bayon, R., J.S. Lovink, and W.J. Veening. 2000. Financing biodiversity conservation. Sustainable Development Department Technical Papers Series No. ENV–134. Inter-American Development Bank,Washington, D.C.

Bazett, M. 2000. Long-term Assessment of the Location and Structure of Forest Industries. Global Vision 2050 for Forestry. World Bank/WWF Project, Washington, D.C. January.

Beek, aus der R., C. Rai, and K. Schuler. 1997. *Community Forestry and Biodiversity: Experiences from Dolakha and Ramechhap Districts (Nepal)*. Nepal–Swiss Community Forest Project, Kathmandu.

Beer, J., M. Ibrahim, and A. Schlonvoigt. 2000. Timber production in tropical agro-forestry systems of Latin America. In B. Krishnapillay et. al., eds., *XXI IUFRO World Congress, 7–12 August 2000, Kuala Lumpur, Malaysia, Sub-Plenary Sessions*. Volume I, International Union of Forestry Research Organizations (IUFRO), Kuala Lumpur, pp. 761–776.

Bekkers,V., and J.Verschuren. 1996. Integration of environmental objectives into agricultural policy and law in the Netherlands. In W. Lockeretz, ed., *op. cit.*, pp. 295–304.

Belcher, B. 2000. Notes on rattan policy research in East Kalimantan, Indonesia. Center for International Forestry Research, Bogor, Indonesia.

Bellwood, Peter. 1985. *Prehistory of the Indo-Malaysian Archipelago*. Academic Press, London.

Ben-Ari, E.T. 1998.A new wrinkle in wildlife management. *BioScience* 48(9):667–673.

Bernardes,Aline Tristao. 1999. Some mechanisms for biodiversity protection in Brazil, with emphasis on their application in the State of Minas Gerais. Prepared for the Brazil Global Overlay Project, Development Research Group. The World Bank, Washington, D.C. Draft.

Blackburn, Harvey W. and Cornelius de Haan. 1999. Livestock and biodiversity. In Wanda W. Collins and Calvin O. Qualset, eds., *Biodiversity in Agroecosystems*. CRC Press,Washington, D.C., pp. 85–99.

Board on Agriculture and Natural Resources. 1991. *Towards Sustainability: A Plan for Collaborative Research on Agriculture and Natural Resource Management*. National Academy Press,Washington, D.C.

Boffa, J.M., L. Petri, and W.A.N. do Amaral. 2000. *In situ* conservation, genetic management, and sustainable use of tropical forests: IPGRI's research agenda. In B. Rishnapillay et al., eds., *XXI IUFRO World Congress, 7–12 August 2000, Kuala Lumpur, Malaysia, Sub-Plenary Sessions*. Volume I, International Union of Forestry Research Organizations (IUFRO), Kuala Lumpur, pp. 120–132.

Bojö, J. 1996. *The economics of wildlife: Case studies from Ghana, Kenya, Namibia and Zimbabwe*. Environmentally Sustainable Development of the African Technical Department Working Paper 19. The World Bank, Washington, D.C.

Borrini-Feyerabend, G., ed. 1997. *Beyond Fences: Seeking Social Sustainability in Conservation*. IUCN (World Conservation Union), Gland, Switzerland.

Bourn, David, and Roger Blench, eds. 1999. *Can Livestock and Wildlife Co-exist? An Interdiscplinary Approach*. Overseas Development Institute, London.

Boyd, Charlotte, Roger Blench, David Bourn, Liz Drake, and Peter Stevenson. 1999. Reconciling interests among wildlife, livestock, and people in eastern Africa: A sustainable livelihoods approach. *Natural Resource Perspectives* 45, June.

Boyle, Tim. 1997. Impacts of disturbance on genetic diversity. *Center for International Forestry Research News* 17:5.

Brand, David. 2000. Notes concerning a proposed approach to calculating a biodiversity credit for native forests or planted forests. May. Unpublished.

Brandon, K., and M. Wells. 1992. Planning for people and parks: Design dilemmas. *World Development* 20(4):557–570.

Bridges, Mike, Ian Hannam, Fritz Penning de Vries, Roel Oldeman, Sara J. Scherr, and Samran Sombatpanit, eds. 2001. *Response to Land Degradation*. Science Publishers, Enfield, N.J.

Bright, C. 1999. Invasive species: Pathogens and globalization. *Foreign Policy* 1999: 51–64.

Brown, G.G., A. Pasini., N.P. Benito, A.M. de Aquino, and M.E. Correia. 2001. Diversity and functional role of soil macrofauna communities in Brazilian no-tillage agroecosystems. In United Nations University et al., *op. cit.,* p. 22.

Brown, L. 1995. *Who Will Feed China? Wake-up Call for a Small Planet*. Worldwatch Institute, Washington, D.C.

Brown, Lester. 1994. How China could starve the world: Its boon is consuming global food supplies. *Washington Post,* 28 August.

Bruner, A.G., R.E. Gullison, R.E. Rise, and G.A.B. da Fonseca. 2001. Effectiveness of parks in protecting tropical biodiversity. *Science* 2981:125–128.

Buchmann, Stephan L., and Gary Paul Nabhan. 1996. *The Forgotten Pollinators.* Island Press, Washington, D.C.

Buck, Louise E., C.C. Geisler, J. Schelhas, E. Wollenberg, eds. 2001. *Biological Diversity: Balancing Interests through Adaptive Collaborative Management.* CRC Publishers, Chelsea, Mich.

Buck, Louise E., James P. Lassoie, and Erick C.M. Fernandes, eds. 1998. *Agroforestry in Sustainable Agricultural Systems.* Advances in Agroecology Series. Lewis Publishers, Boca Raton, Florida.

Buckles, D., B. Triomphe, and G. Sain. 1998. *Cover Crops in Hillside Agriculture, Farmer Innovation with Mucuna.* IDRC, Ottowa, Ont., Canada; and IMWIC, Mexico DF, Mexico.

Buresh, R.J., and P.J.M. Cooper. 1999. The science and practice of short-term improved fallows: Symposium synthesis and recommendations. *Agroforestry Systems* 47(1–3):305–321.

Burke, L., Y. Kura, K. Kassem, M. Spalding, and C. Revenga. 2000. *Pilot Analysis of Global Ecosystems: Coastal Ecosystems.* World Resources Institute, Washington, D.C.

Cairns, M., and D.P. Garrity. 1999. Improving shifting cultivation in Southeast Asia by building on indigenous fallow management strategies. *Agroforestry Systems* 47(1–3):37–48.

Caldecott, J. and E. Lutz. 1996. Conclusions. In E. Lutz and J. Caldecott. *Decentralization and Biodiversity Conservation*. A World Bank Symposium. The World Bank, Washington, D.C., pp. 155–164.

Caldwell, J. 2001. Multilateral environmental agreements and the GATT/WTO regime. In Heinrich Boll Foundation, ed., *Tree and Environment, the WTO and MEAs: Facets of a Complex Relationship*. Heinrich Boll Foundation, Woodrow Wilson International Center for Scholars, and the National Wildlife Foundation. Heinrich Boll Foundation, Washington, D.C., pp. 39–56.

Calegari, A. 2001. No-tillage system in Paraná State, south Brazil. In M. Bridges et al., eds., *op. cit.,* p. 344.

California Wilderness Coalition. 2002. *Wild Harvest: Farming for Wildlife and Profitability*. A report on Private Land Stewardship. California Wilderness Coalition, endorsed by Wild Farm Alliance, Defenders of Wildlife, Community Alliance with Family Farmers, and California Sustainable Agriculture Working Group. July.

Campbell, A. 1991. *Planning for Sustainable Farming: The Potter Farmland Plan Story*. Lothian Books, Port Melbourne, Australia.

———. 1994. *Landcare: Communities Shaping the Land and the Future*. Allen and Unwin, St. Leonards, Australia.

Campbell, F.A. 1998. The day farmers let a leopard go free. Global Environment Facility, Washington, D.C.

Cardinale, Bradley, M.A. Palmer and S.L. Collins. 2002. Species diversity enhances ecosystem functioning through interspecific facilitation. *Nature* 415:426–429.

Carew-Reid, J. 1990. Conservation and protected areas on South Pacific Islands: Importance of tradition. *Environmental Conservation* 17(1):29–38.

Carlton, James T., and Jonathan B. Geller. 1993. Ecological roulette: The global transport of nonindigenous marine organisms. *Science* 261:78–82.

Carson, Rachel. 1962. *Silent Spring*. Houghton Mifflin Company, New York.

Catacutan, D.C., and J. Mercado. 2001. *Technical innovations and institution-building for sustainable upland development: Landcare in the Philippines*. Paper presented to the International Conference on Sustaining Upland Development in Southeast Asia: Issues, Tools and Institutions for Local Natural Resource Management. Makati, Philippines, May 27–30.

Center for Applied Biodiversity Science. 2000. *Designing Sustainable Landscapes: The Brazilian Atlantic Forest*. Center for Applied Biodiversity Science at Conservation International and the Instituto de Estudios Socio-Ambientales do Sul da Bahia: Washington., D.C.

CGIAR. 2000. *Annual Report*. Consultative Group on International Agricultural Research, Washington, D.C.

Chatre, A., P. Rao, G. Utkrsh, P. Pramod, A. Ganguli, and M. Gadgil. 1998. *Srishtigyaan: A Methodology Manual for People's Biodiversity Registers*. Centre for Ecological Sciences, Indian Institute of Science, Bangalore, India.

Cheema, G.S., F. Hartvelt, J. Rabinovitch, R. Work, J.Smit, A. Ratta, and J. Nasr. 1996. *Urban Agriculture: Food, Jobs, and Sustainable Cities*. United Nations Development Programme, New York.

Child, S. 2000. Pearl-farming technology could help save coral reefs and generate new wealth. Available at http://www.futureharvest.org/growth/pearl.bkgnd.shtml.

Chomitz, Kenneth M. 1999. Transferable development rights and forest protection: An exploratory analysis. Prepared for a workshop, Market-Based Instruments for Environmental Protection, 18–20 July, John F. Kennedy School of Government, Harvard University. The World Bank, Washington, D.C.

Christensen, D., E. Jacobsen, and H. Nohr. 1996. A comparative study of bird faunas in conventionally and organically farmed areas. *Dansk Orn. Foren Tidsskr* 90:21–28.

Chung, Beth Ritchie. 1999. *Community-Based Land Use Planning in Conservation Areas: Lessons from Local Participatory Processes That Seek to Balance Economic Uses with Ecosystem Protection.* Training Manual, América Verde No. 3. The Nature Conservancy, Arlington, Va.

CIFOR. 2000. *Integrated Natural Resource Management Research in the CGIAR (Consultative Group for International Agricultural Research).* A Report on the INRM Workshop Held in Penang, Malaysia, 21–25 August 2000. Center for International Forestry Research, Bogor, Indonesia.

Cincotta, Richard P,. and Robert Engelman. 2000. *Nature's Place: Human Population and the Future of Biological Diversity.* Population Action International, Washington, D.C.

Clay, Jason. 2002. *Community-based Natural Resource Management within the New Global Economy: Challenges and Opportunities.* A report prepared by the Ford Foundation. World Wildlife Fund, Washington, D.C.

Coedes, G. 1968. *The Indianized States of Southeast Asia.* University of Malaya Press, Kuala Lumpur.

Cohen, Joel E. 1995a. Population growth and Earth's human carrying capacity. *Science* 269:341–346.

———. 1995b. *How Many People Can the Earth Support?* W.W. Norton, New York.

Colborn, Theo, D. Dumanoski, and J.P. Myers. 1996. *Our Stolen Future.* Dutton Signet, New York.

Compagnoni, Antonio. 2000. Organic agriculture and agroecology in regional parks. In Sue Stolten, Bernward Geier, and Jeffrey A. McNeely, eds. *The Relationship between Nature Conservation, Biodiversity and Organic Agriculture.* IFOAM, Berlin, pp. 87–91.

Considine, A., J. Roe, and K. Willard. 1999. Protecting important natural areas, wildlife habitat, and water quality on Vermont dairy farms through the Vermont Farmland Protection Program. In W. Lockeretz, eds., *op. cit.*, pp. 255–264.

Conway, G., and E. Barbier. 1990. *After the Green Revolution: Sustainable Agriculture for Development.* Earthscan Publishers, London.

Conway, Gordon. 1997. *The Doubly Green Revolution: Food for All in the 21st Century.* Penguin Books, London.

Conway, Gordon and Jules N. Pretty. 1991. *Unwelcome Harvest: Agriculture and Pollution.* Earthscan Publications, London.

Conway, W. 1997. The changing role of zoos in international conservation and the Wildlife Conservation Society. *Society for Conservation Biology Newsletter* 4(2):1, 3.

Convention on Biological Diversity. 2002. Text of the Convention on Biological Diversity. Available at www.biodiv.org.

Coombe, Richard I. 1996. Watershed protection: A better way. In W. Lockeretz, ed., *op. cit.*, pp. 25–34.

Costanza, R., R. d'Arge, R. de Groot, S. Farber, M. Grasso, B. Hannon, K. Limburg, S. Naeem, R.V. O'Neill, J. Paruelo, R.G. Raskin, P. Sutton, and M. van den Belt. 1997. The value of the world's ecosystem services and natural capital. *Nature* 387:253–260.

Coughenour, Michael, Robin Reid, and Philip Thornton. 2000. *The SAVANNA Model: Providing Solutions for Wildlife Preservation and Human Development in East Africa and the Western United States.* http://www.futureharvest.org/news/savannarelease.pdf.

Coward, E. Walter, Jr., Melvin L. Oliver, and Michael E. Conroy. 1999. Building natural assets: Rethinking the Centers' [referring to the Future Harvest Centers supported by the CGIAR] natural resources agenda and its links to poverty alleviation. Paper presented to workshop, Assessing the Impact of Agricultural Research on Poverty Alleviation, San José, Costa Rica, September 18–20.

Cowlishaw, G. 1999. Predicting the pattern of decline of African primate diversity: An extinction debt from historical deforestation. *Conservation Biology* 13(5):1183–1193.

Cox, Paul A., and Thomas Elmqvist. 2000. Pollinator extinction in the Pacific Islands. *Conservation Biology* 14(5):1237–1239.

Cracraft, Joel. 1999. Regional and global patterns of biodiversity loss and conservation capacity: Predicting future trends and identifying needs. In Joel Cracraft and Francesca T. Grifo, eds., *The Living Planet in Crisis: Biodiversity Science and Policy.* Columbia University Press, New York.

Current, Dean. 1995. Economic and institutional analysis of projects promoting on-farm tree planting in Costa Rica. In D. Current, E. Lutz and S.J. Scherr, eds., *Costs, Benefits, and Farmer Adoption of Agroforestry: Project Experience in Central America and the Caribbean.* A CATIE-IFPRI-World Bank Project. World Bank Environment Paper Number 14. The World Bank, Washington, D.C., pp. 45–80.

Current, D., E. Lutz, and S.J. Scherr, eds. 1995. *Costs, Benefits, and Farmer Adoption of Agroforestry: Project Experience in Central America and the Caribbean.* A CATIE-IFPRI-World Bank Project. World Bank Environment Paper Number 14. The World Bank, Washington, D.C.

Daily, G. 1997. *Nature's Services: Societal Dependence on Natural Ecosystems.* Island Press, Washington, D.C.

Daily, Gretchen C. 2000. Developing a scientific basis for managing Earth's life support systems. Department of Biological Sciences, Stanford University, Stanford, Calif. Draft.

Daily, Gretchen C. and Paul R. Ehrlich. 1995. Preservation of biodiversity in small rainforest patches: Rapid evaluations using butterfly trapping. *Biodiversity and Conservation* 4:35–55.

Daily, Gretchen C., Paul R. Ehrlich, and G. Arturo Sánchez-Azofeifa. 2000. Countryside biogeography: Use of human-dominated habitats by the avifauna of southern Costa Rica. *Ecological Applications* 11:1–13.

Daily, G.C., and K. Ellison. 2002. *The New Economy of Nature: The Quest to Make Conservation Profitable.* Island Press/Shearwater Books: Washington.

Daily, Gretchen C., and Brian H. Walker. 2000. Seeking the great transition. *Nature* 403:243–245.

Dakora, F.D. 2001. The case of symbiotic and non-symbiotic microbes and their associated host plants. In United Nations University et al., *op. cit.,* pp. 29.

Daszak, Peter, A. Cunningham, and A.D. Hyatt. 2000. Emerging infectious diseases of wildlife: Threats to biodiversity and human health. *Science* 287:443–449.

Debinski, D.M., and R.D. Holt. 2000. A survey and overview of habitat fragmentation experiments. *Conservation Biology* 14(2):342–355.

Deem, S.L., W.B. Karesh, and W. Weisman. 2001. Putting theory into practice: Wildlife health in conservation. *Conservation Biology* 15(5):1224–1233.

De Foresta, H., and G. Michón. 1994. Agroforests in Sumatra: Where ecology meets economy. *Agroforestry Today* 6:12–13.

———. 1997. The agroforest alternative to Imperata grasslands: when smallholder agriculture and forestry reach sustainability. *Agroforestry Systems* 36:105–120.

de Haan, Cees, Henning Steinfeld, and Harvey Blackburn. 1999. *Livestock and the Environment: Finding a Balance.* Report of a study sponsored by the Commission of the European Communities, The World Bank, and the governments of Denmark, France, Germany, the Netherlands, the United Kingdom, and the United States. WRENmedia, Suffolk, U.K.

de Klemm, C. 1992. *Biological Diversity Conservation and the Law: Legal Mechanisms for Conserving Species and Ecosystems.* IUCN, Gland, Switzerland.

Delgado, C.J., J. Hopkins, and V.A. Kelly (with P. Hazell, A.A. McKenna, P. Gruhn, B. Hojjati, J. Sil, and C. Courbois). 1998. *Agricultural Growth Linkages in Sub-Saharan Africa.* IFPRI Research Report 107. International Food Policy Research Institute, Washington, D.C.

de Moor, I.J., and M.N. Bruton. 1988. *Atlas of Alien and Translocated Indigenous Aquatic Animals in Southern Africa.* South African National Scientific Programmes Report No. 144. CSIR, Pretoria.

De Moraes, C.M., W. Lewis, P. Pare, H. Alborn, and J. Tumlinson. 1998. Herbivore-infested plants selectively attract parasitoids. *Nature* 393:570–573.

Deutsch, W.G., and J.L. Orprecio. 2001. Watershed Data from the Grassroots: Is it Enough to Capture the Trends and Turn the Tide? Paper presented at the Research Synthesis Conference of the Sustainable Agriculture and Natural Resource Management Collaborative Research Support Program, Athens, Georgia, November 28–30, 2001.

Dewees, P., and S. Scherr. 1996. Policies and markets for non-timber tree products. Environment and Production Technology Divison Discussion Paper No. 16. International Food Policy Research Institute, Washington, D.C.

Diegues, A.C. 1992. *The Social Dynamics of Deforestation in the Brazilian Amazon: An Overview.* Discussion Paper 36. United Nations Research Institute for Social Development (UNRISD), Geneva.

Dinerstein, E., A. Rijal, M. Bookbinder, B. Kattel, and A. Rajuria. 1999. Tigers as neighbours: Efforts to promote local guardianship of endangered species in lowland Nepal. In J. Seidensticker, S. Christie, and P. Jackson, eds., *Riding the Tiger.* Cambridge University Press, Cambridge, U.K., pp. 316–333.

DIVERSITAS. 2001. *DIVERSITAS Science Plan.* DIVERSITAS Global Change Programme, Paris. Draft.

Diversity. 2000. Cutting-edge conservation techniques are tested in the cradle of ancient agriculture: GEF Turkish project is a global model for *in situ* conservation of wild crop relatives. *Diversity* 16(4):15–18.

Donaldson, J. 2001. What do we need to conserve in agricultural landscapes and what is possible? Conservation farming in biodiversity hotspots in South Africa. In United Nations University et al., *op. cit.*, p. 33.

Döös, B.R., and A. Nilsson. 1992. *Greenhouse Earth.* John Wiley and Sons, New York.

dos Furtado, J. I., N. Kishor, G. V. Rao, and C. Wood. 1999. *Global Climate Change and Biodiversity: Challenges for the Future and the Way Ahead*. WBI Working Papers. World Bank Institute, The World Bank, Washington, D.C.

Dover, N., and Lee M. Talbot. 1987. *To Feed the Earth: Agro-ecology for Sustainable Development*. World Resources Institute, Washington, D.C.

Duelli, P., M. Obriast, and D. Schmatz. 1999. Biodiversity evaluation in agricultural landscapes: Above-ground insects. *Agriculture, Ecosystems and Environment* 74:33–64.

Durning, A. T. 1994. Redesigning the forest economy. In L.R. Brown, C. Falvin, and H. French, eds., *State of the World 1994*. W. W. Norton, New York, pp. 22–40.

DuToit, J.T., and D.H.M. Cumming. 1999. Functional significance of ungulate diversity in African savannahs and the ecological implications of the spread of pastoralism. *Biodiversity and Conservation* 8:1643–1661.

Eaglesfield, Robert. 2000. Scientists invent a "green muscle" to defeat biblical plague of locusts. Future Harvest Foundation, Washington, D.C.

Eaton, P. 1985. Tenure and taboo: Customary rights and conservation in the South Pacific. In *Third South Pacific National Parks and Reserves Conference: Conference Report Volume 2*. South Pacific Regional Environment Programme, Noumea, New Caledonia, pp. 164–175.

Ebregt, Arthur, and Pol de Greve. 2000. *Bufferzones and their Management*. Theme Study Series No. 5. International Agricultural Centre and the National Reference Centre for Nature Management, Wageningen, the Netherlands.

The Economist. 2001. Marine conservation: Net benefits. *The Economist*. February 24, p. 83.

Elphick, Chris S. 2000. Functional equivalency between rice fields and semi-natural wetland habitats. *Conservation Biology* 14(1):181–191.

Erhlich, Paul R., and Gretchen C. Daily. 1993. Population extinction and saving biodiversity. *Ambio* 22(2–3):64–68.

Ehrlich, Paul R., and John P. Holdren. 1971. Impact of population growth. *Science* 171 (March):1212–1217.

Enters, T. 2001. Incentives for soil conservation. In M. Bridges, et al., eds., *op cit.*, pp. 351–360.

Evans, L.T. 1998. *Feeding the Ten Billion: Plants and Population Growth*. Cambridge University Press, Cambridge, U.K.

European Commission. 1999. *Agriculture, Environment, Rural Development: Facts and Figures*. Office for Official Publications of the European Communities, Luxembourg. Available at http://www.Europa.eu.int/comm/dg06/envir/report/ en/index.htm.

Fairhead, J., and M. Leach. 1998. *Reframing Deforestation: Global Analyses and Local Realities—Studies in West Africa*. Routledge, London.

Fan, Shenggen, Peter Hazell, and Sukhadeo Thorat. 1999. *Linkages between Government Spending, Growth, and Poverty in Rural India*. International Food Policy Research Institute, Washington, D.C. Research Report 110:1–81.

FAO (Food and Agriculture Organization). 1993. *FAO Production Yearbook*. Food and Agriculture Organization of the United Nations, Rome.

———. 1996. *The Sixth World Food Survey*. Food and Agriculture Organization of the United Nations, Rome.

———. 1997. *Rome Declaration on World Food Security and World Food Summit Plan of Action.* FAO Conference Resolution 2/95. Food and Agriculture Organization of the United Nations, Rome. October.

———. 1999a. *Food and Agricultural Organization Statistical Database.* Food and Agriculture Organization of the United Nations, Rome.

———. 1999b. *The Global Strategy for the Management of Farm Animal Genetic Resources.* Food and Agriculture Organization of the United Nations, Rome.

———. 2001. *State of the World's Forests.* 2001. Food and Agriculture Organization of the United Nations, Rome.

FAOSTAT. 1996, 1999, 2001. *FAO Statistical Database.* Food and Agriculture Organization of the United Nations, Rome.

Faris, Robert. 2000. *Deforestation and Land Use on the Evolving Frontier: An Empirical Assessment.* Harvard Institute for International Development, Cambridge, Mass.

Ferrari, S.F., C. Emidio-Silva, M. Aparecida Lopes, and U.L. Bobadilla. 1999. Bearded sakis in south-eastern Amazonia—back from the brink? *Oryx* 33(4):346–351.

Ferraro, P.J., and R.D. Simpson. 2000. *The Cost-Effectiveness of Conservation Payments.* Discussion Paper 00–31. Resources for the Future, Washington, D.C.

Fischer, G., M. Shah, H. van Velthuizen, and F.O. Nachtergaele. 2001. *Global Agroecological Assessment for Agriculture in the 21st Century.* International Institute for Applied Systems Analysis and the Food and Agriculture Organization of the United Nations, Laxenburg, Austria.

Flannery, Tim. 1995. *The Future Eaters: An Ecological History of the Australasian Lands and People.* George Braziller, New York.

Foreign Policy. 2001. Prisoners of geography. *Foreign Policy* (January/February): p. 48.

Forman, Richard T. 1995. *Land Mosaics: The Ecology of Landscapes and Regions.* Cambridge University Press, Cambridge, U.K.

Fox, Jefferson, Dao Minh Truong, A. Terry Rambo, Nghiem Phuong Tuyen, le Trong Cuc, and Stephen Leisz. 2000. Shifting cultivation: A new paradigm for managing tropical forests. *BioScience* 50(6):521–527.

Frankel, O.H., and M.E. Soulé. 1981. *Conservation and Evolution.* Cambridge University Press, Cambridge, U.K.

Franzel, Steven, D. Phiri and Freddy Kwesiga. 2002. Assessing the adoption potential of improved fallows in Easter Zambia. In Franzel and Scherr, eds., *op. cit.,* pp. 37–64.

Franzel, Steven and Sara J.Scherr, eds. 2002. *Trees on the Farm: Assessing the Adoption Potential of Agroforestry Practice in Africa.* CAB International, Wallingford, U.K.

Franzel, S., H. Jaenicke, and W. Janssen. 1996. *Choosing the Right Trees: Setting Priorities for Multipurpose Tree Improvement.* ISNAR Research Report No. 8. International Service to National Agricultural Research, The Hague.

Franzluebbeers, Kathrin, Lloyd R. Hossner, and Anthony S.R. Juo. 1998. *Integrated Nutrient Management for Sustained Crop Production in Sub-Saharan Africa.* (A Review). TropSoils/TAMU Technical Bulletin 98–03, Texas A&M University, Dept. of Soil and Crop Sciences, College Station, Tex.

Frazier, Scott. 1999. *Ramsar Sites Overview: A Synopsis of the World's Wetlands of International Importance.* Wetlands International, Wageningen, the Netherlands.

Fritschel, Heidi, and Uday Mohan. 1999. Are we ready for a meat revolution? *IFPRI News and Views* (March): 1–8.

Fuller, R., R. Gregory, D. Gibbons, J. Marchant, J. Wilson, S. Baillie, and N. Carter. 1998. Population declines and range contractions among lowland farmland birds in Britain. *Conservation Biology* 12(9):1425–1441.

Future Harvest. 1999. Scientists discover previously unknown source of pollution that is killing the world's second largest freshwater lake. Future Harvest Foundation press release, November 5.

Gadgil, M., N. S. Hemam, and D. M. Reddy. 1996. People, refugia, and resilience. In F. Berkes and C. Folke, eds., *Linking Social and Ecological Systems: Management Practices and Social Mechanisms for Building Resilience.* Cambridge University Press, Cambridge, U.K.

Gadgil, M. *et al.* 1998. Where are the people? In S.N. Ravijed, *Hindu Survey of the Environment 1998.* S. Rangarajan: Chennai, India, pp. 107–137.

Gadgil, M., P. Rao, G. Utkarsh, P. Pramod, A. Chatre. 2000. New meanings for old knowledge: the people's biodiversity registers programme. *Ecological Applications* 10(5):1307–1317.

Galindo-Gonzalez, Jorge, Sergio Guevara, and Vinicio Sosa. 2000. Bat- and bird-generated seed rains at isolated trees in pastures in a tropical rainforest. *Conservation Biology* 14(6):1693–1703.

Gallagher, R.S., E.C.M. Fernandes and E.L. McCallie. 1999. Weed management through short-term improved fallows in tropical agroecosystems. *Agroforestry Systems* 47(1/3):197–221.

Gallup, J.L., J.D. Sachs, and A.D. Mellinger. 1999. Geography and economic development. *International Regional Science Review* 22(2):179–232.

Gamez, Luis. 1999. Biodiversity and the Del Oro orange plantation. Ministry of the Environment, San José, Costa Rica. Unpublished memo.

Gari, J.A. 1997. *The Role of Democracy in the Biodiversity Issue: The Case of Quinoa.* Center of Latin American Research, Amsterdam.

GEF (Global Environment Fund). 2000. *Operational Program 13.* Global Environment Fund, Washington, D.C.

Gemmill, B. 2002. *Managing Agricultural Resources for Biodiversity Conservation: A Guide to Best Practices.* Produced with the Support of the UNEP/UNDP GEF Biodiversity Planning Support Programme. Environment Liaison Centre International, Nairobi, Kenya. Draft.

Giovannucci, Daniele. 2001. *Sustainable Coffee Survey of the North American Specialty Coffee Industry.* The World Bank, Washington, D.C.

Glanznig, Andreas, and Michael Kennedy. 2001. Addressing biodiversity loss in Australia: Land degradation and native vegetation clearance in the 1990s. In M. Bridges et al., eds., *op cit,* pp. 395–408.

Glass, Edward. 1988. Biological control of cassava pests in Africa. In *CGIAR 1987/88 Annual Report.* Consultative Group on International Agricultural Research, Washington, D.C., pp. 23–38.

Gleick, Peter H. 1993. Water and conflict: Freshwater resources and international security. *International Security* 18(1):79–112.

Gliessman, S.R. 2001. *Agroecosystem Sustainability: Developing Practical Strategies.* CRC Press, Boca Raton, Fla.

Global Environment Facility. 2000. GEF Operational Program #13 on conservation and sustainable use of biological diversity important to agriculture. Global Environment Facility, Washington, D.C.

Gockowski, James. 2000. *Biodiversity and Agriculture: Reconciling Development, Basic Needs, and the Environment in the Moist Forests of West and Central Africa.* IITA Humid Forest Ecoregional Center, International Institute for Tropical Agriculture, Ibadan, Nigeria.

Goklany, Indur M. 1998. Saving habitat and conserving biodiversity on a crowded planet. *BioScience* 48(11):941–953.

———. 1999. Meeting global food needs: The environmental trade-offs between increasing land conversion and land productivity. *Technology* 6:107–130.

Gomez-Pompa, Arturo, and Andrea Kaus. 1992. Taming the wilderness myth. *BioScience* 42(4):271–279.

Grifo, F., and J. Rosenthal, eds. 1997. *Biodiversity and Human Health.* Island Press, Washington, D.C.

Grumbine, R.E. 1994. What is ecosystem management? *Conservation Biology* 8:27–38.

Guarino, L., A. Jarvis, R.J. Hijmans, and N. Maxted. 2000. *Geographic Information Systems (GIS) and the Conservation and Use of Plant Genetic Resources.* International Plant Genetic Resources Institute, Cali, Colombia.

Guerin-McManus, M., and D. Hill. 2001. *Integrating Conservation into Debt Relief: The HIPC Initiative.* Conservation International, Washington, D.C.

Haila, Yrjö. 2002. A conceptual genealogy of fragmentation research: from island biogeography to landscape ecology. *Ecological Applications* 12(2):321–334.

Hannam, I.D. 2001. A global view of the law and policy to manage land degradation. In M. Bridges, et al., eds., *op. cit.*, pp.385–395.

Hannam, I.D. and B. Boer. 2001. Land degradation and international environmental law. In M. Bridges et al., eds., *op. cit.*

Harvey, Celia A. 2000. Windbreaks enhance seed dispersal into agricultural landscapes in Monteverde, Costa Rica. *Ecological Applications* 10(1):155–173.

Harvey, Celia A., and W.A. Haber. 1999. Remnant trees and the conservation of biodiversity in Costa Rican pastures. *Agroforestry Systems* 44:37–69.

Harvey, Celia A., Carlos F. Guindon, William A. Haber, Deborah Hamilton DeRosier, and K. Greg Murray. 2000. The importance of forest patches, isolated trees, and agricultural windbreaks for local and regional biodiversity: The case of Monteverde, Costa Rica. In B. Krishnapillay et al., eds., *XXI IUFRO World Congress, 7–12 August 2000, Kuala Lumpur, Malaysia, Sub-Plenary Sessions.* Volume I, International Union of Forestry Research Organizations, Kuala Lumpur, pp. 787–798.

Hatton, T.J., and Nulsen, R.A. 1999. Towards achieving functional ecosystem mimicry with respect to water cycling in southern Australian agriculture. *Agroforestry Systems* 445(1-3): 203–214.

Heal, G. 2000. *Nature and the Marketplace.* Island Press, Washington, D.C.

Heath, M.F., and M.I. Evans. 2000. *Important Bird Areas in Europe: Priority Sites for Conservation.* BirdLife International, Cambridge, U.K.

Hecht, S. 1993. *Of Fates, Forests and Futures: Myths, Epistemes and Policy in Tropical Conservation.* XXXII. The Horace M. Albright Conservation Lectureship. University of California, Berkeley, California, February 17.

Heid, Petra. 1998. The weakest go to the wall. *LEISA: ILEIA (Center for Research and Information on Low External Input and Sustainable Agriculture) Newsletter* 14(4):14–15.

Helmuth, Laura. 1999. A shifting equation links modern farming and forests. *Science* 286:1283.

Hess, George R. 1994. Conservation corridors and contagious disease: A cautionary note. *Conservation Biology* 8(1):256–262.

Heywood, V.H., and R.T. Watson, eds. 1995. *Global Biodiversity Assessment*. Cambridge University Press, Cambridge, U.K.

Heywood, Vernon. 1999. Use and potential of wild plants in farm households. *FAO Farm Systems Management Series* 15:1–113.

Hillel, Daniel. 1991. *Out of the Earth: Civilization and the Life of the Soil*. Free Press, New York.

Hobbs, Richard J., and Harold A. Mooney. 1998. Broadening the extinction debate: Population deletions and additions in California and Western Australia. *Conservation Biology* 12(2):271–283.

Hodgkin, T., and R.K. Arora. 2001. Developing conservation strategies and activities for crop wild relatives. International Plant Genetic Resources Institute, Rome. Draft.

Holden, S. 2001. A century of technological change and deforestation in the miombo woodlands of northern Zambia. In A. Angelsen and D. Kaimowitz, eds., *op.cit.*, pp. 251–270.

Homewood, K.M., and W.A. Rodgers. *1991. Maasailand Ecology: Pastoralist Development and Wildlife Conservation in Ngorongoro, Tanzania.* Cambridge University Press, Cambridge, U.K.

Honey, M. 1999. *Ecotourism and Sustainable Development: Who Owns Paradise?* Island Press, Washington, D.C.

Hopkins, Jane C., Sara J. Scherr, and Peter Gruhn. 1994. *Food Security and the Commons: Evidence from Niger.* Report to the United States Agency for International Development, Niamey, Niger. International Food Policy Research Institute, Washington, D.C., November.

Houghton, J.T., L.G. Meira Filho, B.A. Callander, N. Harris, A. Kattenberg, and K. Maskell, eds. 1995. *Climate Change 1995: The Science of Climate Change.* Cambridge University Press, Cambridge, UK.

Howe, H.F. and J. Smallwood. 1982. Ecology of seed dispersal. *Annual Review of Ecology and Systematics* 13:201–228.

Hoyt, E. 1992. *Conserving the Wild Relatives of Crops.* 2nd ed. IBPGR (International Board for Plant Genetic Resources), IUCN, and WWF.

Hughes, Jennifer B., Gretchen C. Daily, and Paul R. Ehrlich. 1997. Population diversity: Its extent and extinction. *Science* 278:689

Hutchings, J.A. 2000. Collapse and recovery of marine fishes. *Nature* 406:882–885.

ICRAF (International Centre for Research in Agroforestry). 1999. *Paths to Prosperity.* International Centre for Research in Agroforestry, Nairobi, Kenya.

ICWDM. 2001. International centre for wildlife damage management: Problems and solutions. Available at http://deal.unl.edu/icwdm/.

IFAD (International Fund for Agricultural Development). 2000. *Issues and Options in Rangelands Development: IFAD's Experience.* Final Draft Report of the Thematic Group on Community-Based Management of Natural Resources (I-Rangelands), International Fund for Agricultural Development, Rome, August.

———. 2001. *Rural Poverty Report 2001: The Challenge of Ending Rural Poverty.* Oxford University Press, Oxford, U.K.

Iisaak Forest Resources. 2000. Iisaak Forest Resources: From conflict to a new economic model for conservation-based forestry. Submission to the Green Economy Secretariat, Government of British Columbia, Canada.

ILEIA (Institute for Low-External Input Agriculture). 1989. *Participatory Technology Development in Sustainable Agriculture.* ILEIA, Leusden, Netherlands.

IMF (International Monetary Fund). 1987. *International Financial Statistics Yearbook.* International Monetary Fund, Washington, D.C.

———. 1998. *International Financial Statistics Yearbook.* International Monetary Fund, Washington, D.C.

International Federation of Organic Agriculture Movements, World Conservation Union, World Wildlife Federation. 2000. *The Relationships between Nature Conservation, Biodiversity, and Organic Agriculture.* IFOAM, IUCN, WWF: Tholey-Theley, Germany.

IPGRI (International Plant Genetic Resources Institute). 2000. IPGRI policy on intellectual property. Available at www.ipgri.org/policy/. Rome.

IRRI (International Rice Research Institute). 2000. *Something to Laugh About.* Los Baños, the Philippines.

IUCN (World Conservation Union). 1994. *Guidelines for Protected Area Management Categories.* IUCN, Gland, Switzerland.

———. 1996. *IUCN Red List of Threatened Animals.* IUCN, Gland, Switzerland.

———. 1997. *IUCN Red List of Threatened Plants.* IUCN, Gland, Switzerland.

———. 2000a. The ecosystem approach. Agenda item 4.2.1, Fifth Meeting of the Subsidiary Body on Scientific, Technical and Technological Advice. Montreal, Canada, 31 January–4 February.

———. 2000b. *IUCN Red List of Threatened Species.* IUCN, Gland, Switzerland.

IUCN and UNEP. 1992. *Caring for the Earth.* IUCN, Gland, Switzerland.

IWMI (International Water Management Institute). 2002. *Dialogue on Water, Food and Environment: Proposal.* Consortium of: FAO, Global Water Partnership, International Commission on Irrigation and Drainage, International Federation of Agricultural Producers, IWMI, IUCN, UNEP, World Health Organization, World Water Council and WWF. Colombo, Sri Lanka. March.

Jackson, D., and G. Boody. 1996. Sustainable farming practices benefit Minnesota landscapes. In W. Lockeretz, *op. cit.*, pp. 35–44.

Jackson, D,. and L. Jackson. 2002. *The Farm as Natural Habitat: Reconnecting Food Systems with Ecosystems.* Island Press, Washington, D.C.

Jackson, Wes. 1985. *New Roots for Agriculture.* University of Nebraska.

Jayasuriya, A.H.M. 1995. National conservation review: The discovery of extinct plants in Sri Lanka. *Ambio* 24:313–316.

Joffre, R., S. Rambal, and P. Ratte. 1999. The dehesa system of southern Spain and Portugal as a natural ecosystem mimic. *Agroforestry Systems* 45:57–79.

Johnson, Brian. 2000. Genetically modified crops and other organisms: Implications for agricultural sustainability and biodiversity. In G.J. Persley and M.M. Lantin, eds. *Agricultural Biotechnology and the Poor: Proceedings of an International Conference, Washington, D.C. 21–22 October 1999*. Consultative Group on International Agricultural Research, Washington, D.C., pp. 131–138.

Johnson, N., A. White, and D. Perrot-Maître. 2001. *Developing Markets for Water Services from Forests: Issues and Lessons for Innovation*. Forest Trends, World Resources Institute, and the Katoomba Group, Washington, D.C.

Jones, G.A., K.E. Sieving, and S.K. Jacobson. 2001. An assessment of bird faunas utilizing conventional and organic farmlands of north-central Florida. Paper presented to the International Symposium on Managing Biodiversity in Agricultural Ecosystems, 8–10 November, Montreal, Canada.

Kaimowitz, D., G. Flores, J. Johnson, P. Pacheco, I. Pavez, J. Montgomery Roper, C. Vallejos, and R. Velez. 2000. *Local Government and Biodiversity Conservation: A Case from the Bolivian Lowlands*. A case study for *Shifting the Power: Decentralization and Biodiversity Conservation*. Biodiversity Support Program, Washington, D.C.

Kaimowitz, David. 1996. *Livestock and Deforestation in Central America in the 1980s and 1990s: A Policy Perspective*. CIFOR Special Publication. Center for International Forestry Research, Jakarta, Indonesia.

Karesh, W.B., S.A. Osofsky, T.E. Rocke, and P. Barrows. 2002. Joining forces to improve our world. *Conservation Biology*. In press.

Karr, J.R., and D.R. Dudley. 1981. Ecological perspective on water quality goals. *Environmental Management* 5:55–68.

Kendall, H.W,. and D. Pimentel. 1994. Constraints on the expansion of the global food supply. *Ambio* 23:198–205.

Kiara, J.K. 2001. The catchment approach to soil and water conservation in Kenya. In M. Bridges et al., eds., *op. cit.*, pp. 370–371.

Kingsolver, Barbara. 1999. *The Poisonwood Bible*. Harper Perennial Library, New York.

Kirk, D., A. Matthew, D. Evenden, and P. Mineau. 1996. Past and current attempts to evaluate the role of birds as predators of insect pests in temperate agriculture. *Current Ornithology* 13:175–269.

Kiss, Agnes, ed. 1990. *Living with Wildlife: Wildlife Resource Management with Local Participation in Africa*. World Bank Technical Paper No. 130, Africa Technical Department Series. The World Bank, Washington, D.C.

Klaffke, O. 1999. The company of wolves. *New Scientist* 6 (February):18–19.

Kleijn D., F. Berendse, R. Smit, and N. Gilissen. 2001. Agri-environment schemes do not effectively protect biodiversity in Dutch agricultural landscapes. *Nature* 413, 723–725.

Kleymeyer, C.D. 1994. *Cultural Expression and Grassroot Development: Cases from Latin America and the Caribbean*. Lynne Rienner Publishers, Boulder, Colo.

Koskela, J., and W.A.N. do Amaral. 2001. Conservation of tropical forest genetic resources: IPGRI's efforts and experiences. Paper presented at the SE-Asian Moving Workshop on Conservation, Management, and Utilisation of Forest Genetic Resources, 26 February–9 March, Thailand.

Kowal, T.M. 1999. Improved fallows and dispersed tree systems. Cornell University MULCH list, Ithaca, New York.

Kremen, Claire, and Taylor Ricketts. 2000. Global perspectives on pollination disruptions. *Conservation Biology* 14(5):1226–1228.

Lacher, Thomas, R. Slack, L. Coburn, and M. Goldstein. 1999. The role of agroecosystems in wild biodiversity. In Wanda W. Collins and Calvin O. Qualset, eds., *Biodiversity in Agroecosystems.* CRC Press, Washington D.C., pp. 147–165.

Land Stewardship Project. 2001. Land Stewardship Project: Working to keep the land and people together. Available at http://www.landstewardshipproject.org/

Lande, R. 1988. Genetics and demography in biological conservation. *Science* 241:1455–1460.

Landers, J.N. 2002. How and why the Brazilian zero tillage explosion occurred. In D.E. Stott, R.H. Mohtar, G.C. Steinhardt, eds., *Sustaining the Global Farm.* Selected Papers for the Tenth International Soil Conservation Organization meeting, 24–29 May 1999, West Lafayette, Indiana. Published by International Soil Conservation Organization in cooperation with USDA Agricultural Resarch Service, National Soil Erosion Research Laboratory, and Purdue University. CD-ROM available from USDA-ARS.

Langewald, J., M.B. Thomas, C.J. Lomer, and O.K. Douro-Kpindou. 1997. Use of *Metarhizium flavoviride* for control of Zonocerus variegatus: A model, relating mortality in caged field samples with disease development in the field. *Entomologia Experimentalis et Applicata* 82:1–8.

Laurance, William F., P. Delamonica, S. Laurance, H. Vasconcelos, and T. Lovejoy. 2000. Rainforest fragmentation kills big trees. *Nature* 404:836.

Lawrence, Debora C., Mark Leighton, and David R. Peart. 1995. Availability and extraction of forest products in managed and primary forest around a Dayak village in West Kalimantan, Indonesia. *Conservation Biology* 9(1):76–88.

Lawton, J.H., and R.M. May, eds. 1995. *Extinction Rates.* Oxford University Press, Oxford, U.K.

Leader-Williams, N., J.A. Kayera, and G.L. Overton, eds. 1996. *Community-based Conservation in Tanzania.* IUCN, Gland, Switzerland.

Leakey, Roger R.B. 1997. Domestication potential of *Prunus africana* ("pygeum") in sub-Saharan Africa. In A. M. Kinuya, W. M. Kofi-Tsekpo, and L.B. Dangana, eds., *Conservation and Utilization of Indigenous Medicinal Plants and Wild Relatives of Food Crops.* UNESCO, Nairobi, Kenya, pp. 99–107.

———. 1999a. Agroforestry for biodiversity in farming systems. In Wanda W. Collins and Calvin O. Qualset, eds. *Biodiversity in Agroecosystems.*: CRC Press, New York, pp. 127–145.

———. 1999b. Potential for novel food products from agroforestry trees: A review. *Food Chemistry* 66:1–14.

———. 2000. Agroforestry and the motor industry. In K. ten Kate and S. Lairds, *Commercial Biodiversity Access and Benefit Sharing Issues.* Earthscan Publications, London.

———. 2001a. Win:win land-use strategies for Africa: 1. Building on experience elsewhere and capitalizing on the value of indigenous tree products. *International Forestry Review* 3:1–10.

———. 2001b. Win:win land use strategies for Africa: 2. Matching economic development with environmental benefits through agroforestry. *International Forestry Review* 3:11–18.

Leakey, Roger R.B., and A-M.N. Izac. 1996. Linkages between domestication and commercialization of non-timber forest products: Implications for agroforestry. In R.R.B. Leakey, A.B. Temu, M. Melnyk, and P. Vantomme, eds. *Domestication and commercialization of non-timber forest products in agroforestry systems.* Non-Wood Forest Products 9. Food and Agriculture Organization of the United Nations: Rome.

Leakey, Roger R.B., and A.C. Newton, eds. 1994. *Tropical Trees: The Potential for Domestication and the Rebuilding of Forest Resources.* Her Majesty's Stationery Office, London.

Leakey, Roger R.B, and Tony Simons. 1998. The domestication and commercialization of indigenous trees in agroforestry for the alleviation of poverty. *Agroforestry Systems* 38:165–176.

Lee, D.R., and C.B. Barrett, eds. 2001. *Trade-offs or Synergies? Agricultural Intensification, Economic Development, and the Environment.* CAB International Publishing, New York.

Lefroy, E.C. 1992. The importance of biodiversity to sustainable agriculture. *Land Management Society Newsletter,* Winter 1992, Perth, Australia.

Lefroy, E.C., and R.J. Stirzaker. 1999. Agroforestry for water management in the cropping zone of southern Australia. *Agroforestry Systems* 45(103):277–302.

Lefroy, E.C., J. Salerian, and R.J. Hobbs. 1992. Integrating economic and ecological considerations: A theoretical framework. In R.J. Hobbs and D.A. Saunders, eds. *Reintegrating Fragmented Landscapes: Towards Sustainable Production and Nature Conservation.* Springer-Verlag, New York.

Lefroy, R., E. Lefroy, R. Hobbs, M. O'Connor, and J. Pate. 1999. *Agriculture as a Mimic of Natural Ecosystems.* Current Plant Science and Biotechnology in Agriculture. Volume 37. Kluwer Academic Publishers, Dortrecht, the Netherlands.

Le Houerou, H.N. 1991. The shrublands of Africa. In *The Biology of the Utilization of Shrubs.* Academic Press, New York, pp. 119–143.

Lehmkuhl, J.F., R.K. Upreti, and U. R. Sharma. 1998. National parks and local development: Grass and people in Royal Chitwan National Park, Nepal. *Environmental Conservation* 15 (2):143–148.

Levin, Simon. 1999. *Fragile Dominion: Complexity and the Commons.* Perseus, New York.

Li, W,. and H. Wang. 1999. Wildlife trade in Yunnan Province, China, at the border with Vietnam. *TRAFFIC Bulletin* 18(1):21–30.

Lichtenberg, Erik. 1996. Using soil and water conservation practices to reduce bay nutrients: How has agriculture done? *Maryland Cooperative Extension Economic Viewpoints* 1(2): 4–8.

———. 2000. Soil and water conservation on Maryland farms: A 1998 update. *Maryland Cooperative Extension Economic Viewpoints* 4(1):5–7.

Lockeretz, W., ed. *Environmental Enhancement through Agriculture.* Proceedings of a conference held in Boston, 15–17 November, 1995, organized by the Tufts University School of Nutrition Science and Policy, the American Farmland Trust, and the Henry A. Wallace Institute for Alternative Agriculture. Center for Agriculture, Food, and Environment, Tufts University, Medford, Mass.

Lomer, C.J., C. Prior, and C. Kooyman. 1997. Development of Metarhizium spp. for the control of grasshoppers and locusts. *Biocontrol Science and Technology* 9:199–214.

Loreau, M., S. Naeem, P. Inchausti, J. Bengtsson, J.P. Grime, A. Hector, D.U. Hooper, M.A. Huston, D. Raffaelli, B. Schmid, D. Tilman, and D.A. Wardle. 2001. Biodiversity and ecosystem functioning: current knowledge and future challenges. *Science* 294:804–808.

Lutz, E., and J. Caldecott, eds. 1996. *Decentralization and Biodiversity Conservation.* The World Bank, Washington, D.C.

Lutz, W., C. Prinz, and J. Langgassmer. 1993. World population projections and possible ecological feedbacks. *POPNET* 12:1–11.

Lynch, Lori. 1999. The enhanced conservation reserve program. *Maryland Cooperative Extension Economic Viewpoints* 3(3):9–11.

Mac, M.J., P.A. Opler, C. Haecker, and P. Doran. 1998. *Status and Trends of the Nation's Biological Resources.* United States Department of the Interior, Reston, Va.

MacArthur, R.H., and E.O. Wilson. 1967. *The Theory of Island Biogeography.* Princeton University Press, Princeton, N.J.

MacKinnon, John, Kathy MacKinnon, Graham Child, and Jim Thorsell. 1984. *Managing Protected Areas in the Tropics.* IUCN, Gland. Also available in Indonesian (Gadjah Mada Press, 1990). Gadjah Mada Press, Yogyakarta, Indonesia.

Mader, P., A. Fliebach, D. Dubois, L. Gunst, P. Fried, and U. Niggli. 2002. Soil fertility and biodiversity in organic farming. *Science* 296(5573):1694.

Main, Martin B., Fritz M. Roka, and Reed F. Noss. 1999. Evaluating costs of conservation. *Conservation Biology* 13(6):1262–1272.

Malik, S. 1998. Rural poverty and land degradation: What does the available literature suggest for priority setting for the Consultative Group on International Agricultural Research? Report prepared for the Technical Advisory Committee of the CGIAR, February.

Maltby, E., M. Holdgate, M. Acreman, and A. Weir. 1999. *Ecosystem Management: Questions for Science and Society.* Royal Holloway Institute for Environmental Research, Virginia Water, U.K.

Mannion, Antoinette M. 1995. Biodiversity, biotechnology, and business. *Environmental Conservation* 22(3):201–210.

Maredia, M.K. 2000. Environmental consequences of intensification: Evidence from the literature and implications for impact assessment. Standing Panel on Impact Assessment, Technical Advisory Committee to the Consultative Group on International Agricultural Research, Rome. Draft.

Margoluis, R., and N. Salafsky. 2000. *Is Our Project Succeeding? A Guide to Threat Reduction for Conservation.* Biodiversity Support Program, Washington, D.C.

Margoluis, R., C. Margoluis, K. Brandon, and N. Salafsky. 2000. *In Good Company: Effective Alliances for Conservation.* Biodiversity Support Program, Washington, D.C.

Margoluis, R., S. Myers, J. Allen, J. Roca, M. Melnyk, and J. Swanson. 2001. *An Ounce of Prevention: Making the Link between Health and Conservation.* Biodiversity Support Program, Washington, D.C.

Margoluis, R., V. Russell, M. Gonzalez, O. Rojas, J. Magdaleno, G. Madrid, and D. Kaimowitz. 2000. *Maximum Yield? Sustainable Agriculture as a Tool for Conservation.* Biodiversity Support Program, Washington, D.C.

Martin, Paul S., and Richard G. Klein, eds. 1994. *Quaternary Extinctions: A Prehistoric Revolution.* University of Arizona Press, Tucson.

Mather, A. 2001. The transition from deforestation to reforestation in Europe. In A. Angelsen and D. Kaimowitz, eds., *op. cit.*, pp. 35–52.

Matthews, Emily, Richard Payne, Mark Rohweder, and Siobhan Murray. 2000. *Pilot Analysis of Global Ecosystems: Forest Ecosystems.* World Resources Institute, Washington, D.C.

Mazzola, M., M.E. Graipel, and N. Dunstone. 2002. Mountain lion depredation in southern Brazil. *Biological Conservation* 105:43–51.

McCalla, Alex F. 2000. *Agriculture in the 21st Century.* CIMMYT Economics Program Fourth Distinguished Economist Lecture. CIMMYT (International Maize and Wheat Improvement Center), Mexico City, March.

McClafferty, Bonnie. 2000. Summary of the state of global malnutrition. International Food Policy Research Institute, Washington, D.C. Unpublished memo.

McKaye, K.R., et al. 1995. African tilapia in Lake Nicaragua. *BioScience* 45(6):406–411.

McMichael, Anthony J., et al. 1999. Globalization and the sustainability of human health: An ecological perspective. *BioScience* 49(3):205–210.

McNeely, J.A. 1978. Dynamics of extinction in Southeast Asia. In J.A. McNeely, D.S. Rabor, and E. A. Sumardja, eds., *Wildlife Management in Southeast Asia.* Biotrop, Bogor, Indonesia, pp. 137–160.

———. 1988. *Economics and Biological Diversity: Developing and Using Economic Incentives to Conserve Biological Diversity.* IUCN, Gland, Switzerland.

———. 1990. How wild relatives of livestock contribute to a balanced environment. *Asian Livestock* 14(10):128–137.

———. 1992. *Parks for Life: Proceedings of the IV World Parks Congress.* IUCN, Gland, Switzerland.

———. 1994. Lessons from the past: Forests and biodiversity. *Biodiversity and Conservation* 3:3–20.

———. 1997. Assessing methods for setting conservation priorities. In OECD, *Investing in Biological Diversity.* Organization for Economic Cooperation and Development, Paris, pp. 25–55.

———. 1999. *Mobilizing Broader Support for Asia's Biodiversity: How Civil Society Can Contribute to Protected Area Management.* Asian Development Bank, Manila.

McNeely, J.A., and G. Ness. 1994. *People, Parks and Biodiversity: Issues in Population and Environment Dynamics.* IUCN (World Conservation Union), Gland.

McNeely, J.A., and P.S. Wachtel. 1988. *Soul of the Tiger: The Relationship between People and Wildlife in Southeast Asia.* Doubleday, New York.

McNeely, J.A., K.R. Miller, W.V. Reid, R.A. Mittermeier, and T.B. Werner. 1990. *Conserving the World's Biological Diversity.* IUCN, World Resources Institute, Conservation International, World Wildlife Fund–U.S., and The World Bank, Washington, D.C.

McNeely, J.A., H.A. Mooney, L.E. Neville, P. Schei, and J.K. Waae, eds. 2001. *A Global Strategy on Invasive Alien Species.* IUCN, Gland, Switzerland.

McNeely, J.A. and S.J. Scherr. 2001. *Common Ground, Common Future: How Ecoagriculture Can Help Feed the World and Save Wild Biodiversity.* World Conservation Union (IUCN) and Future Harvest, Washington, D.C. May.

Meinzen-Dick, R.S., and B.R. Bruns. 2000. Negotiating water rights: Introduction. In B.R. Brun and R.S. Meinzen-Dick, eds., *Negotiating Water Rights.* Vistaar, New Delhi, pp. 23–55.

Meinzen-Dick, Ruth, and Godswill Makombe. 1999. Dambo irrigation systems: Indigenous water management for food security in Zimbabwe. In Anna Knox McCulloch, Suresh Babu, and Peter Hazell, eds. *Strategies for Poverty Alleviation and Sustainable Resource Management in the Fragile Lands of Sub-Saharan Africa*. Proceedings of the International Conference, 25–29 May 1998, Entebbe, Uganda. Deutsche Stiftung für Internationale Entwicklung, Feldalfing, Germany, and International Food Policy Research Institute, Washington, D.C. pp. 279–287.

Menke, J., and G.E. Bradford. 1992. Rangelands. *Agriculture, Ecosystems and Environment* 42(1–2):141–163.

Mercado, Agustin, Jr., Marco Stark, and Dennis P. Garrity. 1997. Enhancing sloping land management technology adoption and dissemination. Paper presented at the International Board for Soil Research and Management Sloping Land Management workshop, 15–21 September, Bogor, Indonesia.

Mew, T. 2000. Research initiatives in cross ecosystem: Exploiting biodiversity for pest management. International Rice Research Institute: Los Baños, the Philippines.

Meyer, Carrie A. 1997. Public-nonprofit partnerships and North-South green finance. *Journal of Environment and Development* 6(2):123–146.

MFP. 2000. *People's Protected Areas (PPA): Sustainable Livelihood Approach with Biodiversity Conservation (SLAB)*. M.P.M.F.P. (Madhya Pradesh State Minor Forest Province) (T&D) Cooperative Federation Ltd., Bhopal, India. October.

Millar J. 2001. Listening to landholders: Approaches to community nature conservation in Queensland. In United Nations University et al., eds., *op. cit.*, pp. 103–112.

Miller, K.R., A.E. Chang, and N. Johnson. 2001. *Defining Common Ground for the Mesoamerican Biological Corridor*. World Resources Institute, Washington, D.C.

Miller, Kenton R. 1996. *Balancing the Scales: Guidelines for Increasing Biodiversity's Chances through Bioregional Management*. World Resources Institute, Washington, D.C.

Mishra, H.R. 1982. Balancing human needs and conservation in Nepal's Royal Chitwan National Park. *Ambio* 11:246–251.

Mitschein, T., and P. Miranda. 1998. POEMA: A proposal for sustainable development in Amazonia. In *The World Bank–DaimlerChrysler Environment Forum,* 12–14 July 1999, Magdeburg, Germany.

Mollison, Bill. 1990. *Permaculture: A Practical Guide for a Sustainable Future*. Island Press, Washington, D.C.

Mooney, H., and J.A. Drake. 1987. The ecology of biological invasions. *Environmentalist* 19(5):10–37.

Murphy, W., J. Silman, L. McCrory, S. Flack, J. Winsten, D. Hoke, A. Schmitt, and B. Pillsbury. 1996. Environmental, economic, and social benefits of feeding livestock on a well-managed pasture. In W. Lockeretz, ed., *op. cit.*, pp. 125–134.

Musters, C.J.M., M. Kruk, H. J. de Graaf, and W. Keurs. 2001. Breeding birds as a farm product. *Conservation Biology* 15(2):363–369.

Myers, Norman. 1987. Tackling mass extinction of species: A great creative challenge. *The Horace M. Albright Lecture in Conservation*. University of California, Berkeley.

———. 1988. Threatened biotas: "Hotspots" in Tropical Forests. *Environmentalist* 8(3):1–20.

———. 1999. The green revolution: Its environmental underpinnings. *Current Science* (4):507–513.

Myers, Norman, and Jennifer Kent. 2001. *Perverse Subsidies: How Tax Dollars Undercut the Environment and the Economy.* Island Press, Washington, D.C.

Myers, Norman, Russ Mittermeier, Cristina G. Mittermeier, Gustavo A.B. da Fonseca, and Jennifer Kent. 2000. Biodiversity hotspots for conservation priorities. *Nature* 403:842–843.

Nabhan, G.P,. and S. Buchmann. 1997. Services provided by pollinators. In G.C. Daily, ed., *Nature's Services: Societal Dependence on Natural Ecosystems.* Island Press, Washington, D.C., pp. 133–150.

Nabhan, G.P., A.M. Rea, K.L. Hardt, E. Mellink, and C.F. Hutchinson. 1982. Papago influences on habitat and biotic diversity: Quitovac Oasis ethno-ecology. *Journal of Ethno-Biology* 2:124–143.

Naeem, S. 1998. Species redundancy and ecosystem reliability. *Conservation Biology* 12(1):39–45.

National Audubon Society. 2000. National Audubon Society Agriculture Policy Program. Available at http://www.audubon.org/campaign/agriculture/.

National Research Council. 1983. *Little-Known Asian Animals with a Promising Economic Future.* National Academy Press, Washington, D.C.

———. 1991. *Microlivestock: Little-Known Small Animals with a Promising Economic Future.* National Academy Press, Washington, D.C.

Nature. 2000. Biodiversity. Nature Insight. Reprinted from *Nature* 405:6783: 207–254. May.

The Nature Conservancy. 2000a. Biodiversity via economic development strategies. The Nature Conservancy, Arlington, Va. Draft.

———. 2000b. *EcoEnterprises Fund (Fondo Ecoempresas): Program Summary.* The Nature Conservancy, Arlington, Va.

Naylor, Rosamond L., Rebecca J. Goldberg, Jurgenne H. Primavera, Nils Kautsky, Malcolm C.M. Beveridge, Jason Clay, Carl Folke, Jane Lubchenco, Harold Mooney, and Max Troell. 2000. Effect of aquaculture on world fish supplies. *Nature* 405:1017–1024.

Nelson, M., and M. Maredia. 1999. *Environmental Impacts of the CGIAR: An Initial Assessment.* Impact Assessment and Evaluation Group. Document ICW/99/08/d. Consultative Group for International Agricultural Research, Washington, D.C.

Nelson, M., R. Dudal, H. Gregersen, N. Jodha, D. Nyamia, J.-P. Groenewold, F. Torres, and A. Kassam. 1997. *Report of the Study on CGIAR Research Priorities for Marginal Lands.* Technical Advisory Committee, Consultative Group on International Research, and FAO, Rome.

Nepal, S.K., and K.E. Weber. 1992. *Struggle for Existence: Park-People Conflict in the Royal Chitwan National Park, Nepal.* Asian Institute of Technology, Bangkok.

Nepstad, D. 2001. Pro-Ambiente: Farmer Credit for Environmental Services in the Amazon. Presentation to The Katoomba Group workshop III on "Markets for Environmental Services of Forests," Teresopolis, Brazil.

Neuenschwander, Peter, and Richard Markham. 2001. Biological control in Africa and its possible effects on biodiversity. In E. Wajnberg, J.K. Scott, and P.C. Quimby, eds., *Evaluating Indirect Ecological Effects of Biological Control.* CAB International, pp. 127–146.

Nietschmann, Bernard. 1992. *The Interdependence of Biological and Cultural Diversity.* Center for World Indigenous Studies, Kenmore, Wash.

Niiler, E. 2001. Fund picks sweet target for first investment. *EcoAmericas* 3(4):5. Published by Fourth Street Press, Santa Monica, Calif.

Nkuba, R.M. 1997. A preliminary assessment of the economic impact of water hyacinth in Uganda. *15th [Fifteenth] East African Biennial Weed Science Conference Proceedings,* Uganda, pp. 279–286.

Noble, Ian R., and Dirzo, Rodolfo. 1997. Forests as human-dominated ecosystems. *Science* 277:522–525.

Nowell, K., and P. Jackson, eds. 1996. *Wild Cats.* IUCN/SSC Cat Specialist Group. Cambridge, U.K.

O'Connor, R.J. 2001. Agricultural regimes and the conservation of farmland biodiversity. Paper presented to the International Symposium on Managing Biodiversity in Agricultural Ecosystems, 8–10 November, Montreal, Canada.

OECD (Organization for Economic Cooperation and Development). 1997a. *Environmental Benefits from Agriculture: Issues and Policies.* The Helsinki Seminar. Organization for Economic Cooperation and Development, Paris.

———. 1997b. *The Environmental Effects of Agricultural Land Diversion Schemes.* Organization for Economic Cooperation and Development, Paris.

———. 1998. *The Environmental Effects of Reforming Agricultural Policies.* Organization for Economic Cooperation and Development, Paris.

Oelhaf, R.C. 1979. *Organic Agriculture: Economic and Ecological Comparisons with Conventional Methods.* Halstead, New York.

Oldfield, S., C. Lusty, and A. MacKinven. 1998. *The World List of Threatened Trees.* IUCN, Cambridge, U.K.

Ong, C., and R.R.B. Leakey. 1999. Why tree-crop interactions in agroforestry appear at odds with tree-grass interactions in tropical savannahs. In Lefroy et al., eds., *op. cit.,* pp. 109–129.

Ong, C., and R.R.B. Leakey 2002. Tree crop intensification in savannahs. International Centre for Research in Agroforestry, Nairobi. Draft.

Ontario Soil and Crop Improvement Association. 2001. *Wildlife Impact Assessment for Ontario Agriculture.* Ontario Soil and Crop Improvement Association, Ontario.

Ostrom, Elinor. 1998. Scales, polycentricity, and incentives: Designing complexity to govern complexity. In Lakshman Guruswamy and J.A. McNeely, eds., *Protection of Global Diversity: Converging Strategies.* Duke University Press, Durham, N.C., pp. 145–163.

Owen, M. 1990. The damage-conservation interface illustrated by geese. *Ibis* 132:238–252.

Paarlberg, R.L. 1996. Rice bowls and dust bowls: Africa, not China, faces a food crisis. Review. *Foreign Affairs.* May/June.

Pacchioli, David. 1996. Critters damage crops. *Research/Penn State* 17(2), June. Available at www.rps.edu/jun96/critters.html.

Pagiola, S., J. Bishop, and N. Landell-Mills, eds. 2002. *Selling Forest Environmental Services: Market-Based Mechanisms for Conservation and Development.* Earthscan Publications, London.

Palm, C.A., R.J.K. Myers, and S.M.Nandwa. 1997. Combined use of organic and inorganic nutrient sources for soil fertility maintenance and replenishment. In R.J. Buresh and P.A. Sanchez, eds., *Replenishing Soil Fertility in Africa.* SSSA Special Publication. Soil Science Society of America and American Society of Agronomy, Madison, Wisc., pp. 193–217.

Palmer, M.A., et al. 2000. Linkages between aquatic sediment biota and life above sediments as potential drivers of biodiversity and ecological processes. *BioScience* 50(12):1062–1075.

Panayotou, T., and P. Ashton. 1992. *Not by Timber Alone: Economics and Ecology for Sustaining Tropical Forests.* Island Press, Washington, D.C.

Paoletti, M.G., and D. Pimentel, eds. 1992. *Biotic Diversity in Agroecosystems.* Elsevier Science, Amsterdam.

Pardey, Philip G., Julian M. Alston, Jason E. Christian, and Shenggen Fan. 1996. *Hidden Harvest: U.S. Benefits from International Research Aid.* Food Policy Report. International Food Policy Research Institute, Washington, D.C.

Payne, J.M., M.A. Bias, and R.G. Kempka. 1996. Valley care: Bringing conservation and agriculture together in California's Central Valley. In W. Lockeretz, ed., *op. cit.*, pp. 79–88.

Pearce, David, F. Putz, and J.K. Vanclay. 1999. A sustainable forest future. Center for Social and Economic Research on the Global Environment (CSERGE). Working paper GEC99–15, University of East Anglia, UK.

Pender, J., S.J. Scherr, and G. Durón. 1999. Pathways of development in the hillsides of Honduras: Causes and implications for agricultural production, poverty, and sustainable resource use. EPTD Discussion Paper No 45. International Food Policy Research Institute, Washington, D.C.

Penning de Vries, F.W.T., H. Acquay, D. Molden, S.J. Scherr, C. Valentin and O. Cofie. 2002. *Integrated Land and Water Management for Food and Environmental Security.* Comprehensive Assessment of Water for Agriculture Working Paper. International Water Management Institute, Colombo, Sri Lanka.

Perfecto, Ivette, et al. 1996. Arthropod biodiversity loss and the transformation of a tropical agro-ecosystem. *Biodiversity and Conservation* 6:935–945.

Persley, G.J., and M.M. Lantin, eds. 2000. *Agricultural Biotechnology and the Poor.* Proceedings of an international conference, Washington, D.C., 21–22 October 1999. Consultative Group on International Agricultural Research, Washington, D.C.

Petrucci, B.T. 1996. Environmental quality in rural and urban settings. In W. Lockeretz, ed., *op. cit.*, pp. 145–150.

Pfiffner, Lukas. 2000. Significance of organic farming for invertebrate diversity: Enhancing beneficial organisms with field margins in combination with organic farming. In Sue Stolten, et al., eds., *The Relationship between Nature Conservation, Biodiversity, and Organic Agriculture.* International Federation of Organic Agricultural Movements, Berlin, pp. 52–66.

Pfiffner, L., and U. Niggli. 1996. Effects of bio-dynamic, organic, and conventional farming on ground beetles and other epigeic arthropods in winter wheat. *Biological Agriculture and Horticulture* 12:353–364.

Phillips, Adrian. 1997. Protected areas and organic agriculture. *Ecology and Farming* (September): 26–27.

Pimbert, M.P., and V. Toledo. 1994. Indigenous people and biodiversity conservation: Myth or reality? *EthnoEcologica* 2(3):1–96.

Pimentel, David, et al. 2001. Economic and environmental threats of alien plant, animal, and microbe Invasions. *Agriculture, Ecosystems, and Environment* 84:1–20.

Pimm, Stuart. 1996. Lessons from a kill. *Biodiversity and Conservation* 5:1059–1067.

———. 2000. In search of perennial solutions. *Nature* 389:126–127.

Pimm, Stuart, and Peter Raven. 2000. Extinction by numbers. *Nature* 403:843–858.

Pinedo-Vasquez, M., C. Padoch, and K. Coffey. 2001. Valuing and promoting smallholder agricultural practices: The approach of the People, Land Management, and Environment Change project. Paper presented to the International Symposium on Managing Biodiversity in Agricultural Ecosystems, 8–10 November, Montreal, Canada.

Pinstrup-Andersen, P,. and Marc Cohen. 1998. Aid to developing-country agriculture: Investing in poverty reduction and new export opportunities. 2020 Brief 56. International Food Policy Research Institute, Washington, D.C.

Pinstrup-Andersen, P., R. Pandya-Lorch, and M.W. Rosegrant. 1997. *The World Food Situation: Recent Developments, Emerging Issues, and Long-Term Prospects.* 2020 Food Policy Report. International Food Policy Research Institute, Washington, D.C.

Piper, Jon K. 1999. Natural systems agriculture. In Wanda W. Collins and Calvin O. Qualset, eds., *Biodiversity in Agroecosystems.* CRC Press, New York, pp. 167–195.

Pirot, J-Y., P.-J. Meynell, and D. Elder. 2000. *Ecosystem Management: Lessons from Around the World. A Guide for Development and Conservation Practitioners.* IUCN, Gland, Switzerland.

Place, F., and A. Waruhiu. 2000. *Options for Biodiversity in Eastern and Southern Africa.* A report on a regional workshop, Mainstreaming Agriculture into Forestry: Towards Systemic Biodiversity Policies, Nairobi, Kenya, 21–22 November 1999. International Centre for Research in Agroforestry, Nairobi.

Plieninger, T,. and C. Wilbrand. 2001. Land use, biodiversity conservation, and rural development in the dehesas of Cuatro Lugares, Spain. *Agroforestry Systems* 51(1):23–34.

Pockley, P. 2000. Global warming identified as main threat to coral reefs. *Nature* 407:932.

Poffenberger, Mark, and Betsy McGean, eds. 1996. *Village Voices, Forest Choices: Joint Forest Management in India.* Oxford University Press, Delhi.

Poffenberger, M., ed. 2001. *Communities and Climate Change: The Clean Development Mechanism and Village-Based Forest Restoration in Central India.* (with N.H. Ravindranath, D.N. Pandey, I.K. Murthy, R. Bist, and D. Jain). Community Forestry International, Inc. and The Indian Institute of Forest Management, Santa Barbara, California.

Postel, Sandra. 1994. Carrying capacity: Earth's bottom line. In L.R. Brown, et al., eds., *State of the World 1994.* W.W. Norton, New York, pp. 3–21.

———. 1998. Water for food production: Will there be enough in 2025? *BioScience* 48(8):629–637.

———. 1999. *Pillar of Sand: Can the Irrigation Miracle Last?* W.W. Norton, New York.

Postel, Sandra L., Gretchen C. Daily, and Paul R. Erhlich. 1996. Human appropriation of renewable freshwater. *Science* 271:785–788.

Potts, G.R. 1991. The environmental and ecological importance of cereal fields. In L.G. Firbank, N. Carter, J. Derbyshire, and G. Potts, eds., *The Ecology of Temperate Cereal Fields.* Blackwell Scientific Publications, Oxford, U.K., pp. 3–21.

Powell, I., A. White, and N. Landell-Mills. 2001. *Developing Markets for the Ecosystem Services of Forests.* Forest Trends, Washington, D.C.

Prescott-Allen, Robert, and Christine Prescott-Allen. 1990. How many plants feed the world? *Conservation Biology* 4(4):365–374.

Pretty, J.N. 2002. *Agri-Culture: Reconnecting People, Land and Nature.* Earthscan Publications, London.

———. 1995. *Regenerating Agriculture: Policies and Practices for Sustainability and Self-Reliance.* National Academy Press, Washington, D.C.

———. 1999. Can sustainable agriculture feed Africa? New evidence on progress, process, and impact. *Environment, Development, and Sustainability* 1:253–274.

Pretty, J.N., and R. Hine. 2000. The promising spread of sustainable agriculture in Asia. *Natural Resources Forum* 24:107–121.

Pretty, J.N., C. Brett, D.Gee, R.E. Hine, C.F. Mason, J.I.L. Morison, H. Raven, M.D. Rayment, and G. van der Bijl. 2000. An assessment of the total external costs of UK agriculture. *Agricultural Systems* 65(2):113–136.

Primack, R.B. 1993. *Essentials of Conservation Biology.* Sinauer Associates, Sunderland, Mass.

Pulliam, H.R., and B.R. Johnson. 2002. In B. Johnson. and K. Hill, eds., *Ecology and Design: Frameworks for Learning.* Island Press, Washington, D.C., pp. 51–84.

Pursel, Vernon. G., C.A. Pinkert, K.F. Miller, D.J. Bolt, R.G. Campbell, R.D. Palmiter, R.L. Brinster, and R.E. Hammer. 1989. Genetic engineering of livestock. *Science* 244:1281–1288.

Putz, Francis E., Kent H. Redford, John G. Robinson, Robert Fimbel, and Geoffrey M. Blate. 2000. Biodiversity conservation in the context of tropical forest management. Environment Department Paper No. 75. The World Bank, Washington, D.C. September.

Queblatin, E.E., E.C. Catacutan, and D.P. Garrity. 2001. *Managing Natural Resources Locally: An Overview of Innovations and Ten Initial Steps for Local Governments.* International Centre for Research in Agroforestry and International Fund for Agricultural Development, Philippines.

Raintree, J.B., and Hermina A. Francisco. 1994. *Marketing of Multipurpose Tree Products in Asia.* Proceedings of an international workshop held in Baguio City, the Philippines, 6–9 December 1993. Winrock International, Bangkok, Thailand.

Rametsteiner, E., and M. Simula. 2001. *Forging Novel Incentives for Environment and Sustainable Forest Management: Background Paper.* For Workshop on Forest Certification, Brussels, September 6–7. European Commission DG Environment. October.

Ramiaramanana, D.M. 1993. Crop-hedgerow interactions with natural vegetative filter strips on sloping acidic land. Master of Science thesis, University of the Philippines, Los Baños.

Redman, Charles L. 1999. *Human Impact on Ancient Environments*. University of Arizona Press, Tucson.

Reijtntjes, C., B. Haverkort, and A. Waters-Bayer. 1992. *Farming for the Future: An Introduction to Low-External-Input and Sustainable Agriculture*. Institute for Low-External-Input Agriculture. MacMillan, London/Basingstoke.

Reijtntjes, Coen, Marilyn Minderhoud-Jones, and Peter Laban, eds. 1999. *LEISA in Perspective: 15 Years of ILEIA*. Institute for Low External Input Agriculture, Leusden, the Netherlands.

Revenga, C., J. Brunner, N. Henninger, K. Kassem, and R. Payne. 2000. *Pilot Analysis of Global Ecosystems: Freshwater Systems*. World Resources Institute, Washington, D.C.

Revkin, Andrew C. 2000. Extinction turns out to be a slow, slow process. *New York Times*, 24 October.

Reynolds, J.D., G.M. Mace, K.H. Redford, and J.G. Robinson, eds. 2001. *Conservation of Exploited Species*. Cambridge University Press, Cambridge, UK.

Rhoades, R.E., and J. Stallings, eds. 2001. *Integrated Conservation and Development in Tropical America: Experiences and Lessons in Linking Communities, Projects, and Policies*. SANREM-CRSP and CARE-SUBIR, Athens, Georgia.

Richards, Michael. 1999. Internalizing the externalities of tropical forestry: A review of innovative financing and incentive mechanisms. European Union Tropical Forestry Paper 1. Overseas Development Institute and European Commission, London.

Ricketts, Taylor H., G.C. Daily, P.R. Ehrlich, and J.P. Fay. 2001. Countryside biogeography of moths in a fragmented landscape: Biodiversity in native and agricultural habitats. *Conservation Biology*. 15(2):378–388.

Robinson, John, and Kent Redford, eds. 1991. *Neotropical Wildlife Use and Conservation*. University of Chicago Press, Chicago.

Rodale Institute. 1999. *Farming Systems Trial*. Rodale Institute, Rodale, Pennsylvania. www.rodaleinstitute.org

Rodriguez Becerra, Manuel, and Eugenia Ponce de León. 1999. Financing the green plan ("Plan Verde") in Colombia: Challenges and opportunities. Presented at the workshop, Financing of Sustainable Forest Management, UNDP, Programme on Forests, 11–13 October, London.

Roosevelt, Anna, ed. 1994. *Amazonian Indians: From Prehistory to the Present*. University of Arizona Press, Tucson.

Rosegrant, Mark W., and Renato G. Schleyer. 1995. Reforming water allocation policy through markets in tradable water rights: Lessons from Chile, Mexico, and California. *Cuadernos de Economía* 32(97):291–315.

RSPB (Royal Society for the Protection of Birds). 2001. Arable stewardship scheme. Available at http://www.rspb.org.uk/wildlife/farming/agrienv/ arblstew.asp.

Rozanov, B.G., V. Targulian, and D.S. Orlov. 1990. Soils. In B.L. Turner II, et al., eds., *The Earth as Transformed by Human Action: Global and Regional Changes in the Biosphere Over the Past 30 Years*. Cambridge University Press with Clark University, Cambridge, pp. 203–214.

Rudel, T.K. 2001. Did a green revolution restore the forests of the American South? In A. Angelsen and D. Kaimowitz, eds., *op. cit.*, pp. 53–68.

Ruitenbeek, J., and C. Cartier. 2001. The invisible wand: Adaptive co-management as an emergent strategy in complex bioeconomic systems. CIFOR Occasional Paper No. 34. Center for International Forestry Research, Bogor, Indonesia.

Rukuni, M., et al., eds. 1994. *Irrigation Performance in Zimbabwe.* Proceedings of two workshops sponsored by [the] Faculty of Agriculture, University of Zimbabwe/AGRITEX/IFPRI Irrigation Performance in Zimbabwe Research Project, held in Harare and Juliasdale, Zimbabwe, 3–6 August 1993.

Russell, Nathan, and Peter Jones. 2000. FloraMap: A new tool for finding and conserving biodiversity. Press release, Centro International de Agricultural Tropical, Cali, Colombia.

Ruthenberg, H. 1980. *Farming Systems of the Tropics.* 3rd ed. Clarendon Press, Oxford, U.K.

Saddler, Gerry, Jeff Waage, and Gabrielle J. Persley. 2001. Impact of the applications of modern biotechnology on biodiversity. In Gunnar Platais and Gabrielle J. Persley, eds., *Biodiversity and Biotechnology: Contributions to and Consequences for Agriculture and the Environment.* The World Bank, Washington, D.C.

Sala, Osvaldo, et al. 2000. Global biodiversity scenarios for the year 2100. *Science* 287:1770–1774.

Salafsky, N., and E. Wollenberg. 2000. Linking livelihoods and conservation: A conceptual framework and scale for assessing the integration of human needs and bio-diversity. *World Development* 28(8):1421–1438.

Salafsky, N., R. Margoluis, and K. Redford. 2001. *Adaptive Management: A Tool for Conservation Practitioners.* Biodiversity Support Program, Washington, D.C.

Sale, Kirkpatrick. 1984. Bioregionalism: A new way to treat the land. *Ecologist* 14(4):167–173.

Sanchez, Pedro A., R.J. Buresh, and R.R.B. Leakey. 1998. In D.J. Greenland, P.J. Gregory, and P.H. Nye, eds., Trees, soils, and food security. *Land Resources: On the edge of the Malthusian precipice?* CAB International and The Royal Society, Wallingford, pp. 89–99.

Sankaram, A. 1993. *Global Agriculture: Perceptions, Prerequisites, Prescriptions.* M.S. Swaminathan Research Foundation, Madras.

Sauwasser, H. 1990. Conserving biological diversity. *Forests and Ecological Management* 35:79–90.

Savina, Gail C., and Alan T. White. 1986. A tale of two islands: Some lessons for marine resource management. *Environmental Conservation* 13(2):107–113.

Savory, A. 1989. *Holistic Resource Management.* Island Press, Washington, D.C.

Sayer, Jeffrey, Natarajan Ishwaran, James Thorseil, and Todd Sigaty. 2000. Tropical forest biodiversity and the World Heritage Convention. *Ambio* 29(6):302–309.

Scheffer, Marten, S. Carpenter, J. Foley, C. Folke and B. Walker. 2001. Catastrophic shifts in ecosystems. *Nature* 413:591–596.

Scherr, Sara J.. 1999a. *Soil Degradation: A Threat to Developing Country Food Security by 2020?* IFPRI Food, Agriculture, and the Environment Discussion Paper 27. International Food Policy Reserach Institute, Washington, D.C.

———. 1999b. *Poverty-Environment Interactions in Agriculture: Key Factors and Policy Implications.* Poverty and Environment Issues Series No. 3. United Nations Development Program and the European Commission, New York.

———. 2000a. A downward spiral? Research evidence on the relationship between poverty and natural resource degradation. *Food Policy* 25:479–498.

————. 2000b. *Hillsides Research in the CGIAR: Towards an Impact Assessment*. Report prepared for the Special Program on Impact Assessment of the Technical Advisory Committee of the Consultative Group for International Agricultural Research. CGIAR, Washington, D.C. Draft. March.

————. 2000c. The evolving role of forestry with agricultural intensification: Forest use and management in the Central Hillsides of Honduras, 1970's to 1990's. Paper presented to the International Union of Forestry Research Organizations World Congress, Kuala Lumpur, Malaysia, August.

————. 2001. Future food security risks and economic consequences of soil degradation in developing countries. In M. Bridges, et al., eds., *op. cit.*, pp. 155–170.

Scherr, Sara J. and Peter A. Dewees. 1994. Policies to promote markets for non-timber tree products. Presentation to the IFPRI, CIFOR, ICRAF international workshop Markets for Non-Timber Tree Products, International Food Policy Research Institute, Washington, D.C.

Scherr, Sara J., and S. Franzel, 2002. Promoting new agroforestry technologies: Policy lessons from on-farm research. In S. Franzel and S.J. Scherr, eds., *op. cit.*, pp. 145–168.

Scherr, Sara J. and Scott Templeton. 2000. Impacts of population increase and economic change on mountain forests. In M.F. Price and N. Butt, eds., *Forests in Sustainable Mountain Development: A State of Knowledge Report for 2000*. CAB International, Wallingford, U.K., pp. 90–96.

Scherr, Sara J., Andy White, and David Kaimowitz. 2002. *Strategies to Improve Rural Livelihoods through Markets for Forest Products and Services*. Forest Trends and the Center for International Forestry Research, Washington, D.C.

Schnitger, F.M. 1964. *Forgotten Kingdoms in Sumatra*. E.J. Brill, Leiden.

Sebastian, Kate. 2001, 2002. Calculations of agricultural extent within global protected areas. International Food Policy Research Institute, Washington, D.C. Unpublished memo.

Seidensticker, J. 1984. *Managing Elephant Depredation in Agricultural and Forestry Projects*. World Bank, Washington, D.C.

Serageldin, I. 1996. *Milestones of Renewal—A Journey of Hope and Accomplishment*. Consultative Group for International Agricultural Research (CGIAR), Washington, D.C.

Settle, W.H. 2001. Ecosystem "services" and pest suppression in tropical irrigated rice. In *Abstracts*, International Symposium on Managing Biodiversity in Agricultural Ecosystems, 8-10 November, Montreal, Canada, p.147.

Shah, Anup. 1995. *The Economics of Third World National Parks: Issues of Tourism and Environmental Management*. Edward Elgar, Aldershot, U.K.

Sharma, Narendra P., ed. 1992. *Managing the World's Forests: Looking for Balance Between Conservation and Development*. Kendall/Hunt Publishing, Dubuque, Iowa.

Sharma, Uday R., and Michael P. Wells. 1997. Decentralization and biodiversity conservation in Nepal. In E. Lutz and Julian Caldecott, eds., *Decentralization and Biodiversity Conservation*. The World Bank, Washington D.C., pp. 65–76.

Sharma, M., M. Garcia, A. Quershi, and L. Brown. 1996. *Overcoming Malnutrition: Is There an Ecoregional Dimension?* Food, Agriculture, and the Environment Discussion Paper 10. International Food Policy Research Institute, Washington, D.C.

Sharratt, Steve. 2000. Survey finds farmers good conservationists: Healthy wildlife populations recognized as good indicators of a healthy environment, new national survey says. *Guardian,* Charlottetown, Canada. 9 November.

Shiklomanov, I.A. 1996. *Assessment of Water Resources and Water Availability in the World.* Report for the Comprehensive Global Freshwater Assessment of the United Nations. State Hydrological Institute, St. Petersburg, Russia. February. Draft.

Shiva, Vandana. 1991. *Violence of the Green Revolution.* Third World Network, Panang, Malaysia, and Zed Books, London.

Shively, G., and E. Martinez. 2001. Deforestation, irrigation, employment, and cautious optimism in southern Palawan, the Philippines. In A. Angelsen and D. Kaimowitz, eds., *op. cit.*, pp. 335–346.

Sibuea, T.T., and D. Herdimansyah. 1993. The variety of mammal species in the agroforest areas of Krui (Lampung), Muara Bungo (Jambi), and Maninjau (West Sumatra). Internal report OSTROM/HIMBIO. Universitas Padjajaran, Bandung, Indonesia.

Simmons, I.G. 1993. *Environmental History: A Concise Introduction.* Blackwell Scientific Publishers, Oxford, U.K.

Singh, S., V. Sankaran, H. Mander, and S. Worah. 2000. *Strengthening Conservation Cultures: Local Communities and Biodiversity Conservation.* UNESCO, Paris.

Sisk, Thomas D., A.E. Launer, K.R. Switky, and P.R. Erhlich. 1994. Identifying extinction threats. *BioScience* 44(9):592–604.

Slabe, Anamarija. 2000. The role of organic farming in the rural landscapes and biodiversity protection in Slovenia. In S. Stolton, B. Geier, and J. McNeely, eds., *The Relationship Between Nature Conservation, Biodiversity and Organic Agriculture.* IFOAM, Berlin, Germany, pp. 99–102.

Slocombe, D. Scott. 1991. *An Annotated, Multidisciplinary Bibliography of Ecosystem Approaches.* Cold Regions Research Center, Wilfred Laurier University, Waterloo, Ontario, and IUCN Commission on Environmental Strategy and Planning, Sacramento, Calif.

Smith, J., and S.J. Scherr. 2002. *Forest Carbon and Local Livelihoods: Assessment of Opportunities and Policy Recommendations.* CIFOR Occasional Paper No. 37. Center for International Forestry Research and Forest Trends, Bogor, Indonesia.

Smith, J., B. Finegan, C. Sabogal, M.S. Goncalvez, P. Ferreira, G. Siles Gonzalez, P. van de Kop, and A. Diaz Barba. 2001. Secondary forests and integrated resource management in colonist swidden agriculture in Latin America. In M. Palo, J. Usivuori, and G. Mery, eds., *World Forests, Markets, and Policies.* Kluwer Academic Publishers: Dordrecht/London/Boston.

Smith, J. Russell. 1987 (original in 1958). *Tree Crops: A Permanent Agriculture.* Island Press, Washington, D.C.

Smits, J.E., D.L. Johnson, and C.J. Lomer. 1999. Avian pathological and physiological responses to dietary exposure to the fungus Metarhizium flavoviride, an agent for control of grasshoppers and locusts in Africa. *Journal of Wildlife Diseases* 35:194–203.

Sotherton, N.W. 1998. Landuse changes and the decline of farmland wildlife: An appraisal of the set-aside approach. *Biological Conservation* 83:259–268.

Soule, J.D., and J.K. Piper. 1992. *Farming in Nature's Image: An Ecological Approach to Agriculture.* Island Press, Washington, D.C.

Soulé, M., and B. Wilcox. 1984. *Conservation Biology: An Evolutionary-Ecological Perspective.* Sinauer Associates, Sunderland, Mass.

Southgate, D.D. 1998. *Tropical Forest Conservation: An Economic Assessment of the Alternatives in Latin America.* Oxford University Press, Oxford.

Spampinato, R.G.. 2000. Organic agriculture in Mt. Etna Park. In Stolten, S., B. Geier, and J. McNeely, eds., *The Relationship between Nature Conservation, Biodiversity, and Organic Agriculture.* IFOAM, Berlin, Germany, pp. 95–98.

Spencer, J.E. 1966. *Shifting Cultivation in Southeast Asia.* University of California Press, Berkeley.

Spiller, Gary 1997. Community-based coastal resources management in Indonesia. *Sea Wind* 11(2):13–19.

Srivastava, J.P. 1998. *Integrating Biodiversity in Agricultural Intensification: Towards Sound Practice.* Environmentally Sound and Socially Sustainable Development Series. The World Bank, Washington, D.C.

Stahl, P., J.M. Vandel, V. Herrenschmidt, and P. Migot. 2001. The effect of removing lynx in reducing attacks on sheep in the French Jura Mountains. *Biological Conservation* 101:15–22.

Stein, Bruce, Lynn Kutner, and Jonathan Adams, eds. 2000. *Precious Heritage: The Status of Biodiversity in the United States.* Oxford University Press, New York.

Steppler, H.A., and P.K. Nair, eds. 1987. *Agroforestry: A Decade of Development.* International Centre for Research in Agroforestry, Nairobi.

Stocking, M. 2001. Soil agrobiodiversity—an aspect of farmers' management of biological diversity. Paper presented to the International Symposium on Managing Biodiversity in Agricultural Ecosystems, 8–10 November, Montreal, Canada.

Stolton, Sue, Bernward Geier, and Jeffrey A. McNeely, eds. 2000. *The Relationship between Nature Conservation, Biodiversity, and Organic Agriculture.* International Federation of Organic Agricultural Movements, Berlin.

Styger, E., E.C.M. Fernandes, and H.M. Raktondramasy. 2001. Restoring biodiversity through agricultural intensification in the rainforest region of Madagascar. Paper presented to the International Symposium on Managing Biodiversity in Agricultural Ecosystems, 8–10 November, Montreal, Canada.

Sukumar, R. 1986. Elephant–man conflict in Karnataka. In C.J. Saldanha, ed., *Karnataka State of Environment Report 1984–85.* Center for Taxonomic Studies, Bangalore, India.

Sutherland, Michael, and Brian Scarsbrick. 2001. Conservation of biodiversity through Landcare. In Bridges, et al., eds., *op. cit.*, pp. 372.

Suyanto, Thomas Tomich, Meine van Noordwijk, and Dennis Garrity. 1996. Slash-without-burn techniques in land clearing: Environmental and economic opportunities and constraints. Seminar presentation. International Center for Research in Agroforestry, Bogor, Indonesia. January.

Swift, M. 2001. Managing below-ground biodiversity. Introductory paper presented to the International Symposium on Managing Biodiversity in Agricultural Ecosystems. Convention on Biodiversity, International Plant Genetic Resources Institute and FAO, 8–10 November, Montreal, Canada.

Swift, M. J., J. Vandermeer, P.S. Ramakrishnan, J.M. Anderson, C.K. Ong, and B.A. Hawkins. 1996. Biodiversity and agroecosystem function. In H.A. Mooney, J.H. Cushman, E. Medina, O.E. Sala, and E.-D. Schulze, eds., *Functional Roles of Biodiversity: A Global Perspective.* John Wiley and Sons, New York, pp. 261–298.

Swinkels, Rob A., Steven Franzel, Keith D. Shepherd, Eva Ohlsson, and James Kamiri Ndufa. 2002. The adoption potential for short rotation improved tree fallows: Evidence from western Kenya. In Steven Franzel and Sara J. Scherr, eds., *Trees on the Farm: Assessing the Adoption Potential of Agroforestry Practice in Africa.* CAB International, Wallingford, U.K., pp. 65–88.

Taylor, R.D. 2000. *A Review of Problem Elephant Policies and Management Options in Southern Africa.* A report to the African Elephant Specialist Group, Human–Elephant Conflict Taskforce, IUCN SSC. Gland, Switzerland.

Tengberg, Anna, and Michael Stocking. 2001. Land degradation, food security, and agro-biodiversity—examining an old problem in a new way. In M. Bridges, et al., eds., *op. cit.*, pp. 171–185.

ten Kate, Kerry, and Sarah A. Laird. 1999. *The Commercial Use of Biodiversity: Access to Genetic Resources and Benefit-Sharing.* Earthscan Publications, London.

Thomas, M.B., and A.J. Willis. 1998. Biocontrol: Risky but necessary? *Trends in Ecology and Evolution* 13(8):325–329.

Thrupp, Lori Ann. 1998. *Cultivating Diversity: Agrobiodiversity and Food Security.* World Resources Institute, Washington, D.C.

Thrupp, Lori Ann, Gilles Bergeron, and William F. Waters. 1995. *Bittersweet Harvests for Global Supermarkets: Challenges in Latin America's Agricultural Export Boom.* World Resources Institute, Washington, D.C.

Thrupp, Lori Ann, Susanna B. Hecht, John O. Browder, Owen J. Lynch, Nabiha Megateli, and William O'Brien. 1997. *The Diversity and Dynamics of Shifting Cultivation: Myths, Realities, and Policy Implications.* World Resources Institute, Washington, D.C.

Tilman, D., J. Fargione, B. Wolff, C. D'Antonio, A. Dobson, R. Howarth, D. Schindler, W.H. Schlesinger, D. Simberloff, and D. Swackhamer. 2001. Forecasting agriculturally driven global environmental change. *Science* 292(13):281–284.

Tilman, David, P.B. Reich, J. Knops, D. Wedin, T. Mielke, and C. Lehman. 2002. Diversity and productivity in a long-term grassland experiment. *Science* 294:843–845.

Tolbert, V.R., and A. Schiller. 1996. Environmental enhancement using short-rotation woody crops and perennial grasses as alternatives to traditional agricultural crops. In W. Lockeretz, ed. Proceedings of the Conference on Environmental Enhancement Through Agriculture. Tufts University, Medford, Mass.

Tomich, T., M. Noordwijk, S. Budidarsono, A. Gillson, T. Kusumanto, D. Murdiyarso, F. Stolle, and A. Fagi. 2001. Agricultural intensification, deforestation, and the environment: Assessing tradeoffs in Sumatra, Indonesia. In D. Lee and C. Barrett, eds., *op. cit.*, pp. 221–244.

Totten, Michael. 1999. *Getting It Right: Emerging Markets for Storing Carbon in Forests.* Forest Trends and World Resources Institute: Washington, D.C.

Trewavas, Anthony J. 2001. The population/biodiversity paradox: Agricultural efficiency to save wilderness. *Plant Physiology* 125:174–179.

Tristao Bernardes, Aline. 1999. Some mechanisms for biodiversity protection in Brazil, with emphasis on their application in the State of Minas Gerais. Prepared for the Brazil Global Overlay Project, Development Research Group, The World Bank, Washington, D.C.

Tucker, G.M., and M.F. Heath. 1994. Birds in Europe: Their conservation status. *Birdlife Conservation Series* 3:366–367.

Turner, I.M., et al. 1996. A century of plant species loss from an isolated fragment of lowland tropical rainforest. *Conservation Biology* 10(4):1229–1244.

Tuxill, John. 1998. Losing strands in the web of life: Vertebrate declines and the conservation of biological diversity. *WorldWatch Paper* 141:1–88.

United Nations. 2002. *World Population Projections by Region.* United Nations, New York. www.un/unpa/

UN Population Division. 1998. *World Population Projections to 2150.* United Nations, New York.

United Nations University, Secretariat of the Convention on Biological Diversity, International Plant Genetic Resources Institute. 2001. *Program Abstracts.* International Symposium: Managing Biodiversity in Agricultural Ecosystems. Montreal. 8–10 November.

USDA (United States Department of Agriculture). 1994. *A Producers' Guide to Preventing Predation of Livestock.* Agriculture Information Bulletin No. 650. USDA Animal and Plant Health Inspection Service, Washington, D.C.

———. 1997. *Agricultural Resources and Environmental Indicators 1996–1997.* Agricultural Handbook No. 712. USDA, Washington, D.C.

USDA (United States Department of Agriculture) Economic Research Service. 2001. *Agri-environmental Policy at the Crossroads. Guideposts on a Changing Landscape* (AER-794). USDA, Washington, D.C.

Uphoff, N., ed. 2002. *Agroecological Innovations: Increasing Food Production with Participatory Development.* Earthscan Publications, London.

Vandel, J.M., and P. Stahl. 1998. Colonisation du massif Jurassien par le lynx (*Lynx lynx*) et impact sure les ongules domestiques. *Givier Faune Sauvage, Game Wild* 15:1161–1169.

Van Eerden, M.R. 1990. The solution of goose damage problems in the Netherlands, with special reference to compensation schemes. *Ibis* 13:253–261.

van Noordwijk, M. 1999. Scale effects in crop-fallow rotations. *Agroforestry Systems* 47(1-3): 239–251.

van Schaik, C.P., and M. van Noordwijk. 2002. Agroforestry and biodiversity: Are they compatible? In. S.M. Sitompul and S.R. Utami, eds. *Akar Pertanian Sehat. Proc. Seminar Ilmiah,* 28 June 2002, Brawijaya University, Malang, Indonesia, pp. 147–156.

Vergara, Napoleon. 1997. *Wood Materials from Non-Forest Areas.* Asia-Pacific Forestry Sector Outlook Study Working Paper Series No. APFSOS/WP/19. Forestry Policy and Planning Division, Rome and Regional Office for Asia and the Pacific, Bangkok. September.

Voegel, Rudi. 2000. Nature protection areas and agriculture in Brandenberg, Germany. In Stolten, et al., eds., *op. cit.*, pp. 92–94.

Vosti, S.A., J. Witcover, C.L. Carpentier, S.J. Magalhaes de Oliveria, and J. Carvalho dos Santos. 2001. Intensifying small-scale agriculture in the western Brazilian Amazon: Issues, implications, and implementation. In D. Lee and C. Barret, eds., *op. cit.,* pp. 245–266.

Waage, J.K. 1991. Biodiversity as a resource for biological control. In D.L. Hawksworth, ed., *The Biodiversity of Micro-organisms and Invertebrates: Its Role in Sustainable Agriculture.* CAB International, Oxford, U.K., pp. 149–163.

Wall, Diana H., and John C. Moore. 1999. Interactions underground: Soil biodiversity, mutualism, and ecosystem processes. *BioScience* 49(2):109–117.

Washburn, S.P., R.J. Knook, J.T. Green Jr., G.D. Jennings, G.A. Benson, J.C. Barker, and M.H. Poore. 1996. Enhancement of communities with pasture-based dairy production systems. In W. Lockeretz, ed., *op. cit.*, pp. 135–144.

Watson, A.J., D. Bakker, A. Ridgewell, P. Boyd, and C. Law. 2000. Effect of iron supply on southern ocean CO_2 uptake and implications for glacial atmospheric CO_2. *Nature* 407:730–733.

Watson, R., et al. 2000. *Land Use, Land Use Change, and Forestry.* Special Report of the International Panel on Climate Change. Established by the World Meteorological Organization and the United Nations Environment Programme. Oxford University Press, Oxford.

Wells, M.P., and K. Brandon. 1992. *People and Parks: Linking Protected Area Management with Local Communities.* World Bank, World Wildlife Fund, and U.S. Agency for International Development, Washington, D.C.

Welsh, Rick. 1999. *The Economics of Organic Grain and Soybean Production in the Midwestern United States.* Policy Report. Henry A. Wallace Institute for Alternative Agriculture, Md.

Western, David, and Mary Pearl. 1989. *Conservation for the 21st Century.* Oxford University Press, New York.

Western, David, and R. Michael Wright, eds. 1994. *Natural Connections: Perspectives in Community-Based Conservation.* Island Press, Washington, D.C.

Wharton, C. 1968. Man, fire, and wild cattle in Southeast Asia. *Proceedings of the Annual Tall Timbers Fire Ecology Conference.* 7:107–167.

White, A.T., and H.P. Vogt. 2000. Philippine coral reefs under threat: Lessons learned after 25 years of community-based reef conservation. *Marine Pollution Bulletin* 40(6):537–550.

White, R., S. Murray, and M. Rohweder. 2000. *Pilot Analysis of Global Ecosystems: Grassland Ecosystems.* World Resources Institute, Washington, D.C.

White, T.A., and A. Martin. 2002. *Who Owns the World's Forests?* Forest Trends and Center for International Environmental Law, Washington, D.C.

Wilcove, David S., et al. 1998. Quantifying threats to imperiled species in the United States. *BioScience* 48(8):607–615.

Williams, Michael. 1990. Forests. In B. L. Turner, W. Clark, R. Kates, J. Richards, J. Matthews, and W. Myer, eds., *The Earth as Transformed by Human Action.* Cambridge University Press, New York, pp. 179–203.

Williams, S.E., M. van Noordwijk, E. Penot. J.R. Healey, F.L. Sinclair, and G. Wibawa. 2001. On-farm evaluation of the establishment of clonal rubber in multistrata agroforests in Jambi, Indonesia. *Agroforestry Systems* 53(2):227–237.

Wilshusen, P.R., S.R. Brechin, C.L. Fortwangler, and P. West. 2002. Reinventing a square wheel: Critique of a resurgent "Protection Paradigm" in international biodiversity conservation. *Society and Natural Resources* 15:17–40.

Wilson, Charlie, Pedro Moura Costa, and Marc Stuart. 1999. *Transfer Payments for Environmental Services to Local Communities: A Local-Regional Approach.* International Fund for Agricultural Development's Proposed Special Program for Asia, IFAD, Rome.

Wilson, Edward O. 2000. Doomed to early demise. *UNESCO Courier* (May):21–23.

———. 2002. *The Future of Life.* Alfred A. Knopf, New York.

Wilson, E.O. 1985. The biological diversity crisis. *BioScience* 35:700–706.

———. 1992. *The Diversity of Life.* Harvard University Press, Cambridge.

Wilson, E.O., and F.M. Peter. 1988. *Biodiversity.* National Academy Press, Washington, D.C.

Wittenberg, R., and M.J.W. Cock. 2001. *Invasive Alien Species: A Toolkit of Best Prevention and Management Practices.* CAB International, Wallingford, U.K.

Witteveen, S. 2001. *Examples of Biodiversity in Agricultural Situations.* Report from the Centre of Expertise for Agriculture, Nature Management, and Fisheries, Ministry of Agriculture, Nature Management, and Fisheries, Amsterdam, the Netherlands.

Wolfe, Martin S. 2000. Crop strength through diversity. *Nature* 406:681–682.

Wolfenbarger, L.L., and P.R. Phifer. 2000. The ecological risks and benefits of genetically engineered plants. *Science* 290:2088–2093.

Wolfensohn, James D., Peter A. Seligmann, and Mohamed T. El-Ashry. 2000. How biodiversity can be preserved if we get smart together. *International Herald Tribune,* Paris, Tuesday, August 22.

Wood, Henry, Melissa McDaniel, and Katharine Warner, eds. 1995. *Community Development and Conservation of Forest Biodiversity through Community Forestry.* Regional Community Forestry Training Centre, Kasetsart University, Bangkok, Thailand.

Wood, S., K. Sebastian, and S.J. Scherr. 2000. *Pilot Analysis of Global Ecosystems: Agroecosystems.* Report prepared for the Millennium Assessment of the State of the World's Ecosystems. International Food Policy Research Institute and the World Resources Institute, Washington, D.C.

Woodford, Michael H. 2000a. Bovine tuberculoses (BTb) in wildlife. Briefing note. February 3. Future Harvest Foundation, Washington, D.C.

———. 2000b. Rinderpest or cattle plague. Briefing note. January 26. Future Harvest Foundation, Washington, D.C.

Woolf, R., ed. 1977. *Organic Farming.* Rodale Press, Emmaus, Penn.

The World Bank. 1992. *World Development Report 1992.* Oxford University Press, New York.

———. 2000. *World Development Report.* The World Bank, Washington, D.C.

———. 2001. *World Development Report.* The World Bank, Washington, D.C.

World Commission on Dams. 2000. *Dams and Development: A New Framework for Decision-making.* Earthscan, London.

World Commission on Protected Areas. 2002. 1992 World Congress on Parks and Protected Areas. Available at http://www.wcpa.iucn.org.

World Conservation Monitoring Centre. 2000. *Global Biodiversity: Earth's Living Resources in the 21st Century.* World Conservation Press, Cambridge, U.K.

World Food Programme. 2000. *Map of World Hunger.* World Food Programme, United Nations, Rome.

World Resources Institute (WRI). 1998. *World Resources 1997–1998.* World Resources Institute, Washington, D.C.

———. 2000. *World Resources 2000–2001: People and Ecosystems: The Fraying Web of Life.* UNDP, UNEP, The World Bank, and WRI, Washington, D.C.

———. 2001. *World Resources.* World Resources Institute, Washington, D.C.

WRI, IUCN, UNEP (World Resources Institute, World Conservation Union, United Nations Environment Programme). 1992. *Global Biodiversity Strategy: Guidelines for Action to Save, Study, and Use Earth's Biotic Wealth Sustainably and Equitably.* World Resources Institute, Washington, D.C.

World Wildlife Fund. 2000. Lessons from the field: What does it take to make conservation work? Biodiversity Support Program, Washington, D.C. July.

————. 1999. *Ecoregions Database.* World Wildlife Fund, Washington, D.C.

Xie, Yan. 1999. Invasive species in China: an overview. Paper prepared for Biodiversity Working Group of the China Council for International Cooperation on Environment and Development, Beijing.

Yachi, Shigeo, and Michel Loreau. 1999. Biodiversity and ecosystem productivity in a fluctuating environment: the insurance hypothesis. *Ecology* 96:1463–1468.

Yoon, C.K. 2000. Simple method found to vastly increase crop yields. 22 August 2000. Available at http://www.nytimes.com/library/national/science/082200sci-gm-rice.html.

Yudelman, M., A. Ratta, and D. Nygaard. 1998. *Pest Management and Food Production: Looking to the Future. A 2020 Vision for Food, Agriculture and the Environment.* 2020 Discussion Paper 25. International Food Policy Research Institute, Washington, D.C.

Zavaleta, Erika. 1999. The emergence of waterfowl conservation among Yup'ik hunters in the Yukon-Kuskokwim Delta, Alaska. *Human Ecology* 27(2):231–266.

Zeddies, J., R.P. Schaab, P. Neuenschwander, and H.R. Herren. 2001. Economics of biological control of cassava mealybug in Africa. *Agricultural Economics* 24:209–219.

Zhu, Y., H. Chen, J. Fan, Y. Wang, Y. Li, J. Chen, J. Fan, S. Yang, L. Hua, H. Leung, T.W. Mew, P.S. Teng, Z. Weng, and C.C. Mundt. 2000. Genetic diversity and disease control in rice. *Nature* 406 (August 17).

Zhu Youyong. 2001. Cultivating biodiversity for disease control, a case study in China. Paper presented at the International Symposium on Managing Biodiversity in Agricultural Ecosystems, 8–10 November 2001, Montreal, Canada.

About the Authors

Jeffrey A. McNeely worked in Asia for twelve years following his training in anthropology at the University of California at Los Angeles. He spent two years as a Peace Corps volunteer in southern Thailand, five years at the Association for Conservation of Wildlife in Bangkok, and two years studying the relationship between people and nature on the Tibetan border in the Himalayas of eastern Nepal. He spent three years in Indonesia as World Wildlife Fund-IUCN (World Conservation Union) representative, establishing IUCN's first country program and running some thirty-five conservation projects in that country. He joined IUCN proper in 1980, designing conservation programs, advising governments and conservation organizations on conservation policy and practice, and producing a variety of technical and popular publications. As Secretary-General of the IV World Congress on National Parks and Protected Areas (Caracas 1992), McNeely helped develop new concepts for relating people to protected areas. Formerly Director of IUCN's Biodiversity Programme and currently Chief Scientist, he has contributed to all of the major global biodiversity initiatives. He has advised over fifty governments on their biodiversity strategies and action plans and was a founder of the Global Biodiversity Forum. He has published numerous papers and over forty books, including *Soul of the Tiger; Mammals of Thailand; Economics and Biological Diversity; Conserving the World's Biological Diversity; Partnerships for Conservation; Protecting Nature: Regional Reviews of Protected Areas; National Parks Conservation and Development;* and *The Great Reshuffling: Human Dimensions of Invasive Alien Species.*

Sara J. Scherr is an agricultural and natural resource economist, specializing in tropical land and forest policy. While completing her B.A. in economics from Wellesley College, she studied deforestation in Costa Rica. A Fulbright Scholar, she received her M.Sc. and Ph.D. in international economics at Cornell University (1983), with theses on river basin development programs in

Mexico and impacts of the oil boom on Mexican cocoa producers. At the Food Research Institute of Stanford University, she compared oil boom impacts on agriculture in Indonesia, Mexico, and Nigeria. She then joined the International Centre for Research in Agroforestry in Nairobi, Kenya, as a Rockefeller fellow and later Principal Researcher, developing methods for on-farm agroforestry research. In 1992, she joined the International Food Policy Research Institute in Washington, D.C., as a Research Fellow, where she led the "Policies for Sustainable Intensification in Fragile Lands" program, with field studies in the hillsides of Honduras and Ethiopia. She is presently Senior Policy Analyst of Forest Trends, and Advisor to the Future Harvest Foundation. She is also an adjunct professor at the Agricultural and Resource Economics Department of the University of Maryland, College Park. Her current research is on policies to promote ecosystem services in agricultural landscapes, forest markets for low-income producers, economic impacts of land degradation, and community watershed management. She has published over 100 papers and nine books, including *Agriculture and the Oil Syndrome; Costs, Benefits, and Farmer Adoption of Agroforestry; Policy Research for Sustainable Development in Mesoamerican Hillsides; Pilot Assessment of Global Ecosystems: Agroecosystems; Response to Land Degradation;* and *Trees on the Farm: Assessing the Adoption Potential of Agroforestry Practices in Africa.*

Index